Higher GCSE for AQA

David Rayner
Edited by **Brian Gaulter**

The shapes in these domes are all regular polygons.

You can find out about regular polygons on page 258.

OXFORD
UNIVERSITY PRESS

OXFORD
UNIVERSITY PRESS

Great Clarendon Street, Oxford OX2 6DP

Oxford University Press is a department of the University of Oxford.
It furthers the University s objective of excellence in research,
scholarship, and education by publishing worldwide in

Oxford   New York
Auckland Bangkok Buenos Aires Cape Town Chennai
Dar es Salaam Delhi Hong Kong Istanbul Karachi Kolkata
Kuala Lumpur Madrid Melbourne Mexico City Mumbai
Nairobi São Paulo Shanghai Taipei Tokyo Toronto

Oxford is a registered trade mark of Oxford University Press in the
UK and in certain other countries

© Oxford University Press 2003

10 9 8 7 6 5 4 3

The moral rights of the author have been asserted. Database right
Oxford University Press (maker)

First published 2003

All rights reserved. No part of this publication may be reproduced,
stored in a retrieval system, or transmitted, in any form or by any
means, without the prior permission in writing of Oxford University
Press, or as expressly permitted by law, or under terms agreed with
the appropriate reprographics rights organisation. Enquiries
concerning reproduction outside the scope of the above should be
sent to the Rights Department, Oxford University Press, at the
above address

You must not circulate this book in any other binding or cover and
you must impose this same condition on any acquirer

British Library Cataloguing in Publication Data

Data available

ISBN 019 914887 2

The authors would like to thank Paul Metcalf for his authoritative
coursework guidance.

The publishers would like to thank AQA for their kind permission
to reproduce past paper questions. AQA accept no responsibility for
the answers to the past paper questions which are the sole
responsibility of the publishers.

The photograph on the cover is reproduced courtesy of
Apex Picture Agency
Additional artwork by Stephen Hill, Nick Bawden, Oxford
Illustrators and Julian Page

Typeset by TechSet Ltd., Gateshead, Tyne and Wear
Printed and bound in Great Britain by Bell and Bain

About this book

This book is designed to help you get your best possible marks in the **AQA Specification B** Higher modular examinations and coursework modules.

How to use this book

The book is arranged so that the content of each module is clear. You can use the tabs at the edge of the page to find the content you need.

The **content modules** – modules 1, 3 and 5 – are broken down into units of work that increase in difficulty.

Each unit starts with an overview of what you are going to learn and a list of what you should already know.

The **'Before you start'** section will help you to remember the key ideas and skills necessary for the exam. The **Check in** questions will help you see what you already know.

At the end of each unit there is a **summary** page.

The learning outcomes are listed as a quick revision guide and each **Check out** question identifies important content that you should remember.

After the summary page you will find a **revision exercise** with past paper questions from AQA. This will help you to prepare for the style of questions you will see in the exam.

At the end of each module you also have two **Practice module tests** – one calculator and one non-calculator. These tests will help you prepare for the real thing.

The **coursework modules** – modules 2 and 4 – are located in a separate unit at the end of the book. The unit tells you what you have to do for your coursework. Each of the Tasks includes **Moderator comments** to help you get better marks.

The numerical **answers** are given at the end of the book. Use these to check you understand what you are doing.

Good luck in your exams!

MODULE 1 CONTENTS

D1 Data Handling — 1–59
- 1.1 Averages and range — 2
- 1.2 Moving averages and frequency tables — 6
- 1.3 Data presentation — 14
- 1.4 Scatter diagrams — 23
- 1.5 Questionnaires and hypotheses — 27
- 1.6 Sampling and bias — 32
- 1.7 Cumulative frequency — 40
- 1.8 Histograms — 47
- Summary and Checkout D1 — 54
- Revision exercise D1 — 57

D2 Probability — 60–90
- 2.1 Relative probability — 60
- 2.2 Working out probabilities — 63
- 2.3 Exclusive events — 71
- 2.4 Independent events and tree diagrams — 75
- 2.5 Conditional probability — 84
- Summary and Checkout D2 — 86
- Revision exercise D2 — 87

Module 1 Practice calculator test — 91

Module 1 Practice non-calculator test — 92

Module 1 Answers — 483–487

MODULE 2: STATISTICAL COURSEWORK — 467–482

MODULE 3 CONTENTS

N1 Number 1 — 94–126
- 1.1 Properties of numbers — 94
- 1.2 Arithmetic without a calculator — 99
- 1.3 Fractions — 104
- 1.4 Estimating — 107
- 1.5 Ratio and proportion — 112
- 1.6 Negative numbers — 116
- 1.7 Mixed numerical problems — 119
- Summary and Checkout N1 — 125
- Revision exercise N1 — 125

N2 Number 2 — 127–178
- 2.1 Percentages — 128
- 2.2 Using a calculator — 137
- 2.3 Indices — 142
- 2.4 Standard form — 146
- 2.5 Rational and irrational numbers — 150
- 2.6 Recurring decimals — 151
- 2.7 Surds — 153
- 2.8 Substitution — 155
- 2.9 Compound measures — 159
- 2.10 Measurements and bounds — 163
- 2.11 Numerical problems — 171
- Summary and Checkout N2 — 175
- Revision exercise N2 — 177

N3 Algebraic Techniques for Module 3 — 179–198
- 3.1 Conjectures — 179
- 3.2 Logical argument — 183
- 3.3 Proof — 184
- 3.4 Direct and inverse proportion — 186
- 3.5 Graphical solution of equations — 192
- Summary and Checkout N3 — 196
- Revision exercise N3 — 196

Module 3 Practice calculator test — 199

Module 3 Practice non-calculator test — 200

Module 3 Answers — 488–498

MODULE 4: INVESTIGATIVE COURSEWORK — 452–467

Module 5 Contents

AS1 Algebra 1   202–251
1.1   Basic algebra   202
1.2   Linear equations   209
1.3   Trial and improvement   216
1.4   Finding a rule   219
1.5   Linear graphs, $y = mx + c$ 227
1.6   Real-life graphs   237
1.7   Simultaneous linear equations   241
     Summary and Checkout AS1   249
     Revision exercise AS1   250

AS2 Space, Space and Measures 1   252–316
2.1   Angle facts   253
2.2   Congruent triangles   261
2.3   Locus   264
2.4   Pythagoras' theorem   269
2.5   Area   273
2.6   Circles, arcs and sectors   278
2.7   Volume and surface area   290
2.8   Similar shapes   300
     Summary and Checkout AS2   312
     Revision exercise AS2   314

AS3 Algebra 2   317–343
3.1   Changing the subject of a formula   318
3.2   Inequalities and regions   323
3.3   Curved graphs   331
3.4   Algebraic indices   339
     Summary and Checkout AS3   341
     Revision exercise AS3   342

AS4 Shape, Spaces and Measures 2   344–408
4.1   Drawing 3-D shapes, symmetry   345
4.2   Trigonometry   351
4.3   Dimensions of formulae   364
4.4   Transformations   367
4.5   Vectors   374
4.6   Sine, cosine, tangent for any angle   381
4.7   Sine and cosine rules   387
4.8   Circle theorems   395
4.9   Geometric proof   402
     Summary and Checkout AS4   404
     Revision exercise AS4   406

AS5 Algebra 3   409–443
5.1   Gradient of a curve   410
5.2   Area under a curve   412
5.3   Distance, velocity, acceleration   414
5.4   Quadratic equations   419
5.5   Algebraic fractions   430
5.6   Transformations of curves   433
5.7   Simultaneous equations, one linear, one quadratic   440
     Summary and Checkout AS5   441
     Revision exercise AS5   442

Module 5 Practice calculator test   444

Module 5 Practice non-calculator test   448

Module 5 Answers   499–519

Index   520–522

D1 Data Handling

Statistics is about handling information, or data.

This unit will show you how to:

- Find the median, mean and range of a distribution
- Calculate moving averages
- Draw stem and leaf diagrams
- Draw scatter diagrams
- Devise a questionnaire to test a hypothesis
- Conduct a survey
- Select and justify a sampling scheme
- Investigate a population, including random and stratified sampling
- Identify possible sources of bias and minimise it
- Draw and interpret box and whisker diagrams
- Draw and interpret cumulative frequency diagrams
- Draw and interpret histograms

Before you start:

You should know how to...	Check in D1
1. Work with percentages and fractions. For example, find 6% of 150. 6% of 150 = $\frac{6}{100} \times 150 = 9$	1. (a) Find 12% of 250. (b) Find $\frac{3}{4}$ of 860. (c) Express 32 as a percentage of 200.
2. Use ratios. For example, divide 72 in the ratio 3:5. Number of parts is $3 + 5 = 8$ Each part is $\frac{72}{8} = 9$ therefore 3 parts is $3 \times 9 = 27$ 5 parts is $5 \times 9 = 45$	2. (a) Divide 120 in the ratio 2:3. (b) Express the ratio 12:18 in its simplest form.
3. Plot points on a grid. For example, plot the point A($^-$3, 2). 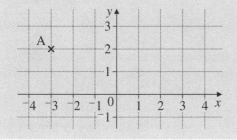	3. Plot the points (1, 5), ($^-$2, 3), ($^-$4, $^-$3) on a set of axes, where x varies from $^-$5 to $^+$5, and y varies from $^-$4 to $^+$4.

1.1 Averages and range

An average can be used to represent the whole range of a set of data, whether the data is discrete (for example, exam marks) or continuous (for example, heights).

The median

The data is arranged in order from the smallest to the largest; the middle number is then selected. This is the central number and is called the **median**.
If there are two 'middle' numbers, the median is in the middle of these two numbers.

The mean

All the data is added up and the total divided by the number of items. This is called the **mean** and is equivalent to sharing out all the data evenly.

The mode

The item which occurs most frequently in a frequency table is selected. This is the most popular value and is called the **mode**.

The word 'mode' comes from the French 'à la mode' meaning 'fashionable'.

Each 'average' has its purpose and sometimes one is preferable to the other.

The median is fairly easy to find and has an advantage in being hardly affected by very large or very small values that occur at the ends of a distribution.

Consider these examination marks.

   20, 21, 21, 22, 23, 23, 25, 27, 27, 27, 29, 98, 98
                              ↑

The median value is 25.
The mean is 35.5. Clearly it does not give a true picture of the centre of distribution of the data.

The mode of this data is 27. It is easy to calculate and it eliminates some of the effects of extreme values. However it does have disadvantages, particularly in data which has two 'most popular' values, and it is not widely used.

Range

In addition to knowing the centre of a distribution, it is useful to know the range or spread of the data.

range = (largest value) − (smallest value)

For the examination marks, range = 98 − 20 = 78.

Example

Find the median, the mean, the mode and the range of this set of 10 numbers: 5, 4, 10, 3, 3, 4, 7, 4, 6, 5.

(a) Arrange the numbers in order of size to find the median.

3,   3,   4,   4,   4,   5,   5,   6,   7,   10
                  ↑

The median is the 'average' of 4 and 5.
∴ median = 4·5

(b) mean = $\frac{(5+4+10+3+3+4+7+4+6+3)}{10} = \frac{51}{10} = 5 \cdot 1$

(c) mode = 4 because there are more 4s than any other number

(d) range = 10 − 3 = 7

Exercise 1A

1. Find the median, the mean, the mode and the range of the following sets of numbers:
   (a) 3, 12, 4, 6, 8, 5, 4
   (b) 7, 21, 2, 17, 3, 13, 7, 4, 9, 7, 9
   (c) 12, 1, 10, 1, 9, 3, 4, 9, 7, 9
   (d) 8, 0, 3, 3, 1, 7, 4, 1, 4, 4.

2. The range for the eight numbers shown is 40. Find the **two** possible values of the missing number.

13	5	27
19	?	
42	11	33

3. The mean weight of ten people in a lift is 70 kg. The weight limit for the lift is 1000 kg. Roughly how many more people can get into the lift?

4. There were ten cowboys in a saloon. The mean age of the men was 25 and the range of their ages was 6. Write each statement below and then write next to it whether it is true, possible or false.
   (a) The youngest man was 18 years old.
   (b) All the men were at least 20 years old.
   (c) The oldest person was 4 years older than the youngest.
   (d) Every man was between 20 and 26 years old.

5. The following are the salaries of 5 employees in a small business:
   Mr A : £22,500   Mr B : £17,900   Mr C : £21,400
   Mr D : £22,500   Mr E : £85,300.
   (a) Find the mean, median and mode of their salaries.
   (b) Which does **not** give a fair 'average'? Explain why in one sentence.

6. A farmer has 32 cattle to sell. Their weights in kg are

81	81	82	82	83	84	84	85
85	86	86	87	87	88	89	91
91	92	93	94	96	150	152	153
154	320	370	375	376	380	381	390

   [Total weight = 5028 kg]

   On the telephone to a potential buyer, the farmer describes the cattle and says the 'average' weight is 'over 157 kg'.
   (a) Find the mean weight and the median weight.
   (b) Which 'average' has the farmer used to describe his animals? Does this average describe the cattle fairly?

7. A gardening magazine sells seedlings of a plant through the post and claims that the average height of the plants after one year's growth will be 85 cm. A sample of 24 of the plants were measured after one year with the following results (in cm)

6	7	7	9	34	56	85	89
89	90	90	91	91	92	93	93
93	94	95	95	96	97	97	99

   [The sum of the heights is 1788 cm.]

   (a) Find the mean and the median height of the sample.
   (b) Is the magazine's claim about average height justified?

8. The mean weight of five men is 76 kg. The weights of four of the men are 72 kg, 74 kg, 75 kg and 81 kg. What is the weight of the fifth man?

9. The mean length of 6 rods is 44·2 cm. The mean length of 5 of them is 46 cm. How long is the sixth rod?

10. (a) The mean of 3, 7, 8, 10 and x is 6. Find x.
    (b) The mean of 3, 3, 7, 8, 10, x and x is 7. Find x.

11. The mean height of 12 men is 1·70 m, and the mean height of 8 women is 1·60 m. Find:
    (a) the total height of the 12 men,
    (b) the total height of the 8 women,
    (c) the mean height of the 20 men and women.

12. The total weight of 6 rugby players is 540 kg and the mean weight of 14 ballet dancers is 40 kg. Find the mean weight of the group of 20 rugby players and ballet dancers.

13. Write down five numbers so that:
    the mean is 6
    the median is 5
    the mode is 4.

14. Find five numbers so that the mean, median, mode and range are all 4.

15. The numbers 3, 5, 7, 8 and N are arranged in ascending order. If the mean of the numbers is equal to the median, find N.

16. The mean of 5 numbers is 11. The numbers are in the ratio $1:2:3:4:5$. Find the smallest number.

17. The median of five consecutive integers is N.
    (a) Find the mean of the five numbers.
    (b) Find the mean and the median of the squares of the integers.
    (c) Find the difference between these values.

1.2 Moving averages and frequency tables

Data that show changes over time are called **time series**.
There are often patterns in data series which can be seen by the use of moving averages.
A **moving average** is the average of consecutive items in a series of data.
Here is a list of the number of children absent from a school for a fifteen-day period.

Day	1	2	3	4	5	6	7	8	9	10	11	12	13	14	15
Number absent	18	12	8	9	8	17	12	8	8	6	15	10	8	6	5

As there are **five** school days in a week you use a **five-point** moving average. This is the average of **five** consecutive numbers.

Thus, at the end of each day, the mean number absent for the last five days is calculated.

At the end of day 5, the mean for the last five days is:
$$\frac{18 + 12 + 8 + 9 + 8}{5} = 11$$

At the end of day 6, the mean for the last five days is:
$$\frac{12 + 8 + 9 + 8 + 17}{5} = 10 \cdot 8$$

At the end of day 7, the mean for the last five days is:
$$\frac{8 + 9 + 8 + 17 + 12}{5} = 10 \cdot 8$$

The averages 11, 10·8, 10·8 are examples of a moving average.

To plot moving averages:

- Draw axes with time along the *x*-axis.
- Plot the original data with crosses (as shown in the diagram).
- Plot the moving averages with ⊙ (as shown in the diagram).

You will notice that the moving averages are plotted along the axis at the mid-point of the days being averaged.

When the moving averages are plotted, a line of best fit is often drawn through the points.
This shows the **trend line**.

In this example, the trend shows a reduction in the number of absences.

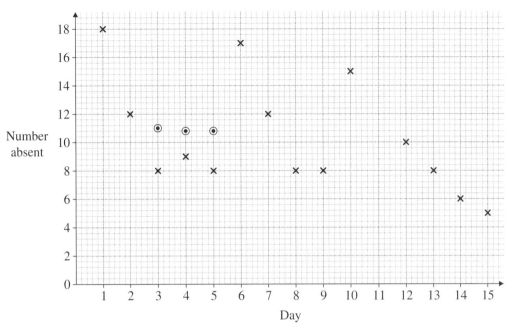

Scatter graph showing absences from school

If the data referred to the sales of a shop which was open six days a week, you would use a **six-point** moving average to include every day of the week the shop was open.

Typical moving averages are three-point, four-point, five-point and six-point moving averages.

Example
The sales, for a period of two weeks, by a farmer at a market which is open three days a week, are plotted together with the moving averages (see page 8).

A line of best fit is drawn through these points.

Assuming that the trend line continues, what would be:

(a) the next value of the moving average,
(b) the sales on the first day of the third week.

The trend line shows a slight increase in the sales.

(a) Extending the line of best fit shows that the next value of the moving average is 121.7.

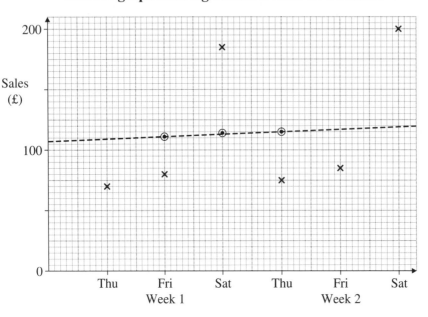

Scatter graph showing farmer's sales at a market

(b) Let S be the sales of the first day of the third week.

121.7 is the average of 86, 201 and S.

$$121.7 = \frac{86 + 201 + S}{3}$$

∴ The sales on the first day of the third week,
$S = 3 \times 121.7 - (86 + 201)$
$= £78.10$

Exercise 1B

1. Traders on the stock market use moving averages as a guide to the performance of a company's share price. Here are the share prices, in pence, of a company over 30 days.

21	24	27	22	25	26	27	23	24	24
28	27	28	26	25	23	25	26	23	25
22	21	19	19	20	19	21	23	24	23

(a) What was the mean price over the first 10 days?
(b) What was the mean price over the 10 day period from day 2 to day 11?
(c) What was the mean price over the 10-day period from day 11 to day 20?

2. Here are the prices, in pence, of shares in 'Tiger Telecom' over a period of 20 days.

| 8 | 7 | 11 | 10 | 9 | 7 | 9 | 11 | 14 | 15 |
| 15 | 14 | 16 | 15 | 13 | 16 | 14 | 10 | 12 | 13 |

At the end of each day the mean price over the last 5 days is calculated. So the mean price at the end of day 5 is

$$\frac{8 + 7 + 11 + 10 + 9}{5} = 9\text{p}$$

The mean price on day 6 $= \dfrac{7 + 11 + 10 + 9 + 7}{5} = 8\cdot 8\text{p}$

Work out the moving average price of the shares in this way up to day 20 and plot the results on a graph.

3. Sue opens her club on Thursday, Friday and Saturday evenings. The attendances over four weeks were:

Week 1			Week 2			Week 3			Week 4		
Thurs	Fri	Sat	Thurs	Fri	Sat	Thurs	Fri	Sat	Thurs	Fri	Sat
76	187	220	75	162	226	72	171	210	69	162	207

Sue uses a three-point moving average to show this data.

(a) Why does Sue use a three-point moving average?
(b) Plot the given data and the moving averages.
(c) What trend does the moving average show?

4. A shop is closed Monday and Tuesday every week. Its sales, rounded to the nearest £500 over a three-week period are:

Week 1					Week 2					Week 3				
Wed	Thurs	Fri	Sat	Sun	Wed	Thurs	Fri	Sat	Sun	Wed	Thurs	Fri	Sat	Sun
500	1000	1500	3500	2500	500	1500	2000	4000	3000	1000	1000	2500	4500	3500

(a) What moving average should be used to show this data?
(b) Plot the given data and the appropriate moving averages.
(c) What trend is shown by your graph?

5. The quarterly phone bills for one family over a three-year period are given in the table below.

Year	1998				1999				2000			
Month	Mar	Jun	Sep	Dec	Mar	Jun	Sep	Dec	Mar	Jun	Sep	Dec
Total bill (£)	98	86	42	102	101	88	47	103	105	91	51	121

(a) Draw a graph for the data.
(b) Calculate the four quarterly moving averages for the data.
(c) Plot the moving averages on the same graph as the original data, joining up the points with a dotted line.
(d) What is the trend in payments over this period?
(e) Do you notice any seasonal variations in the charges?

Frequency tables

A frequency table shows a number x such as a mark or a score, against the frequency f or number of times that x occurs.
The next examples show how these symbols are used in calculating the mean, the median and the mode.
The symbol Σ (or sigma) means 'the sum of'.

Example 1
Discrete data
The marks obtained by 100 students in a test were as follows:

Mark (x)	0	1	2	3	4
Frequency (f)	4	19	25	29	23

Find:
(a) the mean mark   (b) the median mark   (c) the modal mark.

(a) Mean $= \dfrac{\Sigma xf}{\Sigma f}$

where Σxf means 'the sum of the products xf'
i.e. Σ(number × frequency)
and Σf means 'the sum of the frequencies'

$$\text{Mean} = \frac{(0 \times 4) + (1 \times 19) + (2 \times 25) + (3 \times 29) + (4 \times 23)}{100}$$

$$= \frac{248}{100} = 2 \cdot 48$$

(b) The median mark is the number between the 50th and 51st numbers. By inspection, both the 50th and 51st numbers are 3.

∴ Median = 3 marks

(c) The modal mark = 3

Example 2

For **grouped data**, each group can be represented approximately by its mid-point. Suppose the marks of 51 students in a test were:

Mark	30–39	40–49	50–59	60–69
Frequency	7	14	21	9

We say that, for the 30–39 interval, there are 7 marks of 34·5.

Note that the mean calculated by this method is only an estimate because the raw data is not available, whereas in the example on page 10 a true mean could be found.

Exercise 1C

1. A group of 50 people were asked how many books they had read in the previous year; the results are shown in the frequency table below. Calculate the mean number of books read per person.

Number of books	0	1	2	3	4	5	6	7	8
Frequency	5	5	6	9	11	7	4	2	1

2. A teacher conducted a mental arithmetic test for 26 pupils and the marks out of 10 were as follows.

Mark	3	4	5	6	7	8	9	10
Frequency	6	3	1	2	0	5	5	4

(a) Find the mean, median and mode.
(b) The teacher congratulated the class saying that "over three quarters were above average". Which 'average' justifies this statement?

3. The following tables give the distribution of marks obtained by different classes in various tests. For each table, find the mean, median and mode.

(a)

Mark	0	1	2	3	4	5	6
Frequency	3	5	8	9	5	7	3

(b)

Mark	15	16	17	18	19	20
Frequency	1	3	7	1	5	3

4. The table gives the number of words in each sentence of a page of writing.
   (a) Copy and complete the table.
   (b) Work out an estimate for the mean number of words in a sentence.

number of words	frequency f	mid-point x	fx
1–5	6	3	18
6–10	5	8	40
11–15	4		
16–20	2		
21–25	3		
Totals	20	–	

5. The results of 24 students in a test are given.
   (a) Find the mid-point of each group of marks and calculate an estimate of the mean mark.
   (b) Find the modal group.

Mark	Frequency
85–99	4
70–84	7
55–69	8
40–54	5

6. The number of letters delivered to the 26 houses in a street was as follows:

   Calculate an estimate of the mean number of letters delivered per house.

Number of letters delivered	Number of houses (i.e. frequency)
0 – 2	10
3 – 4	8
5 – 7	5
8 –12	3

7. The histogram shows the heights of the 60 athletes in the 1999 British athletics team.
   (a) Calculate an estimate for the mean height of the 60 athletes.
   (b) Explain why your answer is an **estimate** for the mean height.

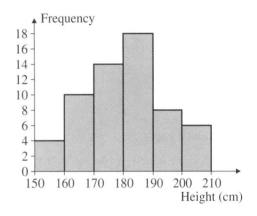

8. The number of goals scored in a series of football matches was as follows:

Number of goals	1	2	3
Number of matches	8	8	x

   (a) If the mean number of goals is 2·04, find x.
   (b) If the modal number of goals is 3, find the smallest possible value of x.
   (c) If the median number of goals is 2, find the largest possible value of x.

9. In a survey of the number of occupants in a number of cars, the following data resulted.

Number of occupants	1	2	3	4
Number of cars	7	11	7	x

   (a) If the mean number of occupants is $2\frac{1}{3}$, find x.
   (b) If the mode is 2, find the largest possible value of x.
   (c) If the median is 2, find the largest possible value of x.

1.3 Data presentation

The results of a statistical investigation are known as data. Data can be presented in a variety of different ways.

Raw data

This is data in the form that it was collected, for example, the number of peas in 40 pods.
Data in this form is difficult to interpret.

```
5 3 6 5 4 6 6 7 4 6
4 7 7 3 7 4 7 5 7 5
6 7 6 7 5 6 6 7 6 6
5 3 6 4 6 5 7 3 6 4
```

Frequency tables

This presentation is in the form of a **tally**; the tally provides a numerical value to the frequency of occurrence.
From the example above, there are 4 pea pods that contain 3 peas, etc.

Number of peas	Tally	Frequency (f)
3	IIII	4
4	IHI I	6
5	IHI II	7
6	IHI IHI III	13
7	IHI IHI	10

Display charts

Three types of display chart are shown; each refers to the initial data about the number of peas in a pod.

Number of peas	Angle on pie chart
3	$\frac{4}{40} \times 360 = 36°$
4	$\frac{6}{40} \times 360 = 54°$
5	$\frac{7}{40} \times 360 = 63°$
6	$\frac{13}{40} \times 360 = 117°$
7	$\frac{10}{40} \times 360 = 90°$

Pie chart

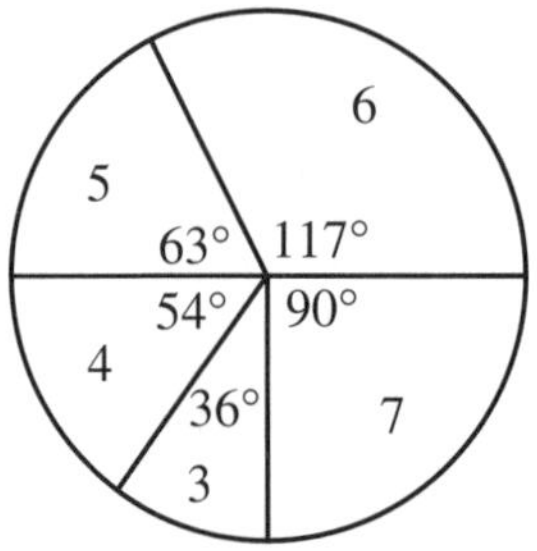

number of peas per pod

Bar chart	Frequency polygon
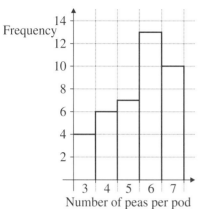	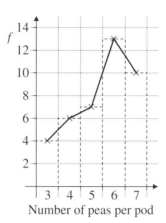

Note that in the bar chart and the frequency polygon, the vertical axis is always used to show frequency. It is the frequency polygon that most clearly shows the mode, that more pods contained 6 peas than any other number of peas. In the frequency polygon, the boxes of the bar chart are replaced by a line joining the tops of their mid-points.

The number of peas in a pod is discrete data; it is concerned solely with individually distinct number quantities. Continuous data is derived from measurements, e.g. height, weight, age, time. It can be handled as follows.

Rounded data

Each measurement can be rounded and the frequency of this rounded quantity then recorded as for discrete data. For example, the lengths of 36 pea pods can be rounded to the nearest mm and then recorded as raw data – a pea pod measuring 59·2 mm is recorded as one that has a length of 59 mm.

Rounded data

```
52  80  65  82  77  60
72  83  63  78  84  75
53  73  70  86  55  88
85  59  76  86  73  89
91  76  92  66  93  84
62  79  90  73  68  71
```

Grouped data

Each measurement can be grouped into classes with defined class boundaries. In the table below, the same data is grouped into the classes shown in the left-hand column.

Grouped frequency table

Length (mm)	Tally	Frequency
50–59	IIII	4
60–69	HHI I	6
70–79	HHI HHI II	12
80–89	HHI HHI	10
90–99	IIII	4

So, the pea pod measuring 59·2 mm is one of 4 recorded in the class 50–59. Because of the rounding of the data, the actual boundary of this first class is from 49·5 to 59·5 (a length of 59·5 mm is rounded up to 60 mm). This is shown below on the horizontal axis of the bar chart for these class boundaries.

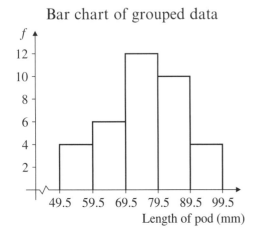

Bar chart of grouped data

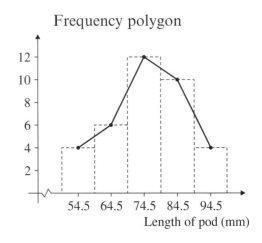

Frequency polygon

The frequency polygon of the same data is also shown. Here the horizontal axis records the mid-points of the classes where the mid-point of the class 50–59 is calculated as $\dfrac{49\cdot5 + 59\cdot5}{2} = 54\cdot5$.

In some cases where there are many small classes, the mid-points of the frequency polygon are joined with a curve. The diagram is then known as a **frequency curve**.

Exercise 1D

1. Karine and Jackie intend to go skiing in February. They have information about the expected snowfall in February for two possible places.

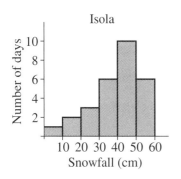
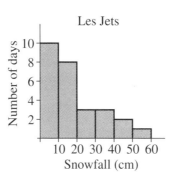

   Decide where you think they should go. It doesn't matter where you decide, but you **must** say why, using the charts above to help you explain.

2. The chart shows information about people who use the internet regularly

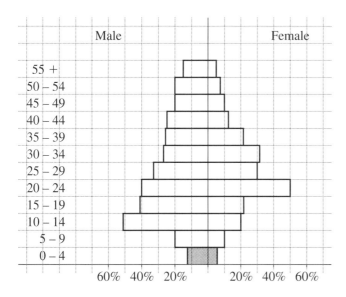

   (a) About what percentage of boys in the 5–9 age group used the internet regularly?
   (b) In what age groups did more women than men use the internet?

3. In a survey, the number of people in 100 cars passing a set of traffic lights was counted. Here are the results:

Number of people in car	0	1	2	3	4	5	6
Frequency	0	10	35	25	20	10	0

(a) Draw a bar chart to illustrate this data.
(b) On the same graph draw the frequency polygon.

Here the bar chart has been started. For frequency, use a scale of 1 cm for 5 units.

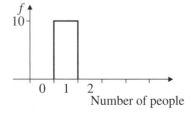

4. Two frequency polygons are shown giving the distribution of the weights of players in two different sports A and B.
   (a) How many people played sport A?
   (b) Comment on two differences between the two frequency polygons.
   (c) Either for A or for B suggest a sport where you would expect the frequency polygon of weights to have this shape. Explain in one sentence why you have chosen that sport.

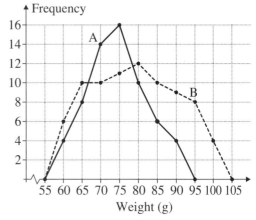

5. A scientist at an agricultural college is studying the effect of a new fertiliser for raspberries. She measures the heights of the plants and also the total weight of fruit collected. She does this for two sets of plants: one with the new fertiliser and one without it. Here are the frequency polygons:

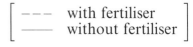

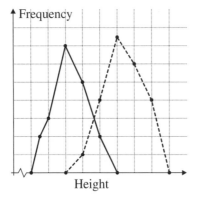
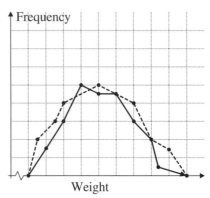

(a) What effect did the fertiliser have on the heights of the plants?
(b) What effect was there on the weights of fruit collected?

6. The diagram illustrates the production of apples in two countries.
   In what way could the pictorial display be regarded as misleading?

UK
470 thousand tonnes

FRANCE
950 thousand tonnes

7. The graph shows the performance of a company in the year in which a new manager was appointed. In what way is the graph misleading?

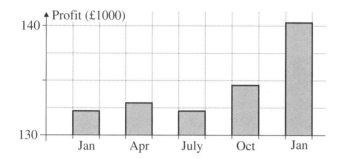

8. The pie chart illustrates the values of various goods sold by a certain shop. If the total value of the sales was £24 000, find the sales value of:
   (a) toys
   (b) grass seed
   (c) records
   (d) food.

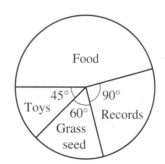

9. A quantity of scrambled eggs is made using the following recipe:

Ingredient	eggs	milk	butter	cheese	salt/pepper
Mass	450 g	20 g	39 g	90 g	1 g

Calculate the angles on a pie chart corresponding to each ingredient.

10. A firm making artificial sand sold its products in four countries.

    5% were sold in Spain
    15% were sold in France
    15% were sold in Germany
    65% were sold in U.K.

What would be the angles on a pie chart drawn to represent this information?

11. The pie chart illustrates the sales of various makes of petrol.
  (a) What percentage of sales does 'Esso' have?
  (b) If 'Jet' accounts for $12\tfrac{1}{2}\%$ of total sales, calculate the angles x and y.

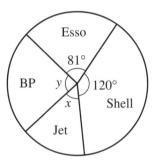

12. The cooking times for meals L, M and N are in the ratio $3:7:x$. On a pie chart, the angle corresponding to L is 60°. Find x.

Stem and leaf diagrams

Data can be displayed in groups in a stem and leaf diagram.

Here are the marks of 20 girls in a science test.

54	42	61	47	24	43	55	62	30	27
28	43	54	46	25	32	49	73	50	45

We will put the marks into groups 20–29, 30–39, … 70–79.
We will choose the tens digit as the 'stem' and the units as the 'leaf'.

The first four marks are shown: [54, 42, 61, 47]

Stem (tens)	Leaf (units)
2	
3	
4	2 7
5	4
6	1
7	

The complete diagram is below ... and then with the leaves in numerical order:

stem	leaf
2	4 7 8 5
3	0 2
4	2 7 3 3 6 9 5
5	4 5 4 0
6	1 2
7	3

stem	leaf
2	4 5 7 8
3	0 2
4	2 3 3 5 6 7 9
5	0 4 4 5
6	1 2
7	3

The diagram shows the shape of the distribution. It is also easy to find the mode, the median and the range.

Back-to-back stem plots

Two sets of data can be compared using a **back-to-back stem plot**. Here are the marks of 20 boys who took the same science test as the girls above.

33	55	63	74	20	35	40	67	21	38
51	64	57	48	46	67	44	59	75	56

These marks are entered onto the back-to-back stem plot shown.
We can see that the boys achieved higher marks than the girls in this test.

It is helpful to have a key.

Boys	stem	Girls
1 0	2	4 5 7 8
8 5 3	3	0 2
8 6 4 0	4	2 3 3 5 6 7 9
9 7 6 5 1	5	0 4 4 5
7 7 4 3	6	1 2
5 4	7	3

key (boys)	key (girls)
1 | 5 means 51	2 | 4 means 24

Exercise 1E

1. The marks of 24 children in a test are shown.

41	23	35	15	40	39	47	29
52	54	45	27	28	36	48	51
59	65	42	32	46	53	66	38

Draw a stem and leaf diagram.
The first three entries are shown.

stem	leaf
1	
2	3
3	5
4	1
5	
6	

22 Data Handling

2. Draw a stem and leaf diagram for each set of data below.

(a)
| 24 | 52 | 31 | 55 | 40 | 37 | 58 | 61 | 25 | 46 |
| 44 | 67 | 68 | 75 | 73 | 28 | 20 | 59 | 65 | 39 |

stem	leaf
2	
3	
4	
5	
6	
7	

(b)
| 30 | 41 | 53 | 22 | 72 | 54 | 35 | 47 |
| 44 | 67 | 46 | 38 | 59 | 29 | 47 | 28 |

3. Here is the stem and leaf diagram showing the masses, in kg, of some people in a lift.
(a) Write down the range of the masses.
(b) How many people were in the lift?
(c) What is the median mass?

stem (tens)	leaf (units)
3	2 5
4	1 1 3 7 8
5	0 2 5 8
6	4 8
7	1
8	2

4. In this question the stem shows the units digit and the leaf shows the first digit after the decimal point. Draw the stem and leaf diagram using the following data:

2·4	3·1	5·2	4·7	1·4	6·2	4·5	3·3
4·0	6·3	3·7	6·7	4·6	4·9	5·1	5·5
1·8	3·8	4·5	2·4	5·8	3·3	4·6	2·8

key
3 | 7 means 3·7

stem	leaf
1	
2	
3	
4	
5	
6	

5. Here is a back-to-back stem plot showing the pulse rates of several people.

(a) How many men were tested?
(b) What was the median pulse rate for the women?
(c) Write a sentence to describe the main features of the data.

```
        Men    | Women
          5 1 | 4 |
        7 4 2 | 5 | 3
        8 2 0 | 6 | 2 1
          5 2 | 7 | 4 4 5 8 9
          2 6 | 8 | 2 5 7
            4 | 9 | 2 8
```

key (men)
1 | 4 means 41

key (women)
5 | 3 means 53

1.4 Scatter diagrams

Sometimes it is interesting to discover if there is a relationship (or **correlation**) between two sets of data. Examples:

- Do tall people weigh more than short people?
- If you spend longer revising for a test, will you get a higher mark?
- Do tall parents have tall children?
- If there is more rain, will there be less sunshine?
- Does the number of Olympic gold medals won by British athletes affect the rate of inflation?

If there is a relationship, it will be easy to spot if your data is plotted on a scatter diagram – that is a graph in which one set of data is plotted on the horizontal axis and the other on the vertical axis.

Example

Each month the average outdoors temperature was recorded together with the number of therms of gas used to heat the house. The results are plotted on the scatter diagram as shown.

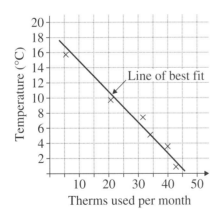

Clearly there is a high degree of **correlation** between these two figures. A **line of best fit** has been drawn 'by eye'.

We can estimate that if the outdoor temperature was 12°C then about 17 therms of gas would be used.

Note: You can only predict within the range of values given.

British Gas do in fact use weather forecasts as their main short-term predictor of future gas consumption over the whole country.

If we extended the line for temperatures below zero the line of best fit predicts that about 60 therms would be used when the temperature is ⁻4°C. But ⁻4°C is well outside the range of the values plotted so the prediction is not valid.

> Perhaps at ⁻4°C a lot of people might stay in bed and the gas consumption would not increase by much. The point is you don't know!

(a) The line in our example has a negative gradient and we say there is **negative correlation**.
(b) If the line of best fit has a positive gradient we say there is **positive correlation**.
(c) Some data when plotted on a scatter diagram does not appear to fit any line at all. In this case there is **no correlation**.

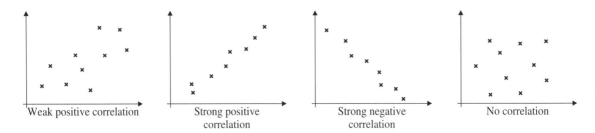

Exercise 1F

1. For this question you need to make some measurements of people in your class.
   (a) Measure everyone's height and 'armspan' to the nearest cm.

   Plot the measurements on a scatter graph. Is there any correlation?

   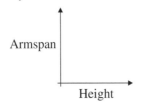
   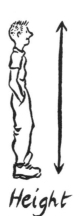

   (b) Now measure everyone's 'head circumference' just above the eyes. Plot head circumference and height on a scatter graph. Is there any correlation?
   (c) Decide as a class which other measurements [e.g. pulse rate] you can (fairly easily) take and plot these to see if any correlation exists.
   (d) Which pair of measurements gave the best correlation?

2. Plot the points given on a scatter graph, with *t* across the page and *z* up the page. Draw axes with values from 0 to 20. Describe the correlation, if any, between the values of *t* and *z*. [i.e. 'strong positive', 'weak negative' etc.]

(a)
t	8	17	5	13	19	7	20	5	11	14
z	9	16	7	13	18	10	19	8	11	15

(b)
t	4	9	13	16	17	6	7	18	10
z	5	3	11	18	6	11	18	12	16

(c)
t	12	2	17	8	3	20	9	5	14	19
z	6	13	8	15	18	2	12	9	12	6

3. Describe the correlation, if any, in these scatter graphs.

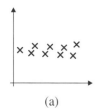 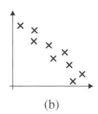 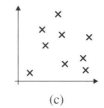 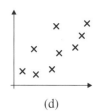

    (a)   (b)   (c)   (d)

4. Plot the points given on a scatter graph, with *s* across the page and *h* up the page. Draw axes with values from 0 to 20.

s	3	13	20	1	9	15	10	17
h	6	13	20	6	12	16	12	17

(a) Draw a line of best fit.
(b) What value would you expect for *h* when *s* is 6?

5. The marks of 7 students in the two papers of a physics examination were as follows.

Paper 1	20	32	40	60	71	80	91
Paper 2	15	25	40	50	64	75	84

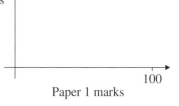

(a) Plot the marks on a scatter diagram, using a scale of 1 cm to 10 marks, and draw a line of best fit.
(b) A student scored a mark of 50 on Paper 1. What would you expect her to get on Paper 2?

6. The table shows (i) the engine size in litres of various cars and (ii) the distance travelled in km on one litre of petrol.

Engine	0·8	1·6	2·6	1·0	2·1	1·3	1·8
Distance	13	10·2	5·4	12	7·8	11·2	8·5

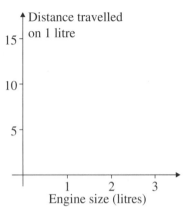

(a) Plot the figures on a scatter graph using a scale of 5 cm to 1 litre across the page and 1 cm to 1 km up the page. Draw a line of best fit.
(b) A car has a 2·3 litre engine. How far would you expect it to go on one litre of petrol?

7. The data shows the latitude of 10 cities in the northern hemisphere and the average high temperatures.

City	Latitude (degrees)	Mean high temperature (°F)
Bogota	5	66
Bombay	19	87
Casablanca	34	72
Dublin	53	56
Hong Kong	22	77
Istanbul	41	64
St. Petersburg	60	46
Manila	15	89
Oslo	60	50
Paris	49	59

(a) Draw a scatter diagram and draw a line of best fit. Plot latitude across the page with a scale of 2 cm to 10°. Plot temperature up the page from 40° F to 90° F with a scale of 2 cm to 10° F.
(b) Which city lies well off the line? Do you know what factor might cause this apparent discrepancy?
(c) The latitude of Shanghai is 31°N. What do you think its mean high temperature is?

8. What sort of pattern would you expect if you took readings of the following and drew a scatter diagram?
   (a) cars on roads; accident rate
   (b) sales of perfume; advertising costs
   (c) birth rate; rate of inflation
   (d) outside temperature; sales of ice cream
   (e) height of adults; age of same adults

1.5 Questionnaires and hypotheses

Hypotheses

Many problems of a statistical nature are made more clear when they are put in the form of a **hypothesis**.
Here are some examples of hypotheses.

- Cats prefer Kit-e-Kat.
- Smoking damages your health.
- A university degree guarantees a higher standard of living.

A hypothesis is generally considered to be a statement which may be true but for which no proof has yet been found.
Statisticians are employed to collect and analyse data in order to obtain information from it which will prove or disprove a hypothesis. Here again are the hypotheses put in the form of questions.

- Is more Kit-e-Kat sold than other brands? — after all, you can't ask the cats!
- Do smokers have a shorter lifespan than non-smokers?
- Do people with a degree earn more than those without a degree?

Several factors need considering when choosing a hypothesis.

Can you test it?

In the smoker's problem, a wide variety of influences affect the lifespan of people: diet, fitness, stress, heredity, etc. How can these be eliminated so that **only** the smoking counts?

Can enough data be collected to give a reasonable result?

Think of where the data will come from. For example in the cat food problem, can you ask your local shop? Will one shop be enough? In the case of the university graduates, how can a lot of graduates (and of course non-graduates) be asked for comparison? Will they each be willing to tell you their income?

How can you know if the hypothesis has been proved or disproved?

Before collecting data, consider the criteria for proof (or disproof). In the undergraduates question, what is meant by more? Do the earnings of the graduates have to be 5% higher (or more) on average than those of the undergraduates?

Can the type of data to be collected be analysed?

Consider the techniques available to you: mean, median, mode, range, scatter diagrams, pie charts, frequency polygons. There are other techniques like histograms and percentiles which will be discussed later.

Is it interesting?

If not, the whole piece of work will be dull and tedious both for you and for others to study.

An excellent way to collect data is by means of a questionnaire. Having decided what you want to know, it is just a matter of asking the right questions.

Questionnaire design

Most surveys are conducted using questionnaires. It is very important to design the questionnaire well so that:

(a) people will cooperate and will answer the questions honestly
(b) the questions are not biased
(c) the answers to the questions can be analysed and presented for ease of understanding.

Checklist

A Provide an introduction to the sheet so that your subject knows the purpose of the questionnaire.

'Proposed new traffic lights'

B Make the questions easy to understand and specific to answer.
Do **not** ask vague questions like this ⟶
The answers could be:

  'Yes, a lot'
  'Not much'
  'Only the best bits'
  'Once or twice a day'

You will find it hard to analyse this sort of data.

Did you see much of the Olympics on TV?

A **better** question is:

'How much of the Olympic coverage did you watch?' Tick one box	
None at all	☐
Up to 1 hour per day	☐
1 to 2 hours per day	☐
More than 2 hours per day	☐

C Make sure that the questions are not **leading** questions. It is human nature not to contradict the questioner. Remember that the survey is to find out opinions of other people, not to support your own.
Do **not** ask
  'Do you agree that BBC has the best sports coverage?'
A better question is:

'Which of the following has the best sports coverage?'					
BBC	ITV	Channel 4	Channel 5	Satellite TV	
☐	☐	☐	☐	☐	

You might ask for one tick or possibly numbers 1, 2, 3, 4, 5 to show an order of preference.

D If you are going to ask sensitive questions (about age or income, for example), design the question with care so as not to offend or embarrass.

Do **not** ask:
    'How old are you?'
or   'Give your date of birth'
A better question is:

'Tick one box for your age group.'

15-17	18-20	21-30	31-50
☐	☐	☐	☐

E   Do not ask more questions than necessary and put the easy questions first.

Exercise 1G

Criticise the following questions and suggest a better question which overcomes the problem involved.
Write some questions with 'yes/no' answers and some questions which involve multiple responses.
Remember to word your questions simply.

1. Do you think it is ridiculous to spend money on food 'mountains' in Europe while people in Africa are starving?

2. What do you think of the new head teacher?

3. How dangerous do you think it is to fly in a single-engined aeroplane?

4. How much would you pay to use the new car park?
    ☐ less than £1   ☐ more than £2·50

5. Do you agree that English and Maths are the most important subjects at school?

6. Do you or your parents often hire videos from a shop?

7. Do you think that we get too much homework?

8. Do you think you would still eat meat if you had been to see the animals killed?

9. Design a questionnaire to test the truth of the statements given.
   (a) 'Most people choose to shop in a supermarket where it is easy to park a car.'
   (b) 'Children of school age watch more television than their parents.'
   (c) 'Boys prefer action films and girls prefer some sort of story.'
   (d) 'Most people who smoke have made at least one serious attempt to give it up.'

Pilot survey

Stop and consider two points.
(a) Will the questionnaire help to prove or disprove the hypothesis?
(b) How many people are needed for the survey?

Both of these questions can be answered by first doing a pilot survey, that is, a quick mini-survey of 5–10 people. Some of the questions may need changing straight away.

Analysis

Having stated clearly what was the hypothesis being tested, display the results clearly. Diagrams like pie charts or bar charts, possibly showing percentages, are a good idea, particularly if drawn using colours. A database or spreadsheet program on a computer may help. Draw conclusions from the results, but make sure they are justified by the evidence.

Your own work

Almost certainly the best hypotheses are your own because you are personally interested.
Here is a list of hypotheses which some students have enjoyed investigating.

(a) Young people are more superstitious than old people.
(b) Given a free choice, most girls would hardly ever choose to wear a dress in preference to something else.
(c) More babies are born in the winter than in the summer.
(d) Year 7 students watch more TV than Year 11 students.
(e) Most cars these days use unleaded petrol.

1.6 Sampling and bias

Statisticians frequently carry out surveys to investigate a characteristic of a population. A population is the set of all possible items to be observed, not necessarily people. Surveys are done for many reasons:

- Newspapers seek to know voting intentions before an election.
- Advertisers try to find out which features of a product appeal most to the public.
- Research workers testing new drugs need to know their effectiveness.
- Government agencies conduct tests for the public's benefit such as the percentage of first class letters arriving the next day or the percentage of '999' calls answered within 5 minutes.

Census

A census is a survey in which data is collected from every member of the population of a country. A census is done in Great Britain every ten years; 1971, 1981, 1991, 2001, ... Data about age, educational qualifications, race, distance travelled to work, etc., is collected from every person in the country on the given 'Census Day'.

Samples

If a full census is not possible or practical, then it is necessary to take a sample from the population. Two reasons for taking samples are:

- it will be much cheaper and quicker than a full census,
- the object being tested may be destroyed in the test, like testing the average lifetime of a new light bulb or the new design of a car tyre.

The choice of a sample prompts two questions.

1. Does the sample truly represent the whole population?
2. How large should the sample be?

After taking a sample, it is assumed that the result for the sample reflects the whole population. For example, if 20% of a sample of 1000 people say they will vote in a particular way, then it is assumed that 20% of the whole electorate would do so likewise, subject to some error.

Simple random sampling

This is a method of sampling in which every item of the population has an equal chance of selection. There are two methods of selection:

(a) Selecting **with** replacement. Here, each item selected is then replaced giving a possibility that it may be reselected.
(b) Selecting **without** replacement. Each item selected is not replaced and is therefore not available for reselection.

For large populations, there is no significant difference between these two methods, but the method chosen should always be stated clearly.

Selection techniques

Out of a school of 758 students, ten are to be selected to take part in an educational experiment in which homework is banned. How would you select these students?
First give each student a three-digit number from 001 to 758.

Method 1

Each person's number is written on a piece of card, the cards placed in a hat and mixed up. Ten cards are then drawn.

Method 2

Use a random number table.
Start anywhere in the table and go either up, down, left, or right to read numbers in groups of three digits.

Random number table

11 74	26 93	81 44	33 93	08 72	32 79	73 31	18 22	64 70	68 50
43 36	12 88	59 11	01 64	56 23	93 00	90 04	99 43	64 07	40 36
93 80	62 04	78 38	26 80	44 91	55 75	11 89	32 58	47 55	25 71
49 54	01 31	81 08	42 98	41 87	69 53	82 96	61 77	73 80	95 27
36 76	87 26	33 37	94 82	15 69	41 95	96 86	70 45	27 48	38 80
07 09	25 23	92 24	62 71	26 07	06 55	84 53	44 67	33 84	53 20
43 31	00 10	81 44	86 38	03 07	52 55	51 61	48 89	74 29	46 47
61 57	00 63	60 06	17 36	37 75	63 14	89 51	23 35	01 74	69 93
31 35	28 37	99 10	77 91	89 41	31 57	97 64	48 62	58 48	69 19
57 04	88 65	26 27	79 59	36 82	90 52	95 65	46 35	06 53	22 54
09 24	34 42	00 68	72 10	71 37	30 72	97 57	56 09	29 82	76 50
97 95	53 50	18 40	89 48	83 29	52 23	08 25	21 22	53 26	15 87
93 73	25 95	70 43	78 19	88 85	56 67	16 68	26 95	99 64	45 69
72 62	11 12	25 00	92 26	82 64	35 66	65 94	34 71	68 75	18 67
61 02	07 44	18 45	37 12	07 94	95 91	73 78	66 99	53 61	93 78
97 83	98 54	74 33	05 59	17 18	45 47	35 41	44 22	03 42	30 00
89 16	09 71	92 22	23 29	06 37	35 05	54 54	89 88	43 81	63 61
25 96	68 82	20 62	87 17	92 65	02 82	35 28	62 84	91 95	48 83
81 44	33 17	19 05	04 95	48 06	74 69	00 75	67 65	01 71	65 45
11 32	25 49	31 42	36 23	43 86	08 62	49 76	67 42	24 52	32 45

This table is published by kind permission of the Department of Statistics
University College, London

Starting on the bottom right-hand corner going up and then down, would give the numbers

553 108 ~~797~~ 049 370 071 605 372 ~~824~~ ~~915~~ 586 670 684

Why were 797, 824, 915 rejected?

So students with the above 10 numbers will take part in the experiment.

Method 3 Random number generator
Use a calculator or a computer.
On the CASIO calculator, random numbers are generated by pressing $\boxed{\text{RAN \#}}$ after using $\boxed{\text{SHIFT}}$ to get this function. The numbers produced are three-digit numbers between .001 and .999. Ignore the decimal point.
Although these are not true random numbers due to their method of generation, they are adequate for school use.

Random two-digit numbers are obtained simply by omitting the last digit.

Example

Find 6 two-digit numbers between 01 and 60.

Ignoring the decimal points, a calculator gave the following:

8̶1̶9̶ 453 480 7̶1̶8̶ 8̶9̶1̶ 326 050 217 9̶9̶0̶ 437

So use: 45   48   32   05   21   43

Exercise 1H

1. One hundred students took a test and their results are given in the table. Each student had a two-digit identification number between 00 and 99.

	\ Second number	0	1	2	3	4	5	6	7	8	9
	0	53	62	48	71	41	78	64	49	82	32
	1	28	66	41	83	49	62	54	67	68	33
	2	75	81	47	35	26	70	85	93	36	35
	3	66	42	55	58	87	38	29	15	41	68
First number	4	84	67	39	62	47	58	62	39	72	86
	5	43	32	49	61	53	80	95	26	44	33
	6	77	62	41	19	26	37	56	57	40	30
	7	38	75	62	41	28	33	62	39	43	26
	8	64	52	53	68	36	42	39	38	65	70
	9	52	45	72	67	90	47	34	55	61	47

For example, Student 31 scored 42% (Row 3, Column 1) and Student 65 scored 37%.

(a) Use your judgement to select a representative sample of ten marks from the 100 given. Find the mean mark of this sample and state its range.

(b) Select several more similar samples of ten marks. Write down the range for each sample mean and find the mean of all the sample means.

2. (a) Use the random number table (page 34) to select a random sample of ten marks from the students' marks of Question 1. Find the sample mean and the range. For example, starting at 07 at the left-hand end of the sixth row of the random number table gives these random numbers and their corresponding marks:

   07   09   25   23   92   24   ...
   49% 32% 70% 35% 72% 26% ...

   (b) Collect other sample means and their range from all the students in your class and find the mean of all the means.
   Compare the results from (a) and (b).

3. This example combines sampling and probability.
   Carefully draw this diagram on graph paper with a scale of 1 cm to 1 unit.
   Points can be selected at random as follows.
   From the random number table read off two pairs, like 75 48.
   For the scale used here, this gives the point (7·5, 4·8).
   Similarly, 83 72 gives the point (8·3, 7·2).

   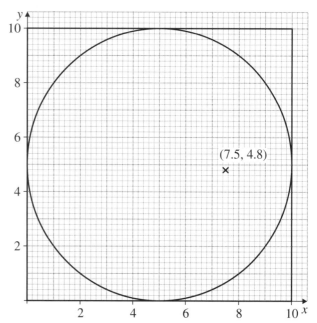

   (a) Select 50 points in this way. Record what fraction of all the points lies inside the circle.
   (b) The expected probability of selecting a point inside the circle can be calculated using areas.

   $$\text{area of circle} = \pi.5^2 = 25\pi$$
   $$\text{area of square} = 10 \times 10 = 100 \text{ cm}^2$$

   $$\frac{\text{Probability of selecting}}{\text{a point inside the circle}} = \frac{25\pi}{100} = \frac{\pi}{4}$$

   (c) Compare the value of the fraction obtained in (a) with the value of the fraction $\frac{\pi}{4}$ by writing the two results in decimals.

Stratified random sampling

Some populations separate naturally into a number of sub-groups or **strata**. Providing that the strata are quite distinct and that every member of the population belongs to one and only one stratum, then the population can be sampled using the method of **stratified random sampling**.

Opinion pollsters like Gallup or N.O.P. use this method when conducting polls on voting intentions. Their claim is that results are accurate to about 3 per cent.

Method
- Separate the population into suitable strata. Find what proportion of the population is in each stratum.
- Select a sample from each stratum proportional to the stratum size, by simple random sampling.

Example

In Year 10 of a school, the 180 students are split into three groups for games; 90 play football, 55 play hockey and 35 play badminton.
Use a stratified random sample of 30 students to estimate the mean weight of the 180 students.

The sample size from each of the three groups (strata) must be proportional to the stratum size. So the 30 students are made up by selecting:

from the football group, $\dfrac{90}{180} \times 30 = 15$,

from the hockey group, $\dfrac{55}{180} \times 30 = 9 \cdot 16$, (say 9)

from the badminton group, $\dfrac{35}{180} \times 30 = 5 \cdot 83$, (say 6)

The three sample means were found to be:

football 53·2 kg   hockey 49·4 kg   badminton 46·4 kg

So the mean for the population of 180 students $= \dfrac{53 \cdot 2 + 49 \cdot 4 + 46 \cdot 4}{3}$

$= 49 \cdot 7 \text{ kg } (3 \text{ s.f.})$

Exercise 1I

1. The table shows the number of pupils in each year group at a new school.
   Naseem takes a stratified sample of 60 pupils from this school.
   Work out how many pupils from each year group should be in Naseem's sample.

Year 7	162
Year 8	161
Year 9	157
Year 10	63
Year 11	58

2. The table gives the number of employees in five firms.
   A stratified sample of 100 people is taken.
   How many people should be selected from each firm?

Firm	Employees
A	398
B	1011
C	409
D	207
E	1985

Bias

Question 1(a) on page 35 required personal judgement to select a representative sample of 10 test marks taken from 100. This provided a non-random judgmental sample.

It is **not** possible to choose a random sample using personal judgement.

Random samples may give results containing errors but these can be predicted and allowed for. The type of error introduced with judgmental sampling is unpredictable and thus corrections for it cannot be made. This type of unpredictable error is called **bias.**

Bias can come from a variety of sources including non-random sampling. For example, for data collected by questionnaire, non-response from particular subjects can introduce bias into the results.

For data collected by a street survey, the time of day and location may well mean that there is a sector or sectors of the population who are not questioned.

A common cause of bias occurs when the questions asked in the survey are not clear or are leading questions.

Exercise 1J

In these questions, decide whether the method of sampling is satisfactory or not. If it is not satisfactory, suggest a better way of obtaining a sample.

1. A teacher, with responsibility for school meals, wants to hear pupils' opinions on the meals currently provided. She waits next to the dinner queue and questions the first 50 pupils as they pass.

2. To find out how satisfied customers are with the service they receive from the telephone company a person telephones 200 people chosen at random from the telephone directory.

3. John, aged 10, wants to find out the average pocket money received by children of his own age. He asks eight of his friends how much they get.

4. An opinion pollster wants to canvas opinion about our European neighbours. He questions drivers as they are waiting to board their ferry at Dover.

5. Jim has a theory that just as many men as women do the shopping in supermarkets. To test his theory he goes to Tesco one Saturday between 09:00 am and 10:00 am and counts the numbers of men and women.

6. A journalist wants to know the views of local people about a new one-way system in the town centre. She takes the electoral roll for the town and selects a random sample of 200 people.

7. A pollster working for the BBC wants to know how many people are watching a new series which is being shown. She questions 200 people as they are leaving a supermarket between 10:00 and 12:00 one Thursday.

8. An electronics company wants to find out what percentage of people have a video at home. In a survey 1000 people are questioned throughout one week as they leave a video hire shop.

1.7 Cumulative frequency

Quartiles, interquartile range

- The range is a simple measure of spread but one extreme (very high or very low) value can have a big effect.
  The **interquartile range** is a better measure of spread.

Example

Find the quartiles and the interquartile range for these numbers:

12 6 4 9 8 4 9 8 5 9 8 10

In order: 4 4 5 | 6 8 8 | 8 9 9 | 9 10 12

↑ lower quartile is 5·5   ↑ median is 8   ↑ upper quartile is 9

Interquartile range = upper quartile − lower quartile
= 9 − 5·5
= 3·5

Box plot

A **box plot** or **box and whisker diagram** shows the spread of a set of data.
Here is a box plot.
It shows the quartiles (Q_1 and Q_3) and the median.
The 'whiskers' extend from the lowest to the highest value and show the range.

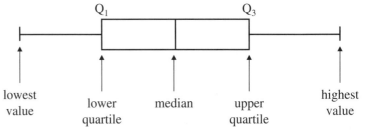

Exercise 1K

For each set of data, work out: (a) the lower and upper quartile
 (b) the inter-quartile range.

1. 1 1 4 4 5 8 8 8 9 10 11 11
2. 5 2 6 4 1 9 3 2 8 4 1 0
3. 7 9 11 15 18 19 23 27
4. 0 0 1 1 2 3 3 4 4 4 6 6 6 7 7

5. Here are three box plots. Estimate: (a) the median
   (b) the interquartile range
   (c) the range.

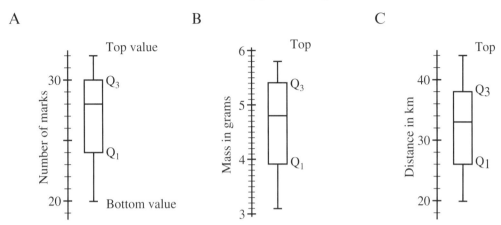

Cumulative frequency

- The total of the frequencies up to a particular value is called the **cumulative frequency**

Data given in a frequency table can be used to calculate cumulative frequencies. These new values, when plotted and joined, form a **cumulative frequency curve**, sometimes called an S-shaped curve.

It is a simple matter to find the median from the halfway point of a cumulative frequency curve.
Other points of location can also be found from this curve. The cumulative frequency axis can be divided into 100 parts.

- The upper quartile is at the 75% point.
- The lower quartile is at the 25% point.

The quartiles are particularly useful in finding the central 50% of the range of the distribution; this is known as the **interquartile range**.

- interquartile range = (upper quartile) − (lower quartile)

The interquartile range is an important measure of spread in that it shows how widely the data is spread.
Half the distribution is in the interquartile range. If the interquartile range is small, then the middle half of the distribution is bunched together.

Example 1

In a survey, 200 people were asked to state their weekly earnings. The results were plotted on the cumulative frequency curve.

(a) How many people earned up to £350 a week?

From the curve, about 170 people earned up to £350 per week.

(b) How many people earned more than £200 a week?

About 40 people earned up to £200 per week. There are 200 people in the survey, so 160 people earned more than £200 per week.

(c) Find the interquartile range.

$$\text{lower quartile (50 people)} = £225$$
$$\text{upper quartile (150 people)} = £325$$
$$\therefore \quad \text{interquartile range} = 325 - 225$$
$$= £100$$

Example 2

A pet shop owner likes to weigh all his mice every week as a check on their state of health. The weights of the 80 mice are shown below.

Weight (g)	Frequency	Cumulative frequency	Weight represented by cumulative frequency
0–10	3	3	$\leqslant 10$ g
10–20	5	8	$\leqslant 20$ g
20–30	5	13	$\leqslant 30$ g
30–40	9	22	$\leqslant 40$ g
40–50	11	33	$\leqslant 50$ g
50–60	15	48	$\leqslant 60$ g
60–70	14	62	$\leqslant 70$ g
70–80	8	70	$\leqslant 80$ g
80–90	6	76	$\leqslant 90$ g
90–100	4	80	$\leqslant 100$ g

The table also shows the cumulative frequency.

Plot a cumulative frequency curve and hence estimate

(a) the median (b) the interquartile range.

- Note: that the points on the graph are plotted at the upper limit of each group of weights.

From the cumulative frequency curve,
$$\text{median} = 55\,\text{g}$$
$$\text{lower quartile} = 36\,\text{g}$$
$$\text{upper quartile} = 68\,\text{g}$$
$$\text{interquartile range} = (68 - 36)\,\text{g} = 32\,\text{g}$$

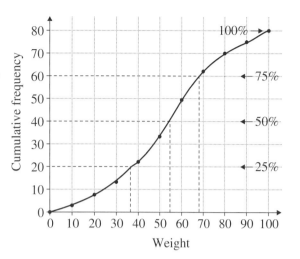

Exercise 1L

1. The graph shows the cumulative frequency curve for the marks of 60 students in an examination. From the graph estimate:
   (a) the median mark
   (b) the mark at the lower quartile and at the upper quartile
   (c) the interquartile range
   (d) the pass mark if three-quarters of the students passed.

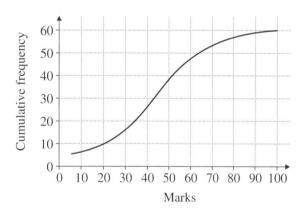

2. The lifetimes of 500 electric light bulbs were measured in a laboratory. The results are shown in the cumulative frequency diagram.
   (a) How many bulbs had a lifetime of 1500 hours or less?
   (b) How many bulbs had a lifetime of between 2000 and 3000 hours?
   (c) After how many hours were 70% of the bulbs dead?
   (d) What was the shortest lifetime of a bulb?

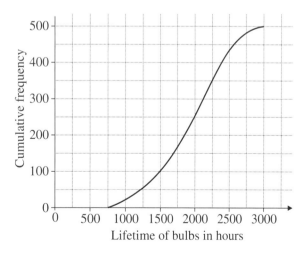

3. A photographer measures all the snakes required for a scene in a film involving a snake pit.

   (a) Draw a cumulative frequency curve for the results below.

Length (cm)	Frequency	Cumulative frequency	Upper limit
0–10	0	0	≤ 10
10–20	2	2	≤ 20
20–30	4	6	≤ 30
30–40	10	16	≤ 40
40–50	17		
50–60	11		
60–70	3		
70–80	3		

   Use a scale of 2 cm for 10 units across the page for the lengths and 2 cm for 10 units up the page for the cumulative frequency. Remember to plot points at the **upper** end of the classes (10, 20, 30 etc.).

   (b) Find: (i) the median   (ii) the interquartile range.

4. As part of a medical inspection, a nurse measures the heights of 48 pupils in a school.

   (a) Draw a cumulative frequency curve for the results below.

Height (cm)	Frequency	Cumulative frequency
$140 \leq h < 145$	2	2 [≤ 145 cm]
$145 \leq h < 150$	4	6 [≤ 150 cm]
$150 \leq h < 155$	8	14 [≤ 155 cm]
$155 \leq h < 160$	9	
$160 \leq h < 165$	12	
$165 \leq h < 170$	7	
$170 \leq h < 175$	4	
$175 \leq h < 180$	2	

   Use a scale of 2 cm for 5 units across the page for the heights and 2 cm for 10 units up the page for the cumulative frequency.

   (b) Find: (i) the median   (ii) the interquartile range.

5. Hugo and Boris are brilliant darts players. They recorded their scores over 60 throws.
Here are Hugo's scores:

Score (x)	$30 < x \leqslant 60$	$60 < x \leqslant 90$	$90 < x \leqslant 120$	$120 < x \leqslant 150$	$150 < x \leqslant 180$
Frequency	10	4	13	23	10

Draw a cumulative frequency curve.

Use a scale of 2 cm for 20 points across the page
and 2 cm for 10 throws up the page.

For Hugo, find: (a) his median score
(b) the interquartile range of his scores.

For his 60 throws, Boris had a median score of 105 and an interquartile range of 20.

Which of the players is more consistent? Give a reason for your answer.

6. The 'life' of a Mickey Mouse photocopying machine is tested by timing how long it works before breaking down. The results for 50 machines are.

Time, t (hours)	$2 < t \leqslant 4$	$4 < t \leqslant 6$	$6 < t \leqslant 8$	$8 < t \leqslant 10$	$10 < t \leqslant 12$
Frequency	9	6	15	15	5

(a) Draw a cumulative frequency graph.
[t across the page, 1 cm = 1 hour; C.F. up the page]
(b) Find: (i) the median life of a machine
(ii) the interquartile range.
(c) The makers claim that their machines will work for at least 10 hours. What percentage of the machines do not match this description?

7. In an international competition 60 children from Britain and France did the same science test.

Marks	Britain frequency	France frequency	Britain cum. freq.	France cum. freq.
1–5	1	2	1 [$\leqslant 5\cdot 5$]	2 [$\leqslant 5\cdot 5$]
6–10	2	5	3 [$\leqslant 10\cdot 5$]	7 [$\leqslant 10\cdot 5$]
11–15	4	11	7 [$\leqslant 15\cdot 5$]	
16–20	8	16		
21–25	16	10		
26–30	19	8		
31–35	10	8		

Note: The upper class boundaries for the marks are 5·5, 10·5, 15·5 etc. The cumulative frequency graph should be plotted for values $\leqslant 5\cdot 5$, $\leqslant 10\cdot 5$, $\leqslant 15\cdot 5$ and so on.

(a) Using the same axes, draw the cumulative frequency curves for the British and French results. Use a scale of 2 cm for 5 marks across the page and 2 cm for 10 people up the page.
(b) Find the median mark for each country.
(c) Find the interquartile range for the British results.
(d) Describe in one sentence the main difference between the two sets of results.

8. A new variety of plant is tested by planting a sample of seeds and measuring the heights of the plants after six weeks. The results were:

Height (cm)	0–	2–	4–	6–	8–	10–	12–	14–	16 and over
Frequency	12	0	2	4	10	12	10	8	0

(a) Draw a cumulative frequency curve for the data. Use a scale of 1 cm to 1 cm across the page and 2 cm to 10 units up the page.
(b) Find the median height of the plants.
(c) The 'average' height of the plants after six weeks is to be printed on the packet.
The actual mean height of the plants is 9·0 cm. Which 'average' height gives the better indication of the likely performance of the plants: the mean or the median? Explain your answer.

Note:
The height interval '0–' means $0 \leqslant$ height < 2. The class boundaries are 0 and 2.

9. The age distribution of the populations of two countries, A and B, is shown below.

Age	Number of people in A (millions)	Number of people in B (millions)
Under 10	15	2
10–19	11	3
20–39	18	5
40–59	7	13
60–79	3	14
80–99	1	7

(a) Copy and complete this cumulative frequency table.

Age	Country A	Country B
Under 10	15	
Under 20	26	
Under 40		
Under 60		
Under 80		
Under 100		

(b) Using the same axes draw cumulative frequency curves for countries A and B.
Use a scale of 2 cm to 20 years across the page and 2 cm to 10 million up the page.

(c) State the population of country B.

(d) State the median age for the two countries.

(e) Describe the main difference in the age distribution of the two countries.

1.8 Histograms

In a histogram, the frequency of the data is shown by the **area** of each bar. Histograms resemble bar charts but are not to be confused with them: in bar charts the frequency is shown by the height of each bar.

Histograms often have bars of varying widths. Because the area of the bar represents frequency, the height must be adjusted to correspond with the width of the bar. The vertical axis is not labelled frequency but **frequency density**.

$$\text{frequency density} = \frac{\text{frequency}}{\text{class width}}$$

Histograms can be used to represent both types of data but their main purpose is for use with continuous data.

Example

Draw a histogram from the table shown for the distribution of ages of passengers travelling on a flight to New York.

Note that the data has been collected into class intervals of different widths.

To draw the histogram, the heights of the bars must be adjusted by calculating frequency density.

Ages	Frequency
$0 \leqslant x < 20$	28
$20 \leqslant x < 40$	36
$40 \leqslant x < 50$	20
$50 \leqslant x < 70$	30
$70 \leqslant x < 100$	18

Ages	Freq.	Freq. density (f.d.)
$0 \leqslant x < 20$	28	$28 \div 20 = 1 \cdot 4$
$20 \leqslant x < 40$	36	$36 \div 20 = 1 \cdot 8$
$40 \leqslant x < 50$	20	$20 \div 10 = 2$
$50 \leqslant x < 70$	30	$30 \div 20 = 1 \cdot 5$
$70 \leqslant x < 100$	18	$18 \div 30 = 0 \cdot 6$

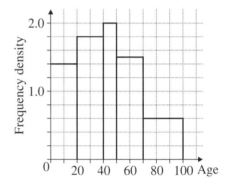

Exercise 1M

1. The lengths of 20 copper nails were measured. The results are shown in the frequency table. Calculate the frequency densities and draw the histogram as started below.

Length l (in mm)	Frequency	Frequency density (f.d.)
$0 \leqslant L < 20$	5	$5 \div 20 = 0 \cdot 25$
$20 \leqslant L < 25$	5	
$25 \leqslant L < 30$	7	
$30 \leqslant L < 40$	3	

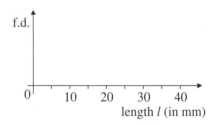

2. The volumes of 55 containers were measured and the results presented in a frequency table as shown.
Calculate the frequency densities and draw the histogram.

Volume (mm^3)	Frequency
$0 \leqslant V < 5$	5
$5 \leqslant V < 10$	3
$10 \leqslant V < 20$	12
$20 \leqslant V < 30$	17
$30 \leqslant V < 40$	13
$40 \leqslant V < 60$	5

3. Thirty students in a class are weighed on the first day of term. Draw a histogram to represent this data.

   Note that the weights do not start at zero. This can be shown on the graph as follows.

Weight (kg)	Frequency
30–40	5
40–45	7
45–50	10
50–55	5
55–70	3

   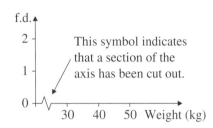

4. The ages of 120 people passing through a turnstyle were recorded and are shown in the frequency table.
   The notation −10 means '0 < age ⩽ 10' and similarly −15 means '10 < age ⩽ 15'.
   The class boundaries are 0, 10, 15, 20, 30, 40.
   Draw the histogram for the data.

Age (yrs)	Frequency
−10	18
−15	46
−20	35
−30	13
−40	8

5. The daily profit made at a beach resort sailing club is shown in the table.

Profit ($)	Frequency
0–40	16
40–80	48
80–100	30
100–120	40
120–160	36
160–240	16

   Draw a histogram for the data.

6. Another common notation is used here for the masses of plums picked in an orchard, shown in the table.

Mass (g)	20–	30–	40–	60–	80–
Frequency	11	18	7	5	0

   The initial 20– means 20 g ⩽ mass < 30 g.
   Draw a histogram with class boundaries at 20, 30, 40, 60, 80.

50 Data Handling

7. The heights of 50 Olympic athletes were measured as shown in the table below.

Height (cm)	170–174	175–179	180–184	185–194
Frequency	8	17	14	11

These values were rounded off to the nearest cm. For example, an athlete whose height h is 181 cm could be entered anywhere in the class $180 \cdot 5 \text{ cm} \leqslant h < 181 \cdot 5 \text{ cm}$. So the table is as follows.

Height	169·5–174·5	174·5–179·5	179·5–184·5	184·5–194·5
Frequency	8	17	14	11

Draw a histogram with class boundaries at 169·5, 174·5, 179·5, ...

Finding frequencies from histograms

If the data is already presented in a histogram, it can be useful to draw up the relevant frequency table. The actual size of the sample tested can then be found as well as the frequencies in each class. However, if a sample size is too small, the results of the analysis may not truly represent the distribution which is being examined.

Example

Given the histogram of children's heights shown, draw up the relevant frequency table and find the size of the sample which was tested.

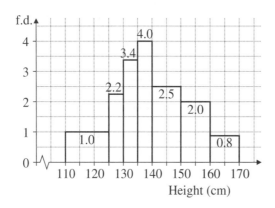

Remember:
The area of each bar represents frequency.

Use   frequency density = $\frac{\text{frequency}}{\text{class width}}$

So   frequency = (frequency density) × (class width)

Height (cm)	f.d.	Frequency
$110 \leqslant x < 125$	1·0	1·0 × 15 = 15
$125 \leqslant x < 130$	2·2	2·2 × 5 = 11
$130 \leqslant x < 135$	3·4	3·4 × 5 = 17
$135 \leqslant x < 140$	4·0	4·0 × 5 = 20
$140 \leqslant x < 150$	2·5	2·5 × 10 = 25
$150 \leqslant x < 160$	2·0	2·0 × 10 = 20
$160 \leqslant x < 170$	0·8	0·8 × 10 = 8
		Total frequency = 116

So there were 116 children in the sample measured.

Exercise 1N

1. For the histogram shown:
   (a) How many of the lengths are in these intervals?
      (i) 40–60 cm   (ii) 30–40 cm
   (b) What is the total frequency?

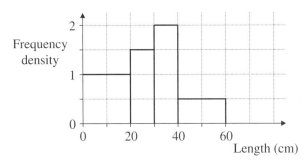

2. For the histogram shown, find the total frequency.

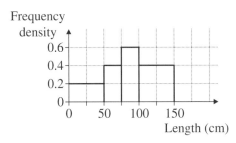

3. The histogram shows the ages of the trees in a small wood. How many trees were in the wood altogether?

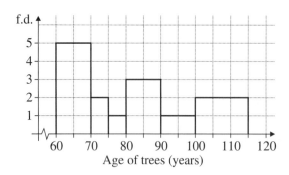

4. One day a farmer weighs all the hens' eggs which he collects. The results are shown in this histogram.
There were 8 eggs in the class 50–60 g.
Work out the numbers on the frequency density axis and hence find the total number of eggs collected.

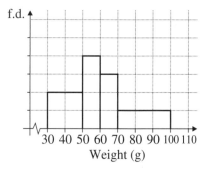

5. This histogram shows the number of letters delivered at some houses in a street. Notice that the data is discrete and that, for example, the class boundaries for 1–2 letters are 0·5–2·5.
Given that 3 letters were delivered to 8 houses, how many letters were delivered altogether?

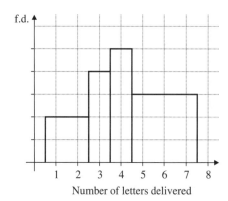

6. Here are the percentage marks obtained by 44 pupils in an exam.

   41 61 11 46 51 21 63 48 25 87 47
   53 52 40 55 15 70 54 74 43 85 38
   47 23 52 18 75 47 73 45 80 59 88
   52 46 76 49 31 57 27 58   5 70 42

   Make a tally chart and a frequency table using groups of marks 0–19, 20–39, 40–49, 50–59, 60–89.
   Work out the frequency densities and hence draw a histogram.

7. The manager of a small company found that the number of sick days taken by his 50 employees in 1999 was as follows:

   10 31 17 22   7   1   9 36   5 13
   26   2   8   6   0 24 30 11 20 28
    8 18 38   3 16 12 23 15   7 21
   12   8 37 22 11 19 40 18   9   6
   13 30 27   3 28   4 29 41 40 14

   (a) Draw up a frequency table and construct a histogram for this set of data after grouping into classes of equal width.
   (b) Repeat using more suitable class widths.
   (c) In 2000, the manager introduced piped music, carpeting and better illumination in the workplace. The number of sick days taken by his employees in 2000 was as follows.

    4   6 11   4 12   5 18   3 10   4
    9   3 13   8 15   0   6   7   1   7
   28   5 23   1 39 14 33   2   7   2
   16 30 13   1 17   7 41   3 22   1
   18 24 38   8   3   9 26 41   0 43

   Draw up a frequency table using the same classes as in part (b) and a histogram. Comment, in one sentence, on the two comparable histograms.

8. The histogram shows the distribution of marks obtained by students in a test.
   To find the number of people with a mark from, say, 50 to 100, use 0.3×50 to get 15. This is really the area of the rectangle with base from 50 to 100.
   The number of people with a mark from 80 to 100 is given by the area of the shaded rectangle which is $0.3 \times 20 = 6$. So 6 people have marks from 80 to 100.
   Similarly, use the areas of rectangles to find the number of people with marks in the following ranges:

   (a) 100–130   (b) 90–150   (c) 90–160   (d) 160–230

Summary

1. You can find the mean, mode and median.

2. You can decide which is the most appropriate moving average to use, you can calculate it, you can draw a trend line and you can extrapolate to the next value.

Check out D1

1. Find (a) the mean (b) the mode (c) the median of:
   21, 23, 25, 27, 19, 18, 21, 30

2. The quarterly electricity bills, in pounds, received by the household are:

2000				2002			
SP	S	A	W	SP	S	A	W
92	71	84	111	95	76	91	115

   (a) What moving average is most appropriate?
   (b) Calculate these moving averages.
   (c) Plot the initial data and the moving average on a graph
   (d) Draw the trend line.
   (e) Extend this trend line to estimate the expected electricity bill for Spring 2003.

3. You can find the mean of a frequency distribution.

3. In a class, each child was asked how many children were in their family. The results were:

Number of children in the family	1	2	3	4	5	6
Number of children	13	8	7	1	0	1

Find the mean number of children per family.

4. You can draw and interpret a stem and leaf diagram.

4. The number of people at an art exhibition was recorded every half hour that the exhibition was open:

7, 15, 25, 26, 21, 11, 4, 1, 19, 25, 17, 18, 25, 29, 14, 15, 5, 8

Using the key of 3|7 to represent 37, show this stem on a stem and leaf diagram.

5. You can draw and interpret a scatter diagram and draw and use a line of best fit.

5. The marks of eight students in two examinations are given in the table below:

Paper 1:   74, 37, 51, 48, 59, 61, 21, 91
Paper 2:   54, 28, 41, 37, 47, 48, 16, 68

(a) Draw a scatter graph to show this data and draw a line of best fit.

(b) One student, absent for Paper 2, acheived 42 marks in Paper 1. Estimate the mark this student should have achieved on Paper 2.

6. You know how to construct a questionnaire and carry out a survey.

6. To test the hypothesis that local people visit a supermarket more often than people who live further away, Ben writes a number of questions.
Give three examples of suitable questions which would help Ben.

7. You know what is meant by bias, and how to reduce it.

7. To see whether commuters bought a newspaper before travelling to work, Hazel asked people as they left a newsagent. She started at 7.30 am.

(a) Was her sample appropriate?

(b) How could it have been improved?

8. You understand the purpose in using, and know how to carry out, quota sampling or stratified random sampling.

8. In a school there are:

   70 students in year 7, 80 students in year 8,
   80 students in year 9, 70 students in year 10,
   100 students in year 11,

   How would you select around 100 students to form a random sample of students in the school?

9. You can draw and interpret:
   (a) a cumulative frequency curve
   (b) a box and whisker diagram
   (c) a histogram.

9. Students in a school are arranged in 64 tutor groups. The number of students in each tutor group is shown below:

Number of students in tutor group	Number of tutor groups
0–15	0
16–20	4
21–25	16
26–30	26
31–35	17
36–40	1

(a) Draw a cumulative frequency curve.

(b) Find (i) the median and
         (ii) the interquartile range.

(c) How many tutor groups have less than 23 students?

(d) Show this data in a box and whisker diagram.

(e) Show this data in a histogram.

Revision exercise D1

1. The table shows the times taken by 50 students to travel to college one morning.

Time taken, t (minutes)	Frequency
$0 < t \leqslant 20$	12
$20 < t \leqslant 40$	20
$40 < t \leqslant 60$	15
$60 < t \leqslant 80$	3

   (a) Calculate an estimate of the mean travel time.

   (b) Complete the cumulative frequency table below.

Time taken, t (minutes)	Cumulative frequency
$t \leqslant 20$	
$t \leqslant 40$	
$t \leqslant 60$	
$t \leqslant 80$	

   (c) Draw the cumulative frequency graph for the times taken by the 50 students.

   (d) Use your graph to estimate how many of the 50 students took **more** than 12 minutes. [SEG]

2. A college records the number of people who sign up for adult education classes each term. The table shows the numbers from Autumn 2000 to Summer 2002.

Term	Autumn 2000	Spring 2001	Summer 2001	Autumn 2001	Spring 2002	Summer 2002
Number of people	520	300	380	640	540	500

   (a) Calculate the first value of the three-point moving average for these data.

   (b) Explain why a three-point moving average was appropriate.

   The time series graph on page 58 shows the original data. The remaining values of the three-point moving average are also plotted (as crosses).

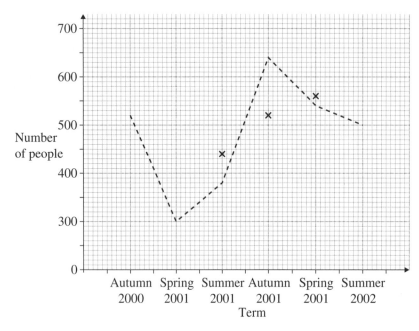

(c) Plot the value from part (a) on the graph.

(d) Use the trend to estimate the three-point moving average that would be plotted at Summer 2002.

(e) Use the value from part (d) to calculate a prediction of the number of people who will sign up for adult education classes in Autumn 2002. [AQA]

3. Barney wants to find out how many burgers children eat.

(a) Write a question with multiple responses that he could use.

Barney also records the number of portions of chips eaten by 60 children in the past week.

Number of portions of chips	0	1–3	4–7	8–14	15+
Number of children	22	17	12	8	1

(b) Which class contains the median? [SEG]

4. Sixty cyclists were asked how many kilometres they had cycled last week.
The cumulative frequency graph opposite shows the results.

(a) How many of the cyclists had cycled less than 50 kilometres last week?

(b) Estimate the median and the interquartile range for the distances. [SEG]

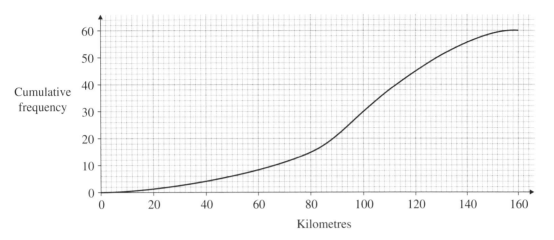

5. The histogram shows the annual snowfall at a ski resort.

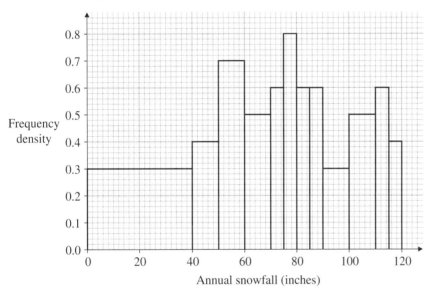

(a) Estimate the number of years in which there were less than 30 inches of snow.

(b) Estimate the most likely level of snowfall next year. [SEG]

6. A glass blower has made 18 wine glasses, 32 ornaments and 7 bottles in one day. He is going to check the quality of the items by looking at a 20% sample.

(a) Explain why he should use a stratified sample rather than a random sample.

(b) How many items of each type should he choose?

(c) Describe how he should choose the wine glasses for the sample from the 18 he has made. [SEG]

D2 Probability

Probability is a measure of how likely a particular event is to occur.

This unit will show you how to:

- Use relative frequency
- Calculate with equally likely outcomes
- Calculate with mutually exclusive and non-exclusive events
- Use a tree diagram

Before you start:

You should know how to...	Check in D2
1. Calculate with fractions. For example, $\frac{2}{3} + \frac{1}{4} = \frac{8}{12} + \frac{3}{12}$ $= \frac{11}{12}$	**1.** Find the value of: (a) $\frac{3}{5} + \frac{2}{11}$ (b) $\frac{2}{5} \times \frac{3}{4}$ (c) $\frac{3}{4} \times \frac{2}{5} + \frac{1}{4} \times \frac{3}{5}$
2. Calculate with decimals. For example, $0.2 + 0.37 = 0.57$ $0.3 \times 0.2 = 0.06$	**2.** Find the value of: (a) $0.4 + 0.08$ (b) $0.4 - 0.21$ (c) 0.3×0.4 (d) 0.1×0.05 (e) $\frac{2}{5} \times 0.15$

2.1 Relative frequency

To work out the probability of a drawing pin landing point up, we can conduct an experiment in which a drawing pin is dropped many times. If the pin lands 'point up' on x occasions out of a total number of N trials, the **relative frequency** of landing 'point up' is $\dfrac{x}{N}$.

When an experiment is repeated many times we can use the relative frequency as an estimate of the probability of the event occurring.

The graph shows how the ratio $\frac{x}{N}$ settles down to a consistent value as N becomes large (say 1000 or 10 000 trials). So the relative frequency is a more reliable guide to the probability of the event occurring when the number of trials is large.

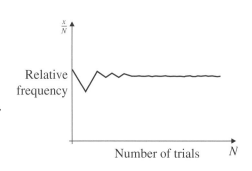

There are four different ways of estimating probabilities.

Method A Use the theoretical probability.
- The probability of rolling a 3 on a fair dice is $\frac{1}{6}$. This is because all the scores, 1, 2, 3, 4, 5, 6 are equally likely.
- Similarly the probability of getting a head when tossing a fair coin is $\frac{1}{2}$.

Method B Conduct an experiment or survey to collect data.
- The dice shown has five faces, not all the same shape. To estimate the probability of the dice landing with a '1' showing, I could conduct an experiment to see what happened in, say, 500 trials.
- I might want to know the probability that the next car going past the school gates is driven by a woman. I could conduct a survey in which the drivers of cars are recorded over a period of time.

Method C Look at past data.
- If I wanted to estimate the probability of my plane crashing as it lands at Heathrow airport I could look at accident records at Heathrow over the last five years or so.

Method D Make a subjective estimate.
We have to use this method when the event is not repeatable. It is not really a 'method' in the same sense as are methods A, B, C.
- We might want to estimate the probability of England beating France in a soccer match next week. We could look at past results but these could be of little value for all sorts of reasons. We might consult 'experts' but even they are notoriously inaccurate in their predictions.

Exercise 2A

In Questions **1** to **10** state which method A, B, C or D you would use to estimate the probability of the event given.

1. The probability that a person chosen at random from a class will be left-handed.

2. The probability that there will be snow in the ski resort to which a school party is going in February next year.

3. The probability of drawing an 'ace' from a pack of playing cards.

4. The probability that you hole a six foot putt when playing golf.

5. The probability that the world record for running 1500 m will be under 3 min 20 seconds by the year 2020.

6. The probability of winning the National Lottery.

7. The probability of rolling a 3 using a dice which is suspected of being biased.

8. The probability that a train will arrive within ten minutes of its scheduled arrival time.

9. The probability that the current Wimbledon Ladies Champion will successfully defend her title next year.

10. Sam rolls a dice and records the number of ones that he gets.

Number of rolls	10	25	50	100	300	1000	2000
Number of ones	1	3	5	12	54	160	338

    Plot a graph of relative frequency of getting one against the number of rolls.

    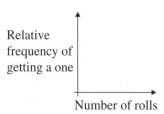

11. Conduct an experiment where you cannot predict the
result. You could roll a dice with a piece of 'blu-tack'
stuck to it. Or make a spinner where the axis is not quite
in the centre. Or drop a drawing pin.
Conduct the experiment many times and work out the
relative frequency of a 'success' after every 10 or 20 trials.
Plot a relative frequency graph to see if the results
'settle down' to a consistent value.

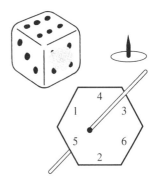

12. The spinner has an equal chance of giving any digit from
0 to 9. Four friends did an experiment when they spun
the pointer a different number of times and recorded the
number of zeros they got. Here are their results.

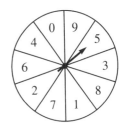

	Number of spins	Number of zeros	Relative frequency
Steve	10	2	0·2
Nick	150	14	0·093
Mike	200	41	0·205
Jason	1000	104	0·104

One of the four recorded his results incorrectly. Say
who you think this was and explain why.

2.2 Working out probabilities

The probability of an event occurring can be calculated
using symmetry. We argue, for example, that when we toss
a coin we have an equal chance of getting a 'head' or a
'tail'. So the probability of spinning a 'head' is a half.
We write 'P (spinning a head) $=\frac{1}{2}$'.

Example

A single card is drawn from a pack of 52 playing cards.
Find the probability of the following results:
(a) the card is a Queen,
(b) the card is a Club,
(c) the card is the Jack of Hearts.

13 Spades
13 Hearts
13 Diamonds
13 Clubs
Total = 52

There are 52 equally likely outcomes of the 'trial' (drawing a card).
(a) P (Queen) $= \frac{4}{52} = \frac{1}{13}$
(b) P (Club) $= \frac{13}{52} = \frac{1}{4}$
(c) P (Jack of Hearts) $= \frac{1}{52}$

Exercise 2B

1. If one card is picked at random from a pack of 52 playing cards, what is the probability that it is:
   (a) a King,
   (b) the Ace of Clubs,
   (c) a Heart?

2. Nine counters numbered 1, 2, 3, 4, 5, 6, 7, 8, 9 are placed in a bag. One is taken out at random. What is the probability that it is:
   (a) a '5',
   (b) divisible by 3,
   (c) less than 5,
   (d) divisible by 4?

3. A bag contains 5 green balls, 2 red balls and 4 yellow balls. One ball is taken out at random. What is the probability that it is:
   (a) green,   (b) red,   (c) yellow?

4. A cash bag contains two 20p coins, four 10p coins, five 5p coins, three 2p coins and three 1p coins. Find the probability that one coin selected at random is:
   (a) a 10p coin,   (b) a 2p coin,   (c) a silver coin.

5. One card is selected at random from those shown. Find the probability of selecting:
   (a) a Heart,
   (b) an Ace,
   (c) the 10 of Clubs,
   (d) a Spade,
   (e) a Heart or a Diamond.

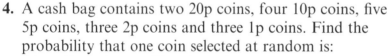

6. (a) A bag contains 5 red balls, 6 green balls and 2 black balls. Find the probability of selecting:
   (i) a red ball   (ii) a green ball   (iii) a ball that is not red.
   (b) One black ball is removed from the bag. Find the new probability of selecting:
   (i) a red ball   (ii) a black ball.

7. One ball is selected at random from a bag containing 12 balls of which x are white.
   (a) What is the probability of selecting a white ball?
   When a further six white balls are added the probability of selecting a white ball is doubled.
   (b) Find x.

8. A large firm employs 3750 people. One person is chosen at random. What is the probability that person's birthday is on a Monday in the year 2000?

9. The numbering on a set of 28 dominoes is as follows:
   (a) What is the probability of drawing a domino from a full set with
      (i) at least one six on it?
      (ii) at least one four on it?
      (iii) at least one two on it?
   (b) What is the probability of drawing a 'double' from a full set?
   (c) If I draw a double five which I do not return to the set, what is the probability of drawing another domino with a five on it?

Expectation

Example

A fair dice is rolled 240 times. How many times would you expect to roll a number greater than 4?

We can roll a 5 or a 6 out of the six equally likely outcomes.
∴ P (number greater than 4) $= \frac{2}{6} = \frac{1}{3}$

Expected number of successes = (probability of a success) × (number of trials)

Expected number of scores greater than 4 $= \frac{1}{3} \times 240 = 80$

Exercise 2C

1. A fair dice is rolled 300 times. How many times would you expect to roll:
   (a) an even number
   (b) a 'six'?

2. The spinner shown has four equal sectors.
   How many 3s would you expect in 100 spins?

3. About one in eight of the population is left-handed.
   How many left-handed people would you expect to find
   in a firm employing 400 people?

4. A bag contains a large number of marbles of which one
   in five is red. If I randomly select one marble on 200
   occasions how many times would I expect to select a
   red marble?

5. The spinner shown is used for a simple game. A player
   pays 10p and then spins the pointer, winning the
   amount indicated.
   (a) What is the probability of winning nothing?
   (b) If the game is played by 200 people how many
       times would you expect the 50p to be won?

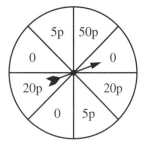

6. These numbered cards are shuffled and put into a pile.

   One card is selected at random and not replaced. A
   second card is then selected.
   (a) If the first card was the '11' find the probability of
       selecting an even number with the second draw.
   (b) If the first card was an odd number, find the
       probability of selecting another odd number.

7. When the ball is passed to Vinnie, the probability
   that he kicks it is 0·2 and the probability that he
   heads it is 0·1. Otherwise he will miss the ball
   completely, fall over and claim a foul.
   In one game his ever-optimistic team mates
   pass the ball to Vinnie 150 times.
   (a) How often would you expect him to head
       the ball?
   (b) How often would you expect him to miss
       the ball?

8. A point is chosen at random from inside the square.
   (a) What is the probability of choosing a point which lies outside the circle? Give your answer as a decimal to 4 s.f.
   (b) In a computer simulation 5000 points are chosen using a random number generator. How many points would you expect to be chosen which lie outside the circle?

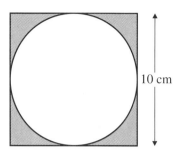

Combined events

When a 10p coin and a 50p coin are tossed together we have two events occurring:

- tossing the 10p coin
- tossing the 50p coin.

We can list all the possible outcomes as shown.

Note that a **sample space** is a list of all the possible outcomes and a **sample space diagram** is a table or diagram which shows all the possible outcomes.

10p	50p
head,	head
head,	tail
tail,	head
tail,	tail

Example

A red dice and a black dice are thrown together. Show all the possible outcomes.

We could list them in pairs with the red dice first:

(1,1), (1,2), (1,2), ...(1,6)
(2,1), (2,2), ...
and so on.

In this example it is easier to see all the possible outcomes when they are shown on a grid.

The X shows 6 on the red dice and 2 on the black.

The ○ shows 3 on the red dice and 5 on the black.

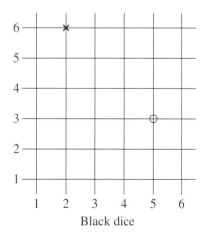

Exercise 2D

1. Three coins (10p, 20p, 50p) are tossed together.
   (a) List all the possible ways in which they could land.
   (b) What is the probability of getting three heads?

2. List all the possible outcomes when four coins are tossed together. How many are there altogether?

3. A black dice and a white dice are thrown together
   (a) Draw a grid to show all the possible outcomes. [See the box in the example.]
   (b) How many ways can you get a total of nine on the two dice?
   (c) What is the probability that you get a total of nine?

4. The two spinners shown are spun together.

   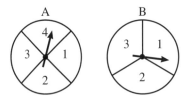

   List all the possible outcomes. Find the probability of obtaining:
   (a) a total of 4,   (b) the same number on each spinner.

5. Four friends, Wayne, Xavier, Yves and Zara, each write their name on a card and the four cards are placed in a hat. Two cards are chosen to decide who does the maths homework that night.
   List all the possible combinations.
   What is the probability that the names drawn are Xavier and Yves?

6. The spinner is spun and the dice is thrown at the same time.
   (a) Draw a grid to show all the possible outcomes.
   (b) A 'win' occurs when the number on the spinner is greater than or equal to the number on the dice.
   Find the probability of a 'win'.

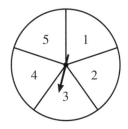

7. A dice is thrown; when the result has been recorded, the dice is thrown a second time. Display all the possible outcomes of the two throws. Find the probability of obtaining:
   (a) a total of 4 from the two throws,
   (b) a total of 8 from the two throws,
   (c) a total between 5 and 9 inclusive from the two throws,
   (d) a number on the second throw which is double the number on the first throw,
   (e) a number on the second throw which is four times the number on the first throw.

8. Shirin and Dipika are playing a game in which three coins are tossed. Shirin wins if there are no heads or one head. Dipika wins if there are either two or three heads. Is the game fair to both players?

9. Pupils X, Y and Z play a game in which four coins are tossed.
   X wins if there is 0 or 1 head.
   Y wins if there are 2 heads.
   Z wins if there are 3 or 4 heads.
   Is the game fair to all three players?

10. Four cards numbered 2, 4, 5 and 7 are mixed up and placed face down.

    In a game you pay 10p to select two cards.
    You win 25p if the total of the two cards is nine.
    How much would you expect to win or lose if you played the game 12 times?

11. Two dice and two coins are thrown at the same time. Find the probability of obtaining:
    (a) two heads and a total of 12 on the dice,
    (b) a head, a tail and a total of 9 on the dice.

Exercise 2E (A practical exercise)

1. The RAN # button on a calculator generates random numbers between ·000 and ·999. It can be used to simulate tossing three coins.
We could say any **odd** digit is a **tail** and any **even** digit is a **head**.
So the number ·568 represents THH
and ·605 represents HHT

```
  0.623
   ↑ ↑ ↑
   H H T
```

Use the RAN # button to simulate the tossing of three coins.
'Toss' the three coins 32 times and work out the relative frequencies of (a) three heads and (b) two heads and a tail. Compare your results with the values that you would expect to get theoretically.

2. A calculator can be used to simulate throwing imaginary dice with 10 faces numbered 0, 1, 2, 3, 4, 5, 6, 7, 8, 9.

The RAN # button gives a random 3 digit number between 0·000 and 0·999. We can use the first 2 digits after the point to represent the numbers on two ten-sided dice.

So 0·763 means 7 on one dice and 6 on the second. [Ignore the '3']

and 0·031 means 0 on one dice and 3 on the second. [Ignore the '1']

We add the numbers on the two 'dice' to give the total.
So for 0·763 the total is 13 and for 0·031 the total is 3.

Press the RAN # button lots of times and record the totals in a tally chart.

Total	0	1	2	...	18
Tally		‖	‖‖‖‖	...	

This grid can be used to predict the result. There are 100 equally likely outcomes like 3,7 or 4,1. The ten results shown each give a total of 9 which is the most likely score. So in 100 throws you could expect to get a total of 9 on ten occasions.
Plot your results on a frequency graph. Compare your results with the values you would expect.

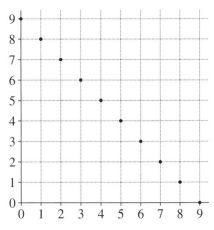

3. In Gibson Academy there are six forms (A, B, C, D, E, F) in Year 10 and the Mathematics Department has been asked to work out a schedule so that, over 5 weeks, each team can play each of the others in a basketball competition. The games are all played at lunch time on Wednesdays and three games have to be played at the same time.
For example in Week 1 we could have

> Form A *v* Form B
> Form C *v* Form D
> Form E *v* Form F

Work out a schedule for the remaining four weeks so that each team plays each of the others. Check carefully that each team plays every other team just once.

2.3 Exclusive events

Events are **mutually exclusive** if they cannot occur at the same time.

Examples
- Selecting an ace } from a
  Selecting a ten } pack of cards

- Tossing a 'head'
  Tossing a 'tail'

- The total sum of the probabilities of mutually exclusive events is 1.
- The probability of something happening is 1 minus the probability of it not happening.

The 'OR' rule: the addition rule

- For exclusive events A and B

    $P(A \text{ or } B) = P(A) + P(B)$

Many questions involving exclusive events can be done without using the addition rule.

Example

A card is selected at random from a pack of 52 playing cards.
What is the probability of selecting any King or Queen?

(a) Count the number of ways in which a King **or** Queen can occur. That is 8 ways.

$$P \text{ (selecting a King or a Queen)} = \frac{8}{52} = \frac{2}{13}$$

(b) Since 'selecting a King' and 'selecting a Queen' are exclusive events, we could use the addition rule.

$$P \text{ (selecting a King)} = \frac{4}{52} \quad P \text{ (selecting a Queen)} = \frac{4}{52}$$

So, P (selecting a King or a Queen) $= \frac{4}{52} + \frac{4}{52} = \frac{8}{52}$ as before.

You can decide for yourself which method is easier in any given question.

Non-exclusive events

If the events are not exclusive, the addition rule **cannot** be used.

Here is a spinner with numbers and colours.
The events 'spinning a red' and 'spinning a 1' are **not** exclusive because a red and a 1 can occur at the same time.

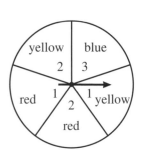

Exercise 2F

1. State whether or not these pairs of events are mutually exclusive.
   (a) Drawing a Club from a pack of cards – Drawing a red card.
   (b) Drawing an Ace from a pack of cards – Drawing a Diamond.
   (c) [Two coins are tossed] Getting at least one head – Getting two tails.
   (d) [Two dice are thrown] Getting the same on both dice – Getting two numbers less than 4.
   (e) [A single dice is thrown] Getting a prime number – Getting an even number.

2. A dice is thrown. Pair the following events so that each pair is mutually exclusive.
   Event A getting an odd number   Event B getting a 6
   Event C getting a prime number   Event D getting a number greater than 4
   Event E getting a factor of 4   Event F getting a 2

3. A bag contains 10 red balls, 5 blue balls and 7 green balls. Find the probability of selecting at random:
   (a) a red ball,
   (b) a green ball,
   (c) a blue **or** a red ball,
   (d) a red **or** a green ball.

4. A roulette wheel has the numbers 1 to 36 once only. What is the probability of spinning either a 10 or a 20?

5. A bag contains a large number of balls coloured red, white, black or green. The probabilities of selecting each colour are as follows:

Colour	red	white	black	green
Probability	0·3	0·1		0·3

   Find the probability of selecting a ball:
   (a) which is black
   (b) which is not white.

6. In an opinion poll people were asked to state the party for which they are going to vote.

   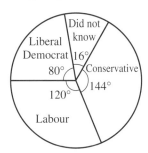

   If one person is chosen at random from the sample, what is the probability that he or she would vote either Conservative or Labour?

7. In a survey the number of people in cars is recorded. When a car passes the school gates the probability of having 1, 2, 3, … occupants is as follows.

Number of people	1	2	3	4	more than 4
Probability	0·42	0·23		0·09	0·02

   (a) Find the probability that the next car past the school gates contains (i) three people (ii) less than four people.
   (b) One day 2500 cars passed the gates. How many of the cars would you expect to have two people inside?

8. Thirty pupils were asked to state the activities they enjoyed from swimming (S), tennis (T) and hockey (H). The numbers in each set are shown.
   One pupil is randomly selected.
   (a) Which of the following pairs of events are exclusive?
      (i) 'selecting a pupil from S', 'selecting a pupil from H'.
      (ii) 'selecting a pupil from S', 'selecting a pupil from T'.
   (b) What is the probability of selecting a pupil who enjoyed either hockey or tennis?

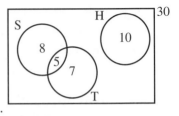

9. The spinner shown has four equal sectors.
   (a) Which of the following pairs of events are exclusive?
      (i) 'spinning 1', 'spinning green',
      (ii) 'spinning 3', 'spinning 2'.
      (iii) 'spinning blue', 'spinning 1'.
   (b) What is the probability of spinning either a 1 or a green?

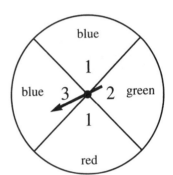

10. Here is a spinner with unequal sectors. When the pointer is spun the probability of getting each colour and number is as follows.

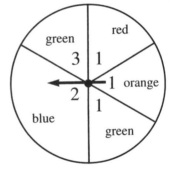

red	0·1
blue	0·3
orange	0·2
green	0·4

1	0·5
2	0·3
3	0·2

   (a) What is the probability of spinning either 1 or 2?
   (b) What is the probability of spinning either blue or green?
   (c) Why is the probability of spinning either a 1 or a green **not** 0·5 + 0·4?

11. According to a weather forecaster in Norway, 'The probability of snow on Saturday is 0·6 and the probability of snow on Sunday is 0·5 so it is certain to snow over the weekend.' Explain why this statement is wrong.

2.4 Independent events and tree diagrams

Two events are **independent** if the occurrence of one event is unaffected by the occurrence of the other.
So, obtaining a head on one coin and a tail on another coin when the coins are tossed at the same time are independent.

The 'AND' rule: the multiplication rule
- For independent events A and B
  $P(A \text{ and } B) = P(A) \times P(B)$

The multiplication rule only works for **independent** events.

Two coins are tossed and the results listed.
  HH   HT   TH   TT
The probability of tossing two heads is $\frac{1}{4}$.

By the multiplication rule for independent events:
$P(\text{two heads}) = P(\text{head on first coin}) \times P(\text{head on second coin})$
∴ $P(\text{two heads}) = \frac{1}{2} \times \frac{1}{2} = \frac{1}{4}$ as before.

Example 1
A fair coin is tossed and a fair dice is rolled. Find the probability of obtaining a 'head' and a 'six'.

The two events are independent.

$P(\text{head and six}) = P(\text{head}) \times P(\text{six}) = \frac{1}{2} \times \frac{1}{6} = \frac{1}{12}$

Example 2
When the two spinners are spun, what is the probability of getting a B on the first and a 3 on the second?

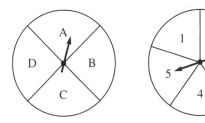

The events 'B on the first spinner' and '3 on the second spinner' are independent.

∴ $P(\text{spinning B and 3}) = P(B) \times P(3) = \frac{1}{4} \times \frac{1}{5} = \frac{1}{20}$

Exercise 2G

1. A card is drawn from a pack of playing cards and a dice is thrown. Events A and B are as follows:
   A: 'a Jack is drawn from the pack'
   B: 'a one is thrown on the dice'.
   (a) Write down the values of P(A), P(B).
   (b) Write down the value of P(A and B).

2. A coin is tossed and a dice is thrown. Write down the probability of obtaining:
   (a) a 'head' on the coin,
   (b) an odd number on the dice,
   (c) a 'head' on the coin and an odd number on the dice.

3. Box *A* contains 3 red balls and 3 white balls. Box *B* contains 1 red and 4 white balls.

   One ball is randomly selected from Box *A* and one from Box *B*. What is the probability that both balls selected are red?

   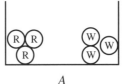 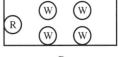

4. In an experiment, a card is drawn from a pack of playing cards and a dice is thrown.
   Find the probability of obtaining:
   (a) a card which is an ace and a six on the dice,
   (b) the king of clubs and an even number on the dice,
   (c) a heart and a 'one' on the dice.

5. A ball is selected at random from a bag containing 3 red balls, 4 black balls and 5 green balls. The first ball is replaced and a second is selected. Find the probability of obtaining:
   (a) two red balls,   (b) two green balls.

6. The letters of the word 'INDEPENDENT' are written on individual cards and the cards are put into a box. A card is selected and then replaced and then a second card is selected. Find the probability of obtaining:
   (a) the letter 'P' twice,   (b) the letter 'E' twice.

7. A fruit machine has three independent reels and pays out a Jackpot of £100 when three apples are obtained.

   Each reel has 15 pictures. The first reel has 3 apples, the second has 4 apples and the third has 2 apples. Find the probability of winning the Jackpot.

8. A coin is biased so that it shows 'Heads' with a probability of $\frac{2}{3}$. The same coin is tossed three times. Find the probability of obtaining:
   (a) two tails on the first two tosses,
   (b) a head, a tail and a head (in that order).

9. A fair dice and a biased dice are thrown together. The probabilities of throwing the numbers 1 to 6 are shown for the biased dice.

   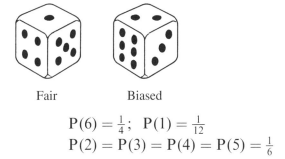

   Fair   Biased

   $P(6) = \frac{1}{4}; \quad P(1) = \frac{1}{12}$
   $P(2) = P(3) = P(4) = P(5) = \frac{1}{6}$

   Find the probability of obtaining a total of 12 on the two dice.

10. Philip and his sister toss a coin to decide who does the washing up. If it's heads Philip does it. If it's tails his sister does it.
    What is the probability that Philip does the washing up every day for a week (7 days)?

11. This spinner has six equal sectors.

    Find the probability of getting:
    (a) 20 Qs in 20 trials.
    (b) No Qs in n trials
    (c) At least one Q in n trials.

    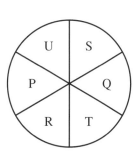

Tree diagrams

Example 1

A bag contains 5 red balls and 3 green balls. A ball is drawn at random and then replaced. Another ball is drawn.

What is the probability that both balls are green?

The branch marked * involves the selection of a green ball twice.
The probability of this event is obtained by simply multiplying the fractions on the two branches.

$$\therefore \quad P(\text{two green balls}) = \tfrac{3}{8} \times \tfrac{3}{8} = \tfrac{9}{64}$$

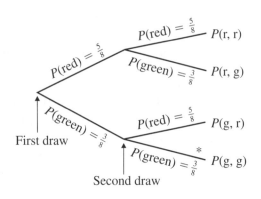

Example 2

A bag contains 5 red balls and 3 green balls. A ball is selected at random and *not* replaced. A second ball is then selected.
Find the probability of selecting
(a) two green balls
(b) one red ball and one green ball.

Notice that since the first ball is not replaced there are only 7 balls in the bag for the second selection.

(a) $P \text{ (two green balls)} = \tfrac{3}{8} \times \tfrac{2}{7}$

$$= \tfrac{3}{28}$$

(b) $P \text{ (one red, one green)} = (\tfrac{5}{8} \times \tfrac{3}{7}) + (\tfrac{3}{8} \times \tfrac{5}{7})$

$$= \tfrac{15}{28}$$

Notice that we can add here because the events 'red then green' and 'green then red' are exclusive.

As a check all the fractions at the ends of the branches should add up to one.

So $(\tfrac{5}{8} \times \tfrac{4}{7}) + (\tfrac{5}{8} \times \tfrac{3}{7}) + (\tfrac{3}{8} \times \tfrac{5}{7}) + (\tfrac{3}{8} \times \tfrac{2}{7}) = \tfrac{20}{56} + \tfrac{15}{56} + \tfrac{15}{56} + \tfrac{6}{56}$

$$= 1$$

Exercise 2H

1. A bag contains 10 discs; 7 are black and 3 white. A disc is selected, and then replaced. A second disc is selected. Copy and complete the tree diagram showing all the probabilities and outcomes.

   Find the probability of the following:
   (a) both discs are black
   (b) both discs are white.

2. A bag contains 5 red balls and 3 green balls. A ball is drawn and then replaced before a ball is drawn again. Draw a tree diagram to show all the possible outcomes. Find the probability that:
   (a) two green balls are drawn
   (b) the first ball is red and the second is green.

3. A bag contains 7 green discs and 3 blue discs. A disc is drawn and **not** replaced. A second disc is drawn. Copy and complete the tree diagram.

   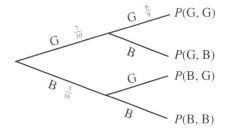

   Find the probability that:
   (a) both discs are green
   (b) both discs are blue.

4. A bag contains 4 red balls, 2 green balls and 3 blue balls. A ball is drawn and not replaced. A second ball is drawn.

   Find the probability of drawing:
   (a) two blue balls
   (b) two red balls
   (c) one red ball and one blue ball (in any order)
   (d) one green ball and one red ball (in any order).

5. A six-sided dice is thrown three times. Complete the tree diagram, showing at each branch the two events: 'three' and 'not three' (written $\overline{3}$).

   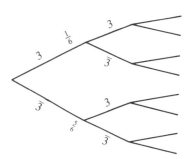

   What is the probability of throwing a total of
   (a) three threes,
   (b) no threes,
   (c) one three,
   (d) at least one three [use part (b)]?

6. A card is drawn at random from a pack of 52 playing cards. The card is replaced and a second card is drawn. This card is replaced and a third card is drawn. What is the probability of drawing:
   (a) three hearts?
   (b) at least two hearts?
   (c) exactly one heart?

7. A bag contains 6 red marbles and 4 yellow marbles. A marble is drawn at random and **not** replaced. Two further draws are made, again without replacement. Find the probability of drawing:
   (a) three red marbles,
   (b) three yellow marbles,
   (c) no red marbles,
   (d) at least one red marble. [Hint: Use part (c).]

8. A dice has its six faces marked 0, 1, 1, 1, 6, 6. Two of these dice are thrown together and the total score is recorded. Draw a tree diagram.

   (a) How many different totals are possible?
   (b) What is the probability of obtaining a total of 7?

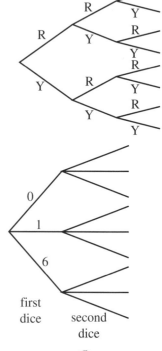

9. Louise loves the phone! When the phone rings $\frac{2}{3}$ of the calls are for Louise, $\frac{1}{4}$ are for her brother Herbert and the rest are for me.
   On Christmas Day I answered the phone twice. Find the probability that:
   (a) both calls were for Louise
   (b) one call was for me and one was for Herbert.

10. Bag A contains 3 red balls and 1 white ball.
    Bag B contains 2 red balls and 3 white balls.

    A ball is chosen at random from each bag in turn.
    Find the probability of taking:
    (a) a white ball from each bag
    (b) two balls of the same colour.

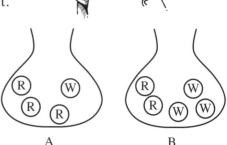

11. A teacher decides to award exam grades A, B or C by a new fairer method. Out of 20 children, three are to receive A's, five B's and the rest C's. She writes the letters A, B and C on 20 pieces of paper and invites the pupils to draw their exam result, going through the class in alphabetical order. Find the probability that:
    (a) the first three pupils all get grade 'A'
    (b) the first three pupils all get grade 'B'
    (c) the first four pupils all get grade 'B'.
    (Do not cancel down the fractions.)

Example

There are 10 boys and 12 girls in a class. Two pupils are chosen at random. What is the probability that one boy and one girl are chosen in either order?

Here two pupils are chosen. It is like choosing one pupil without replacement and then choosing a second.
Here is the tree diagram.

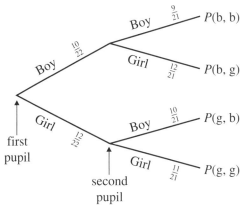

$P(b, g) + P(g, b) = \left(\frac{10}{22} \times \frac{12}{21}\right) + \left(\frac{12}{22} \times \frac{10}{21}\right)$
$= \frac{40}{77}$

Remember:
When a question says '2 balls are drawn' or '3 people are chosen', remember to draw a tree diagram **without** replacement.

Exercise 2I

1. There are 1000 components in a box of which 10 are known to be defective. Two components are selected at random. What is the probability that:
   (a) both are defective
   (b) neither are defective
   (c) just one is defective?
   (Do *not* simplify your answers.)

2. There are 10 boys and 15 girls in a class. Two children are chosen at random. What is the probability that:
   (a) both are boys
   (b) both are girls
   (c) one is a boy and one is a girl?

3. Two similar spinners are spun together.
   (a) What is the most likely total score on the two spinners?
   (b) What is the probability of obtaining this score on three successive spins of the two spinners?

4. A bag contains 3 red, 4 white and 5 green balls. Three balls are selected without replacement.
   Find the probability that the three balls chosen are:
   (a) all red   (b) all green   (c) one of each colour.
   (d) If the selection of the three balls was carried out 1100 times, how often would you expect to choose three red balls?

5. The diagram represents 15 students in a class and shows the following.
   G represents those who are girls
   S represents those who are swimmers
   F represents those who believe in Father Christmas.

   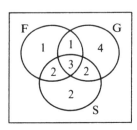

   A student is chosen at random. Find the probability that the student:
   (a) can swim
   (b) is a girl swimmer
   (c) is a boy swimmer who believes in Father Christmas.

   Two students are chosen at random. Find the probability that:
   (d) both are boys,
   (e) neither can swim.

6. There are 500 ball bearings in a box of which 100 are known to be undersize. Three ball bearings are selected at random. What is the probability that:
   (a) all three are undersize   (b) none are undersize?
   Give your answers as decimals correct to three significant figures.

7. There are 9 boys and 15 girls in a class. Three children are chosen at random. What is the probability that:
   (a) all three are boys
   (b) all three are girls
   (c) one is a boy and two are girls?
   Give your answers as fractions.

8. A box contains x milk chocolates and y plain chocolates. Two chocolates are selected at random. Find, in terms of x and y, the probability of choosing:
   (a) a milk chocolate on the first choice
   (b) two milk chocolates
   (c) one of each sort
   (d) two plain chocolates.

9. A pack of z cards contains x 'winning cards'. Two cards are selected at random. Find, in terms of x and z, the probability of choosing:
   (a) a 'winning' card on the first choice
   (b) two 'winning cards' in the two selections
   (c) exactly one 'winning' card in the pair.

10. When a golfer plays any hole, he will take 3, 4, 5, 6, or 7 strokes with probabilities of $\frac{1}{10}, \frac{1}{5}, \frac{2}{5}, \frac{1}{5}$, and $\frac{1}{10}$ respectively. He never takes more than 7 strokes. Find the probability of the following events:
    (a) scoring 4 on each of the first three holes
    (b) scoring 3, 4 and 5 (in that order) on the first three holes
    (c) scoring a total of 28 for the first four holes
    (d) scoring a total of 10 for the first three holes.

11. Assume that births are equally likely on each of the seven days of the week. Two people are selected at random. Find the probability that:
    (a) both were born on a Sunday
    (b) both were born on the same day of the week.

12. (a) A playing card is drawn from a standard pack of 52 and then replaced. A second card is drawn. What is the probability that the second card is the same as the first?
    (b) A card is drawn and *not* replaced. A second card is drawn. What is the probability that the second card is *not* the same as the first?

2.5 Conditional probability

Example
The performance of a climber is affected by the weather. When it rains, the probability that he will fall on a mountain is $\frac{1}{10}$, whereas in dry weather the probability he will fall is only $\frac{1}{50}$. In the climbing seasons, the probability of rain on any day is $\frac{1}{4}$.

What is the probability that he falls on the mountain on a day chosen at random?

Draw a tree diagram.

$$P(\text{he falls}) = P(r, f) + P(\bar{r}, f)$$
$$= \left(\frac{1}{4} \times \frac{1}{10}\right) + \left(\frac{3}{4} \times \frac{1}{50}\right) = \frac{1}{25}$$

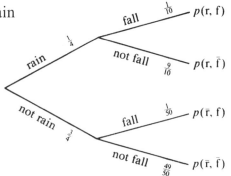

Exercise 2J

1. Bag *A* contains 3 red balls and 3 blue balls. Bag *B* contains 1 red ball and 3 blue balls. A ball is taken at random from Bag *A* and placed in Bag *B*. A ball is then chosen from Bag *B*. What is the probability that the ball taken from *B* is red?

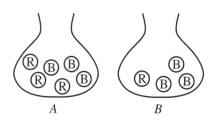

2. On a Monday or a Thursday Mr Gibson paints a 'masterpiece' with a probability of $\frac{1}{5}$. On any other day, the probability of producing a masterpiece is $\frac{1}{100}$. In common with other great painters, Gibson never knows what day it is. Find the probability that on one day chosen at random, he will in fact paint a masterpiece.

3. If a hedgehog crosses a certain road before 7.00 a.m., the probability of being run over is $\frac{1}{10}$. After 7.00 a.m., the corresponding probability is $\frac{3}{4}$. The probability of the hedgehog waking up early enough to cross before 7.00 a.m. is $\frac{4}{5}$.

   What is the probability of the following events:
   (a) the hedgehog waking up too late to reach the road before 7.00 a.m.,
   (b) the hedgehog waking up early and crossing the road in safety,
   (c) the hedgehog waking up late and crossing the road in safety,
   (d) the hedgehog waking up early and being run over,
   (e) the hedgehog crossing the road in safety?

4. In Stockholm the probability of snow falling on a day in January is 0·2. If it snows on one day, the probability of snow on the following day increases to 0·35.
   What is the probability that for two consecutive days:
   (a) it will snow on both days
   (b) it will not snow on either day
   (c) it will snow on just one of the two days?

5. Two boxes are shown containing red and white balls. One ball is selected from Box *A* and placed in Box *B*. A ball is then selected from Box *B* and placed in Box *A*.
   What is the probability that Box *A* now contains 4 red balls?

   Box *A*

   Box *B*

6. Michelle has been revising for a multiple choice exam in history but she had only had time to learn 70% of the facts being tested. If there is a question on any of the facts she has revised she will get that question right. Otherwise she will simply guess one of the five possible answers.
   (a) A question is chosen randomly from the paper. What is the probability that she will get it right?
   (b) If the paper has 50 questions what mark do you expect her to get?

7. Surgeons can operate to cure Pythagoratosis but the success rate at the first attempt is only 65%. If the first operation fails the operation can be repeated but this time the success rate is only 20%. After a second failure there is so little chance of success that surgeons will not operate again.

There is an outbreak of the disease at Roundhay College and 38 students contract the disease.
How many of these students can we expect to be saved after both operations?

Summary

Check out D2

1. You understand and can calculate relative frequency

1. You pick a card from a pack of 52.
   What is the probability that the card is a Queen?

2. You can complete and use a sample space diagram

2. You toss two dice, each marked from 1 to 6.
   By means of a sample space diagram, calculate the probability of getting a total of 5.

3. You understand mutually exclusive events, non-exclusive events and independent events.

3. What is meant by the statement:
   'Two events are mutually exclusive'?

4. You can draw and use a tree diagram.

4. You have an unbiased coin and a biased coin, which when tossed has a probability of $\frac{1}{3}$ of being a head.
   Draw a tree diagram to show the possible results when both coins are tossed and find the probability of getting two heads.

5. You can use a tree diagram to solve questions using conditional probability.

5. The probability of John oversleeping is $\frac{2}{3}$. If he oversleeps the probability that he is late for work is $\frac{3}{4}$. If he does not oversleep the probability that he is late is $\frac{1}{4}$.

Find the probability that John is late for work.

Revision exercise D2

1. Past records show that in a typical week in September it rains on 4 days out of 7.

   (a) Copy the tree diagram and fill in the probabilities for the first and second Saturdays in September.

   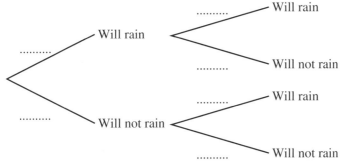

   (b) Calculate the probability that it rains on both Saturdays.
   (c) Calculate the probability that it rains on at least one of the Saturdays. [SEG]

2. Shaz has ten one pound coins.
   Six have a thistle design and four have a leek design.

   She chooses a one pound coin at random.
   If the first coin has a thistle design she replaces it, and chooses again.
   If the first coin has a leek design she does not replace it, but chooses again.
   What is the probability that the second coin has a leek design? [AQA]

3. A manufacturer tested a new cat food on groups of cats. Each cat was given two types of food. One of these was the new food, the other was a different make.
An observer recorded whether each cat ate the new food first (✔) or the other make first (✘).

The results for the first group of 10 cats are as follows.

✔ ✔ ✘ ✔ ✘ ✔ ✔ ✘ ✔ ✔

(a) Use these ten results to calculate an estimate of the probability that a cat will eat the new food first (✔).

The experiment was carried out on three more groups of 10 cats. The results are as follows.

Second group of 10 cats   5 ate the new food first
Third group of 10 cats   7 ate the new food first
Fourth group of 10 cats   4 ate the new food first

(b) Out of 100 cats, how many would you expect to eat the new food first?   [SEG]

4. Sean has a set of 4 cards numbered 1, 2, 3 and 4.
He picks a card at random and does not replace it.
He then picks another card at random from the three remaining cards.

(a) What is the probability that **neither** of the two cards that Sean picked was the 4?

(b) What is the probability that the numbers on the two cards that Sean picked add up to 4?

Sean now picks a third card from the remaining two cards.

(c) Calculate the probability that the numbers on the first two cards add up to the number on third card.

You **must** show all your working.   [SEG]

5. Sanjit enters a competition where he takes part in two events.
The probability that he will win the running event is 0·3.
The probability that he will win the cycling event is 0·6.
The two events are independent.

(a) Calculate the probability that Sanjit will win both events.

(b) Calculate the probability that Sanjit will win the running event and lose the cycling event.

(c) Calculate the probability that Sanjit will win exactly one of the two events.   [SEG]

6. Two fair spinners are numbered as shown.
Both spinners are spun once.

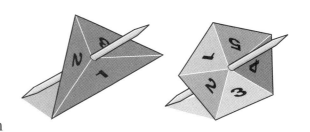

(a) Calculate the probability that both spinners land on a 3.

(b) Calculate the probability that both spinners land on an odd number.

(c) Calculate the probability that neither spinner lands on an odd number. [SEG]

7. The table shows the probability for the delivery time of letters posted first class.

Delivery time (days)	1	2	3 or more
Probability	0·7	0·2	0·1

(a) 100 letters are posted first class.
How many will be delivered in 1 or 2 days?

(b) Two letters are posted first class in different towns.
What is the probability that only one will be delivered in one day? [SEG]

8. A bag contains 3 red balls, 2 yellow balls and 1 blue ball, identical except for their colours.

Harry picks two balls from the bag at random and without replacement.

(a) What is the probability that Harry picks at least one yellow ball?

A second bag also contains 3 red balls, 2 yellow balls and 1 blue ball, identical except for their colours.

Sandi picks one ball from the bag at random.

(b) Given that Sandi does not pick a yellow ball, what is the probability that she picks a red ball?

Sandi does not replace the ball. She then picks another ball from the bag at random.

(c) Given that neither of the balls is yellow, what is the probability that both are red? [SEG]

9. A bag contains 20 coins.
   There are 6 gold coins and the rest are silver.

   A coin is taken at random from the bag.
   The type of coin is recorded and the coin is then **returned** to the bag.
   A second coin is then taken at random from the bag.

   (a) The tree diagram shows all the ways in which two coins can be taken from the bag.
   Write the probabilities on a copy of the diagram.

   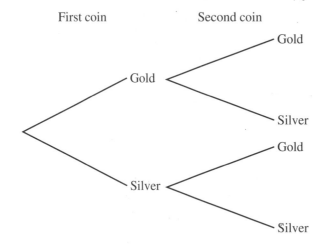

   (b) Use your tree diagram to calculate the probability that one coin is gold and one coin is silver. [SEG]

10. An office has two photocopiers, A and B.
    On any one day,

    the probability that A is working is 0·8,
    the probability that B is working is 0·9.

    (a) Calculate the probability that, on any one day, both photocopiers will be working.

    (b) Calculate the probability that, on any one day, only one of the photocopiers will be working. [SEG]

Module 1 Practice Calculator Test

1. One afternoon a survey was taken of 100 customers at a supermarket.
   The time spent queuing at the checkout was recorded.
   The results are shown below.

Time (t min.)	Number of customers
$0 < t \leqslant 4$	41
$4 < t \leqslant 8$	44
$8 < t \leqslant 12$	12
$12 < t \leqslant 16$	2
$16 < t \leqslant 20$	1

   Calculate an estimate of the mean time these customers had to queue.   **(4 marks)**

2. The table shows the amounts on 10 of Meera's gas bills.

Date of Bill	Mar 1999	Jun 1999	Sept 1999	Dec 1999	Mar 2000	Jun 2000	Sept 2000	Dec 2000	Mar 2001	Jun 2001
Amount (£)	17	21	34	40	25	29	40	42	30	32

   The time series graph shows Meera's original data.

   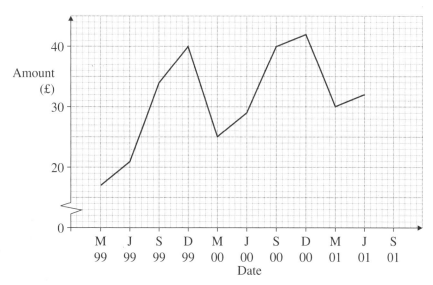

The values of the four-point moving average for the data are

28, 30, 32, 33.5, 34, 35.25, 36

(a) Plot the values of the moving average on the graph.

(b) Use your graph to estimate Meera's gas bill for September 2001. **(6 marks)** [AQA]

3. Louise believes that $\frac{1}{5}$ of all cars are silver in colour.

   (a) Draw a fully labelled probability tree diagram showing all the possibilities of whether each of the next two cars which pass Louise are silver.

   (b) Calculate the probability that **neither** of the next two cars are silver. **(5 marks)**

4. Rob is an estate agent. He currently has 160 houses for sale.

   (a) Describe how you would select a random sample of 25 of the houses.

   Rob has 31 detached houses, 69 semi-detached houses, 40 terraced houses and 20 other house for sale.

   (b) Rob wants to select a stratified sample of 25 of the houses.

   Calculate the number of houses of each type that should be chosen. **(5 marks)** [SEG]

Module 1   Practice Non-Calculator Test

1. The weights of 80 bags of rice are measured. The table summarises the results.

Minimum	480 g
Lower quartile	500 g
Median	540 g
Upper quartile	620 g
Maximum	720 g

   (a) draw a box plot to show this information.

(b) Write down the interquartile range for these data.
(c) How many bags weigh
  (i) less than 480 g
  (ii) less than 500 g? **(6 marks)** [AQA]

2. The table shows the number of cars that crossed a bridge each hour.
   (a) How many cars crossed the bridge in total?
   (b) Draw a cumulative frequency graph for these data.
   (c) Use your graph to estimate the time when half the total number of cars have crossed the bridge.
   (d) Use your graph to estimate how many cars crossed the bridge before 15 50.

Time, t	Number of cars
$10\,00 \leqslant t < 11\,00$	15
$11\,00 \leqslant t < 12\,00$	20
$12\,00 \leqslant t < 13\,00$	25
$13\,00 \leqslant t < 14\,00$	15
$14\,00 \leqslant t < 15\,00$	10
$15\,00 \leqslant t < 16\,00$	10
$16\,00 \leqslant t < 17\,00$	5

(6 marks) [SEG]

3. There are 8 pens in a box, of which 5 are red and 3 are blue.
   Karen selects 2 pens at random.
   What is the probability that Karen has 1 red and 1 blue pen?
   (4 marks)

4. The histograms of two distributions, A and B, are shown below.

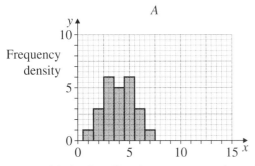
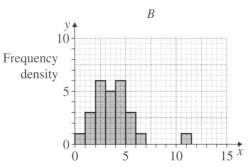

(a) Which distribution, A or B, has the greater mean? Explain your answer.
(b) (i) Which average, the mean, median or mode, is the **most** appropriate as a representative value for distribution A? Explain your answer.
   (ii) Give **two** reasons why the median is the most appropriate average for distribution B.

(4 marks) [AQA]

N1 Number 1

This unit will show you how to:

- Use numerical terms
- Calculate with decimals
- Calculate with fractions
- Use approximation and estimation
- Calculate ratios and proportions
- Calculate with directed numbers

Before you start:

You should know how to...	Check in N1
1. Multiply and divide by 10, 100, 1000. For example, $5 \cdot 34 \times 1000 = 5340$	**1.** (a) Multiply $2 \cdot 41$ by: (i) 100 (ii) 1000. (b) Divide $3 \cdot 1$ by: (i) 100 (ii) 1000.
2. Give answers to an appropriate degree of accuracy.	**2.** Give, to the nearest penny: (a) £74 ÷ 81 (b) £8 ÷ 75

1.1 Properties of numbers

- **Factors**   Any number which divides exactly into 8 is a **factor** of 8. The factors of 8 are 1, 2, 4, 8.
- **Multiples**   Any number in the 8-times table is a **multiple** of 8. The first five multiples of 8 are 8, 16, 24, 32, 40.
- **Prime**   A **prime** number has just two different factors: 1 and itself. The number 1 is **not** prime. [It does not have two different factors.] The first five prime numbers are 2, 3, 5, 7, 11.
- **Prime factor**   The factors of 8 are 1, 2, 4, 8. The only **prime factor** of 8 is 2. It is the only prime number which is a factor of 8.
- **Square numbers**   A square number can be written in the form $n \times n$, where n is an integer. [Examples: 4, 9, 100.]
- **Cube numbers**   A cube number can be written in the form $n \times n \times n$, where n is an integer. [Examples: 1, 8, 27, 64.]

Example

(a) Express 6930 as a product of primes.

$2 \overline{)69^13^10}$ (divide by 2)
$3 \overline{)34^16^15}$ (divide by 3)
$3 \overline{)11^25^15}$ (divide by 3)
$5 \overline{)\ \ 38^35}$ (divide by 5)
$7 \overline{)\ \ \ \ 77}$ (divide by 7)
$\qquad\qquad 11$ (stop because 11 is prime)

∴ $6930 = 2 \times 3 \times 3 \times 5 \times 7 \times 11$

(b) In prime factors,

$90 = 2 \times 3 \times 3 \times 5$

$42 = 2 \times 3 \times 7$

By inspection we see that the **Highest Common Factor** (HCF) of 90 and 42 is 6 [i.e. 2×3]

Exercise 1A

1. Write down all the factors of the following numbers.
   (a) 6 (b) 15 (c) 18 (d) 21 (e) 40

2. Write down all the prime numbers less than 20.

3. Write down two prime numbers which add up to another prime number. Do this in **two** ways.

4. Which of the following are prime numbers?
   3, 11, 15, 19, 21, 23, 27, 29, 31, 37, 39, 47, 51, 59, 61, 67, 72, 73, 87, 99

5. Express each of the following numbers as a product of primes. See the example above.
   (a) 600   (b) 693   (c) 2464
   (d) 3510   (e) 4000   (f) 22 540

6. Copy the grid and then write the numbers 1 to 9, one in each box, so that all the numbers satisfy the conditions for both the row and the column.

	Prime number	Multiple of 3	Factor of 16
Number greater than 5	7	9	8
Odd number	5	3	1
Even number	2	6	4

7. Find each of the mystery numbers below.
   (a) I am an odd number and a prime number. I am a factor of 14.
   (b) I am a two-digit multiple of 50.
   (c) I am one less than a prime number which is even.
   (d) I am odd, greater than one and a factor of both 20 and 30.

8. The first few multiples of 4 are   4, 8, 12, 16, ⓴, 24, 28 …
   The first few multiples of 5 are   5, 10, 15, ⓴, 25, 30, 35 …
   The **Least Common Multiple** (LCM) of 4 and 5 is 20.
   The LCM is usually known as the lowest common multiple.
   It is the lowest number which is in both lists.
   (a) Write down the first four multiples of 6.
   (b) Write down the first four multiples of 9.
   (c) Write down the LCM of 6 and 9.

9. Find the LCM of:
   (a) 6 and 15   (b) 20 and 30   (c) 6 and 10.

10. Write down the first 15 square numbers. **Learn** these numbers.

11. Write down the first 5 cube numbers.

12. Work out which of the following are prime numbers.
    (a) 91   (b) 101   (c) 143   (d) 151   (e) 293
    [Hint: Divide by the prime numbers 2, 3, 5, 7, 11 etc.]

13. Put three different numbers in the circles so that when you add the numbers at the end of each line you always get a square number.

    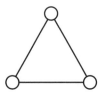

14. (a) Express 1008 and 840 as products of their prime factors.
    (b) Find the HCF (Highest Common Factor) of 1008 and 840.
    (c) Find the smallest number which can be multiplied by 1008 to give a square number.

15. (a) Express 19 800 and 12 870 as products of their prime factors.
    (b) Find the HCF of 19 800 and 12 870.
    (c) Find the smallest number which can be multiplied by 19 800 to give a square number.

16. The first of three consecutive numbers is *n*. Find the smallest value of *n* if the three numbers are all non-prime.

17. The odd numbers are written here in a special way.

    Row 1   1
    Row 2   3   5
    Row 3   7   9   11
    Row 4   13   15   17   19

    (a) (i)   Write down the sum of each of the first four rows.
        (ii) What is the special name of the numbers that you have written down?
        (iii) What is the sum of the numbers in the hundredth row?
    (b) (i)   Write down the sum of all the numbers in the first two rows, in the first three rows, in the first four rows.
        (ii) What is the special name of the numbers that you have written down?
        (iii) What is the sum of all the numbers in the first ten rows?

18. Make six prime numbers using digits 1 to 9 once each.

19. Consider the expression $5^n + 3$.
    Substitute six different values for *n* and find out if the result is a multiple of 4 each time.

20. Substitute six pairs of values of *a* and *n* in the expression $a^n - a$.
    Is the result always a multiple of *n*?

Operations and inverses

Operation	Inverse	Example	
Add n	Subtract n	$9 + 7 = 16,$	$16 - 7 = 9$
Subtract n	Add n	$12 - 5 = 7,$	$7 + 5 = 12$
Multiply by n	Divide by n	$20 \times 4 = 80,$	$80 \div 4 = 20$
Divide by n	Multiply by n	$12 \div 3 = 4,$	$4 \times 3 = 12$
Square	Square root	$7^2 = 49,$	$\sqrt{49} = 7$
Raise to power n	Find nth root $\left[\text{raise to the power } \frac{1}{n}\right]$	$10^4 = 10000,$	$\sqrt[4]{10000} = 10$

[The nth root is discussed more fully on page 143.]

Also: The 'reciprocal' of n is $\frac{1}{n}$. $(n \neq 0)$

The reciprocal of $\frac{1}{n}$ is n.

Note that zero has no reciprocal because division by zero is not defined.
The reciprocal is sometimes called the 'multiplicative inverse'.

Exercise 1B

In the flow charts, the boxes A, B, C and D each contain a single mathematical operation (like $+5$, $\times 4$, -15, $\div 2$).

Look at flow charts (i) and (ii) together and work out what is the same operation which will replace A. Complete the flow chart by replacing B, C and D.

Now copy and complete each flow chart on the right, using the same operations.

1. (i) $2 \to \boxed{A} \to 17 \to \boxed{B} \to 34 \to \boxed{C} \to 12 \to \boxed{D} \to 3$

(ii) $4 \to \boxed{A} \to 19 \to \boxed{B} \to 38 \to \boxed{C} \to 16 \to \boxed{D} \to 4$

(a) $7 \to \boxed{A} \to ? \to \boxed{B} \to ? \to \boxed{C} \to ? \to \boxed{D} \to ?$

(b) $10 \to \boxed{A} \to ? \to \boxed{B} \to ? \to \boxed{C} \to ? \to \boxed{D} \to ?$

(c) $? \to \boxed{A} \to ? \to \boxed{B} \to 62 \to \boxed{C} \to ? \to \boxed{D} \to ?$

(d) $? \to \boxed{A} \to 15\frac{1}{2} \to \boxed{B} \to ? \to \boxed{C} \to ? \to \boxed{D} \to ?$

2.
(i) $\to 2 \to [A] \to 4 \to [B] \to 12 \to [C] \to 2 \to [D] \to 1 \to$

(ii) $\to 3 \to [A] \to 9 \to [B] \to 27 \to [C] \to 17 \to [D] \to 8\frac{1}{2} \to$

(a) $\to 4 \to [A] \to 16 \to [B] \to ? \to [C] \to ? \to [D] \to ? \to$

(b) $\to 5 \to [A] \to ? \to [B] \to ? \to [C] \to ? \to [D] \to ? \to$

(c) $\to ? \to [A] \to ? \to [B] \to 108 \to [C] \to ? \to [D] \to ? \to$

(d) $\to ? \to [A] \to ? \to [B] \to ? \to [C] \to 182 \to [D] \to ? \to$

In questions **3**, **4** and **5** find the operations A, B, C, D.

3. (i) $\to 4 \to [A] \to 16 \to [B] \to 4 \to [C] \to {}^-6 \to [D] \to 12 \to$

(ii) $\to 9 \to [A] \to 36 \to [B] \to 6 \to [C] \to {}^-4 \to [D] \to 8 \to$

4. (i) $\to 5 \to [A] \to \frac{1}{5} \to [B] \to 1\cdot 2 \to [C] \to 1\cdot 44 \to [D] \to 0\cdot 48 \to$

(ii) $\to \frac{1}{2} \to [A] \to 2 \to [B] \to 3 \to [C] \to 9 \to [D] \to 3 \to$

5. (i) $\to {}^-1 \to [A] \to 2 \to [B] \to 8 \to [C] \to {}^-4 \to [D] \to 96 \to$

(ii) $\to 7 \to [A] \to 10 \to [B] \to 1000 \to [C] \to {}^-500 \to [D] \to {}^-400 \to$

1.2 Arithmetic without a calculator

Decimals

Example

(a) $0\cdot 032 \times 100$
$= 3\cdot 2$

(b) $51 \div 1000$
$= 0\cdot 051$

(c) $26\cdot 1 \div 5$
$$\begin{array}{r} 5\cdot 22 \\ 5\overline{)26\cdot {}^11{}^10} \end{array}$$ ← Add a zero
$26\cdot 1 \div 5 = 5\cdot 22$

(d) $\begin{array}{r} 2\cdot 5 \\ \times\ 0\cdot 3 \\ \hline 0\cdot 75 \end{array}$
↖ two figures after the decimal points
↖ two figures after the decimal point

(e) $\times 10 \left(\begin{array}{c} 8\cdot 64 \div 0\cdot 2 \\ 86\cdot 4 \div 2 \\ = 43\cdot 2 \end{array} \right) \times 10$

(f) $\times 1000 \left(\begin{array}{c} 6\cdot 9 \div 0\cdot 003 \\ 6900 \div 3 \\ = 2300 \end{array} \right) \times 1000$

Exercise 1C

Evaluate the following:
1. $7\cdot 6 + 0\cdot 31$
2. $15 + 7\cdot 22$
3. $7\cdot 004 + 0\cdot 368$
4. $0\cdot 06 + 0\cdot 006$
5. $4\cdot 2 + 42 + 420$
6. $3\cdot 84 - 2\cdot 62$
7. $11\cdot 4 - 9\cdot 73$
8. $4\cdot 61 - 3$
9. $17 - 0\cdot 37$

10. $8.7 + 19.2 - 3.8$
11. $25 - 7.8 + 9.5$
12. $3.6 - 8.74 + 9$
13. $20.4 - 20.399$
14. 2.6×0.6
15. 0.72×0.04
16. 27.2×0.08
17. 0.1×0.2
18. $(0.01)^2$

19. Sharon is paid a basic weekly wage of £85 and then a further 30p for each item completed. How many items must be completed in a week when she earns a total of £191·50?

20. James has the same number of 10p and 50p coins. The total value is £12. How many of each coin does he have?

21. A pile of 10p coins is 25·6 cm high. Calculate the total **value** of the coins if each coin is 0·2 cm thick.

25.6 cm

Work out:

22. $0.34 \times 100\,000$
23. $3.6 \div 0.2$
24. $56.9 \div 100$
25. $0.1401 \div 0.06$
26. $3.24 \div 0.002$
27. $0.968 \div 0.11$
28. $600 \div 0.5$
29. $0.007 \div 4$
30. $2640 \div 200$
31. $1100 \div 5.5$
32. $(11 + 2.4) \times 0.06$
33. $(0.4)^2 \div 0.2$
34. $77 \div 1000$
35. $(0.3)^2 \div 100$
36. $(0.1)^4 \div 0.01$

37. (a) The calculation $\dfrac{480 \times 55}{22}$ is easier to perform after cancelling.

$$\dfrac{480 \times \overset{5}{\cancel{55}}}{\underset{2}{\cancel{22}}} = \dfrac{\overset{240}{\cancel{480}} \times 5}{\underset{1}{\cancel{2}}} = 1200$$

(b) Use cancelling to help work out the following.

(i) $\dfrac{180 \times 4}{36}$ (ii) $\dfrac{92 \times 4.6}{2.3}$ (iii) $\dfrac{35 \times 0.81}{45}$

(iv) $\dfrac{72 \times 3000 \times 24}{150 \times 64}$ (v) $\dfrac{63 \times 600 \times 0.2}{360 \times 7}$

38. Fill in the missing numbers so that the answer is always 0·7.

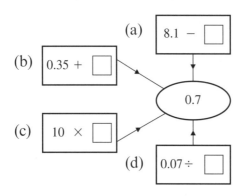

39. When you multiply by a number greater than 1 you make it bigger.

so $5.3 \times 1.03 > 5.3$ and $6.75 \times 0.89 < 6.75$

When you divide by a number greater than 1 you make it smaller.

so $8.92 \div 1.13 < 8.92$ and $11.2 \div 0.73 > 11.2$

State whether these are true or false:
(a) $3.72 \times 1.3 > 3.72$
(b) $253 \times 0.91 < 253$
(c) $0.92 \times 1.04 > 0.92$
(d) $8.5 \div 1.4 > 8.5$
(e) $113 \div 0.73 < 113$
(f) $17.4 \div 2.2 < 17.4$
(g) $0.73 \times 0.73 < 0.73$
(h) $2511 \div 0.042 < 2511$
(i) $614 \times 0.993 < 614$

40. Write down each statement with either $>$ or $<$ in place of the box.

(a) $17.83 \div 0.07 \;\square\; 17.83$
(b) $9.04 \times 1.13 \;\square\; 9.04$

(c) $207 \times 0.96 \;\square\; 207$
(d) $308 \div 2.8 \;\square\; 308$

(e) $1.19 \div 0.06 \;\square\; 1.19$
(f) $0.74 \div 0.95 \;\square\; 0.74$

Long multiplication and division
Example
To work out 327×53 we will use the fact that $327 \times 53 = (327 \times 50) + (327 \times 3)$.

Set out the working like this:

```
    327
     53 ×
  16350  → This is 327 × 50
    981  → This is 327 × 3
  17331  → This is 327 × 53
```

Here is another example:

```
    541
     84 ×
  43280  → This is 541 × 80
   2164  → This is 541 × 4
  45444  → This is 541 × 84
```

Exercise 1D

Work out:
1. 35 × 23
2. 27 × 17
3. 26 × 25
4. 31 × 43
5. 45 × 61
6. 52 × 24
7. 323 × 14
8. 416 × 73
9. 504 × 56
10. 306 × 28
11. 624 × 75
12. 839 × 79
13. 694 × 83
14. 973 × 92
15. 415 × 235

With ordinary 'short' division, we divide and find remainders. The method for 'long' division is really the same but we set it out so that the remainders are easier to find.

Example

Work out 736 ÷ 32.

```
       23
32 ) 736
     64 ↓
      96
      96
       0
```

(a) 32 into 73 goes 2 times
(b) 2 × 32 = 64
(c) 73 − 64 = 9
(d) 'bring down' 6
(e) 32 into 96 goes 3 times

Exercise 1E

Work out:
1. 672 ÷ 21
2. 425 ÷ 17
3. 576 ÷ 32
4. 247 ÷ 19
5. 875 ÷ 25
6. 574 ÷ 26
7. 806 ÷ 34
8. 748 ÷ 41
9. 666 ÷ 24
10. 707 ÷ 52
11. 951 ÷ 27
12. 803 ÷ 31
13. 2917 ÷ 45
14. 2735 ÷ 18
15. 56 274 ÷ 19

Exercise 1F

1. A shop owner buys 56 tins of paint at 84p each. How much does he spend altogether?

2. Eggs are packed eighteen to a box.

How many boxes are needed for 828 eggs?

3. On average a man smokes 146 cigarettes a week. How many does he smoke in a year?

4. Sally wants to buy as many 23p stamps as possible. She has £5 to buy them. How many can she buy and how much change is left?

5. How many 49-seater coaches will be needed for a school trip for a party of 366?

6. An office building has 24 windows on each of eight floors. A window cleaner charges 42p for each window. How much is he paid for the whole building?

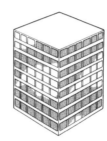

7. A lottery prize of £238 million was won by a syndicate of 17 people who shared the prize equally between them. How much did each person receive?

8. It costs £7905 to hire a plane for a day. A trip is organised for 93 people. How much does each person pay?

9. The headmaster of a school discovers an oil well in the school playground. As is the custom in such cases, he receives all the money from the oil. The oil comes out of the well at a rate of £15 for every minute of the day and night.
How much does the headmaster receive in a 24-hour day?

10. Tins of peaches are packed 24 to a box. How many boxes are needed for 1285 tins?

11. Work out:
    (a) $2 \cdot 1 \times 3 \cdot 2$ (b) $0 \cdot 65 \times 4 \cdot 3$ (c) $0 \cdot 71 \times 5 \cdot 6$

12. Work out the following, giving your answer correct to 1 decimal place.
    (a) $8 \cdot 32 \div 1 \cdot 7$ (b) $0 \cdot 7974 \div 0 \cdot 23$ (c) $1 \cdot 4 \div 0 \cdot 71$

13. Susie breeds gerbils for pet shops. The number of gerbils trebles each year. Susie has 40 gerbils at the end of year 1.
    (a) Copy and complete the table below.

End of year	1	2	3	4	5
Number of gerbils	40				

    (b) A gerbil cage can hold 16 gerbils. Work out the minimum number of cages needed by the end of year 6.

1.3 Fractions

Fraction arithmetic

- Common fractions are added or subtracted from one another directly only when they have a common denominator.

Example 1

(a) $\frac{3}{4} + \frac{2}{5} = \frac{15}{20} + \frac{8}{20}$
$= \frac{23}{20}$
$= 1\frac{3}{20}$

(b) $\frac{3}{7} - \frac{1}{3} = \frac{9}{21} - \frac{7}{21}$
$= \frac{2}{21}$

so $2\frac{3}{8} - 1\frac{5}{12} = \frac{19}{8} - \frac{17}{12}$
$= \frac{57}{24} - \frac{34}{24}$
$= \frac{23}{24}$

- Multiplying and dividing fractions is made easier by 'cancelling down'.

Example 2

(d) $\frac{2}{5} \times \frac{6}{7} = \frac{12}{35}$

(no cancelling here)

(e) $\frac{\cancel{12}^3}{13} \times \frac{5}{\cancel{8}_2} = \frac{15}{26}$

(cancel down)

(f) $2\frac{2}{5} \div 6 = \frac{12}{5} \div 6$
$= \frac{\cancel{12}^2}{5} \times \frac{1}{\cancel{6}_1} = \frac{2}{5}$

- Notice in part (f): $2\frac{2}{5}$ is written as $\frac{12}{5}$;

  'Dividing by 6' is the same as 'multiplying by $\frac{1}{6}$'.

Exercise 1G

1. Copy and complete.

(a) $\frac{1}{4} + \frac{1}{8}$
$= \frac{\square}{8} + \frac{1}{8} =$

(b) $\frac{1}{2} + \frac{2}{5}$
$= \frac{\square}{10} + \frac{\square}{10} =$

(c) $\frac{2}{5} + \frac{1}{3}$
$= \frac{\square}{15} + \frac{\square}{15} =$

2. Work out:
   (a) $\frac{1}{2}+\frac{1}{3}$
   (b) $\frac{1}{4}+\frac{1}{6}$
   (c) $\frac{1}{3}+\frac{1}{4}$
   (d) $\frac{1}{4}+\frac{1}{5}$
   (e) $\frac{3}{4}+\frac{4}{5}$
   (f) $\frac{5}{6}+\frac{2}{3}$
   (g) $1\frac{1}{4}+\frac{1}{3}$
   (h) $2\frac{1}{3}+1\frac{1}{4}$

3. Work out:
   (a) $\frac{2}{3}-\frac{1}{2}$
   (b) $\frac{3}{4}-\frac{1}{3}$
   (c) $\frac{4}{5}-\frac{1}{3}$
   (d) $\frac{1}{2}-\frac{2}{5}$
   (e) $1\frac{3}{4}-\frac{2}{3}$
   (f) $1\frac{2}{3}-1\frac{1}{2}$

4. Copy and complete the addition square:

+			$\frac{1}{3}$
$\frac{1}{2}$	$\frac{3}{4}$		$\frac{5}{6}$
$\frac{1}{8}$	$\frac{3}{8}$	$\frac{1}{2}$	
$\frac{1}{5}$			

5. Work out these multiplications.
   (a) $\frac{2}{3}\times\frac{4}{5}$
   (b) $\frac{1}{7}\times\frac{5}{6}$
   (c) $\frac{5}{8}\times\frac{12}{13}$
   (d) $1\frac{3}{4}\times\frac{2}{3}$
   (e) $2\frac{1}{2}\times\frac{1}{4}$
   (f) $3\frac{2}{3}\times 2\frac{1}{2}$
   (g) $\frac{2}{3}$ of $\frac{2}{5}$
   (h) $\frac{3}{7}$ of $\frac{1}{2}$

6. Work out these divisions.
   (a) $\frac{1}{3}\div\frac{4}{5}$
   (b) $\frac{3}{4}\div\frac{1}{6}$
   (c) $\frac{5}{6}\div\frac{1}{2}$
   (d) $\frac{5}{8}\div 2$
   (e) $\frac{3}{8}\div\frac{1}{5}$
   (f) $1\frac{3}{4}\div\frac{2}{3}$
   (g) $3\frac{1}{2}\div 2\frac{3}{5}$
   (h) $6\div\frac{3}{4}$

7. A rubber ball is dropped from a height of 300 cm. After each bounce, the ball rises to $\frac{4}{5}$ of its previous height. How high, to the nearest cm, will it rise after the fourth bounce?

8. Steve Braindead spends his income as follows:
   (a) $\frac{2}{5}$ of his income goes in tax,
   (b) $\frac{2}{3}$ of what is left goes on food, rent and transport,
   (c) he spends the rest on cigarettes, beer and betting.
   What fraction of his income is spent on cigarettes, beer and betting?

9. Given that $a = \frac{3}{4}$, $b = \frac{2}{5}$, $c = \frac{1}{3}$, work out:

   (a) a^2 (b) $a - b$ (c) $\frac{b}{c}$ (d) $\frac{1}{b} - \frac{1}{a}$

   (e) abc (f) $2b - c$ (g) $\frac{1}{a+b}$ (h) $\frac{b}{ac}$

10. In the equation below all the asterisks stand for the same number. What is the number?

    $$\left[\frac{*}{*} - \frac{*}{6} = \frac{*}{30}\right]$$

11. Find the value of n if $\left(1\frac{1}{3}\right)^n - \left(1\frac{1}{3}\right) = \frac{28}{27}$

12. A formula used by opticians is

    $$\frac{1}{f} = \frac{1}{u} + \frac{1}{v}$$

    Given that $u = 3$ and $v = 5\frac{1}{2}$ find the exact value of f.

13. In each equation find two **positive** integers a and b.

    (a) $\frac{a}{4} + \frac{b}{3} = \frac{11}{12}$ (b) $\frac{a}{6} + \frac{b}{21} = \frac{17}{42}$ (c) $\frac{a}{12} + \frac{b}{18} = \frac{29}{36}$

14. Copy and complete the multiplication square.

×	$\frac{2}{5}$		
$\frac{1}{3}$		$\frac{1}{4}$	
		$\frac{3}{8}$	
$\frac{1}{4}$			$\frac{1}{6}$

15. A cylinder is $\frac{1}{4}$ full of water. After 60 ml of water is added the cylinder is $\frac{2}{3}$ full. Calculate the total volume of the cylinder.

18. (a) Here is a sequence:

$$\frac{1}{2} + \frac{1}{4} = \frac{3}{4}$$

$$\frac{1}{2} + \frac{1}{4} + \frac{1}{8} = \frac{7}{8}$$

$$\frac{1}{2} + \frac{1}{4} + \frac{1}{8} + \frac{1}{16} = \square$$

$$\frac{1}{2} + \frac{1}{4} + \frac{1}{8} + \frac{1}{16} + \frac{1}{32} = \square$$

Fill in the missing answers.

(b) Write down the answer to the nth line of the sequence.

1.4 Estimating

It is always sensible to check that the answer to a calculation is 'about the right size'.

Example

Estimate the value of $\dfrac{57 \cdot 2 \times 110}{2 \cdot 146 \times 46 \cdot 9}$, correct to 1 significant figure.

Round each number to 1 significant figure: $\dfrac{60 \times 100}{2 \times 50}$

$$= \frac{6000}{100}$$

$$= 60$$

On a calculator the value is 62·52 (to 4 significant figures).

62·52 is close to the estimate of 60.

Exercise 1H

In this exercise there are 25 questions, each followed by three possible answers. In each case only one answer is correct. Write down each question and decide (by estimating) which answer is correct.

	Question	Answer A	Answer B	Answer C
1.	7.2×9.8	52.16	98.36	70.56
2.	2.03×58.6	118.958	87.848	141.116
3.	23.4×19.3	213.32	301.52	451.62
4.	313×107.6	3642.8	4281.8	33678.8
5.	6.3×0.098	0.6174	0.0622	5.98
6.	1200×0.89	722	1068	131
7.	0.21×93	41.23	9.03	19.53
8.	88.8×213	18914.4	1693.4	1965.4
9.	0.04×968	38.72	18.52	95.12
10.	0.11×0.089	0.1069	0.0959	0.00979
11.	$13.92 \div 5.8$	0.52	4.2	2.4
12.	$105.6 \div 9.6$	8.9	11	15
13.	$8405 \div 205$	4.6	402	41
14.	$881.1 \div 99$	4.5	8.9	88
15.	$4.183 \div 0.89$	4.7	48	51
16.	$6.72 \div 0.12$	6.32	21.2	56
17.	$20.301 \div 1010$	0.0201	0.211	0.0021
18.	$0.28896 \div 0.0096$	312	102.1	30.1
19.	$0.143 \div 0.11$	2.3	1.3	11.4
20.	$159.65 \div 515$	0.11	3.61	0.31
21.	$(5.6 - 0.21) \times 39$	389.21	210.21	20.51
22.	$\dfrac{17.5 \times 42}{2.5}$	294	504	86
23.	$(906 + 4.1) \times 0.31$	473.21	282.131	29.561
24.	$\dfrac{543 + 472}{18.1 + 10.9}$	65	35	85
25.	$\dfrac{112.2 \times 75.9}{6.9 \times 5.1}$	242	20.4	25.2

Exercise 1I

1. For a wedding the caterers provided food at £39·75 per head. There were 207 guests at the wedding. Estimate the total cost of the food.

2. 985 people share the cost of hiring an ice rink. About how much does each person pay if the total cost is £6017?

3. On a charity walk, Susie walked 31 miles in 11 hours 7 minutes. Estimate the number of minutes it took to walk one mile.

In Questions **4** to **9**, estimate which answer is closest to the actual answer.

4. The height of a double-decker bus:
   A 3 m   B 6 m   C 10 m

5. The mass of a £1 coin:
   A 1 g   B 10 g   C 100 g

6. The volume of your classroom:
   A 20 m³   B 200 m³   C 2000 m³

7. The top speed of a Grand Prix racing car:
   A 600 km/h   B 80 km/h   C 300 km/h

8. The thickness of one page in this book:
   A 0·01 cm   B 0·001 cm   C 0·0001 cm

9. The number of cars in a traffic jam 10 km long on a 3-lane motorway:
   A 3000   B 30 000   C 300 000
   [Assume each car takes up 10 m of road.]

In Questions **10** and **11** there are six calculations and six answers. Write down each calculation and insert the correct answer from the list given. Use estimation.

10. (a) 8·9 × 10·1   (b) 7·98 ÷ 1·9   (c) 112 × 3·2
    (d) 11·6 + 47·2   (e) 2·82 ÷ 9·4   (f) 262 ÷ 100
    Answers: 2·62, 58·8, 0·3, 89·89, 358·4, 4·2

11. (a) 49·5 ÷ 11 (b) 21 × 22 (c) 9·1 × 104
    (d) 86 − 8·2 (e) 2·4 ÷ 12 (f) 651 ÷ 31
    Answers: 21, 946·4, 0·2, 4·5, 462, 77·8

12. Mr Johnson, the famous maths teacher, has won the pools. He decides to give a rather unusual prize for the person who comes top in his next maths test.
    The prize winner receives his or her own weight in coins and they can choose to have either 1p, 2p, 5p, 10p, 20p, 50p or £1 coins. All the coins must be the same.

Approximate masses	
1p	3·6 g
2p	7·2 g
5p	3·2 g
10p	6·5 g
20p	5·0 g
50p	7·5 g
£1	9·0 g

    Shabeza is the winner and she weighs 47 kg.
    Estimate the highest value of her prize.

13. The largest tree in the world has a diameter of 11 m.
    Estimate the number of 'average' 15-year-olds required to circle the tree so that they form an unbroken chain.

14. There are about 7000 cinemas in the U.K. and every day about 300 people visit each one.
    The population of the U.K. is about 60 million.
    A film magazine report said:
    Is the magazine report fair?
    Show the working you did to decide.

    'Over 3% of British people go to the cinema everyday'.

15. The surface area, A, of an object is given by the formula
    $$A = \pi r\left[10r + \sqrt{(r^2 + d^2)}\right].$$
    Estimate the value of A correct to one significant figure if $r = 1·08$ and $d = 5·87$.

Approximations

(a) Reminders: 35·2 | 6 = 35·3 to 3 significant figures
       ↑
      '5 or more'

    0·041 | 2 = 0·041 to 2 significant figures
    ↑ ↑
  Do not count zeros at the beginning.

    15·26 | 66 = 15·27 to 2 decimal places
      ↑

    0·349 | 7 = 0·350 to 3 decimal places
     ↑

(b) Suppose you timed a race with a stopwatch and got 13·2 seconds while your friend with an electronic watch got 13·20 seconds. Is there any difference? Yes! When you write 13·20, the figure is accurate to 2 decimal places even though the last figure is a zero. The figure 13·2 is accurate to only one decimal place.

(c) Similarly a weight, given as 12·00 kg, is accurate to 4 significant figures while '12 kg' is accurate to only 2 significant figures.

(d) Suppose you measure the length of a line in cm and you want to show that it is accurate to one decimal place. The measured length using a ruler might be 7 cm. To show that it is accurate to one decimal place, you must write 'length = 7·0 cm'.

Exercise 1J

1. Round off to the number of decimal places indicated.
 (a) 0·672 (1 d.p.) (b) 8·814 (2 d.p.)
 (c) 0·7255 (3 d.p.) (d) 1·1793 (2 d.p.)
 (e) 0·863 (1 d.p.) (f) 8·2222 (2 d.p.)
 (g) 0·07518 (3 d.p.) (h) 11·7258 (3 d.p.)
 (i) 20·154 (1 d.p.) (j) 6·6666 (2 d.p.)
 (k) 0·342 (1 d.p.) (l) 0·07248 (4 d.p.)

2. Round off to the number of significant figures indicated
 (a) 2·658 (2 s.f.) (b) 188·79 (3 s.f.)
 (c) 2·87 (1 s.f.) (d) 0·3569 (2 s.f.)
 (e) 1·7231 (2 s.f.) (f) 0·041551 (3 s.f.)
 (g) 0·0371 (1 s.f.) (h) 811·1 (1 s.f.)
 (i) 9320 (2 s.f.) (j) 8·051 (2 s.f.)
 (k) 6·0955 (3 s.f.) (l) 8·205 (1 s.f.)

Common sense

When you find the answer to a problem, you should choose the degree of accuracy appropriate for that situation.

(a) Suppose you were calculating how much tax someone should pay in a year and your actual answer was £2153·6752. It would be sensible to give the answer as £2154 to the nearest pound.

(b) In calculating the average speed of a car journey from Bristol to Cardiff, it would not be realistic to give an answer of 38·241 km/h. A more sensible answer would be 38 km/h.

1.5 Ratio and proportion

Ratio

- The word 'ratio' is used to describe a fraction. If the **ratio** of a boy's height to his father's height is $4:5$, then he is $\frac{4}{5}$ as tall as his father.

Example

(a) Change the ratio $2:5$ into the form

(i) $1:n$ (ii) $m:1$

(i) $2:5 = 1:\frac{5}{2}$ (ii) $2:5 = \frac{2}{5}:1$
$= 1:2\cdot5$
$= 0\cdot4:1$

(b) Divide £60 between two people A and B in the ratio $5:7$.

Consider £60 as 12 equal parts (i.e. $5+7$). Then A receives 5 parts and B receives 7 parts.

∴ A receives $\frac{5}{12}$ of £60 = £25

B receives $\frac{7}{12}$ of £60 = £35

Exercise 1K

Express the following ratios in the form $1:n$.

1. $2:6$ 2. $5:30$ 3. $2:100$
4. $5:8$ 5. $4:3$ 6. $8:3$

Express the following ratios in the form $n:1$.

7. $12:5$ 8. $5:2$ 9. $4:5$

In Questions **10** to **13**, divide the quantity in the ratio given.

10. £40; (3 : 5)   **11.** £120; (3 : 7)

12. 180 kg; (1 : 5 : 6)   **13.** 184 minutes; (2 : 3 : 3)

14. When £143 is divided in the ratio 2 : 4 : 5, what is the difference between the largest share and the smallest share?

15. If $\frac{5}{8}$ of the children in a school are boys, what is the ratio of boys to girls?

16. Find the ratio (shaded area) : (unshaded area) for each diagram.

(a) 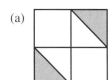   (b) 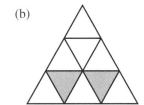   (c)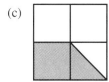

17. A man and a woman share a bingo prize of £1000 between them in the ratio 1 : 4. The woman shares her part between herself, her mother and her daughter in the ratio 2 : 1 : 1.
How much does her daughter receive?

18. A man and his wife share a sum of money in the ratio 3 : 2. If the sum of money is doubled, in what ratio should they divide it so that the man still receives the same amount?

19. In a herd of x cattle, the ratio of the number of bulls to cows is 1 : 6. Find the number of bulls in the herd in terms of x.

20. If $x : 3 = 12 : x$, calculate the positive value of x.

21. £400 is divided between Ann, Brian and Carol so that Ann has twice as much as Brian, and Brian has three times as much as Carol. How much does Brian receive?

22. A cake weighing 550 g has three ingredients: flour, sugar and raisins. There is twice as much flour as sugar and one and a half times as much sugar as raisins. How much flour is here?

Map scales

Example

A map is drawn to a scale of 1 to 50 000.
Calculate the length of a road which appears as 3 cm long on the map.

1 cm on the map is equivalent to 50 000 cm on the earth.

$$\therefore \quad 3\,\text{cm} \equiv 3 \times 50\,000 = 150\,000\,\text{cm}$$
$$= 1500\,\text{m}$$
$$= 1\cdot5\,\text{km}$$

The road is 1·5 km long.

Exercise 1L

1. On a map of scale 1 : 100 000, the distance between Tower Bridge and Hammersmith Bridge is 12·3 cm. What is the actual distance in km?

2. On a map of scale 1 : 15 000, the distance between Buckingham Palace and Brixton Underground Station is 31·4 cm. What is the actual distance in km?

3. If the scale of a map is 1 : 10 000, what will be the length on this map of a road which is 5 km long?

4. The distance from Hertford to St Albans is 32 km. How far apart will they be on a map of scale 1 : 50 000?

5. The 17th hole at the famous St Andrews golf course is 420 m in length. How long will it appear on a plan of the course of scale 1 : 8000?

6. The scale of a map is 1 : 1000. What are the actual dimensions of a rectangle which appears as 4 cm by 3 cm on the map? What is the area on the map in cm^2? What is the actual area in m^2?

7. The scale of a map is 1 : 100. What area does 1 cm^2 on the map represent? What area does 6 cm^2 represent?

8. The scale of a map is 1 : 20 000. What area does 8 cm^2 represent?

Proportion

Example

(a) If 9 litres of petrol costs £5·76, find the cost of 20 litres.

The cost of petrol is **directly** proportional to the quantity bought.

9 litres costs £5·76
∴ 1 litre costs £5·76 ÷ 9 = £0·64
∴ 20 litres costs £0·64 × 20 = £12·80

(b) If five men can paint a bridge in 12 days, how long would it take three men?

The length of time required to paint the bridge is **inversely** proportional to the number of men who paint.

5 men take 12 days
∴ 1 man takes 60 days.
∴ 3 men take 20 days.

In the first example we found the cost of **one** litre of petrol.

In the second example we found the time needed for **one** man.

Exercise 1M

This exercise contains a mixture of questions involving both **direct** and **inverse** proportion.

1. If seven discs cost £1·54, find the cost of five discs.

2. Twelve people are needed to harvest a crop in 2 hours. How long would it take 4 people?

3. Nine milk bottles contain $4\frac{1}{2}$ litres of milk between them. How much do five bottles hold?

4. A car uses 10 litres of petrol in 75 km. How far will it go on 8 litres?

5. A wire 11 cm long has a mass of 187 g. What is the mass of 7 cm of this wire?

6. A train travels 30 km in 120 minutes. How long will it take to travel 65 km at the same speed?

7. A ship has sufficient food to supply 600 passengers for 3 weeks. How long would the food last for 800 people?

8. Usually it takes 12 hours for 5 men to tarmac a road. How many men are needed to tarmac the same road in 10 hours?

9. 80 machines can produce 4800 identical pens in 5 hours. At this rate:
   (a) how many pens would one machine produce in one hour?
   (b) how many pens would 25 machines produce in 7 hours?

10. Three men can build a wall in 10 hours. How many men would be needed to build the wall in $7\frac{1}{2}$ hours?

11. If it takes 6 men 4 days to dig a hole 3 feet deep, how long will it take 10 men to dig a hole 7 feet deep?

12. 5 machines produce 10 000 bottles in 10 hours. How many bottles would 8 machines produce in 6 hours?

13. It takes x ants n hours to build a nest. How long will it take $x + 1000$ ants to build the same size nest?

1.6 Negative numbers

- For adding and subtracting use the number line.

Example Work out: (a) $^-1 + 4$ (b) $^-2 - 3$ (c) $4 - 6$

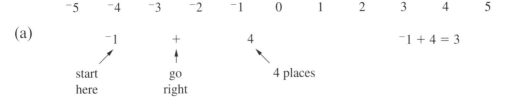

(a)

(b)

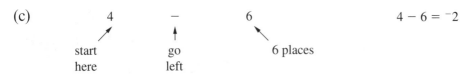

(c)

Exercise 1N
Work out:
1. $^-6 + 2$
2. $^-7 - 5$
3. $^-3 - 8$
4. $^-5 + 2$
5. $^-6 + 1$
6. $8 - 4$
7. $4 - 9$
8. $11 - 19$
9. $4 + 15$
10. $^-7 - 10$
11. $16 - 20$
12. $^-7 + 2$

13. ⁻6 − 5   14. 10 − 4   15. ⁻4 + 0   16. ⁻6 + 12
17. ⁻7 + 7   18. 2 − 20   19. 8 − 11   20. ⁻6 − 5
21. ⁻8 − 4   22. ⁻3 + 7   23. ⁻6 + 10   24. ⁻5 + 5
25. ⁻11 + 3   26. 7 − 10   27. ⁻5 + 8   28. ⁻12 + 0
29. ⁻1 + 19   30. 20 − 25   31. ⁻6 − 60   32. ⁻2 + 100

- When you have two (+) or (−) signs together use this rule:

  ++ = +   +− = −
  −− = +   −+ = −

Example
(a) $3 - (^-6) = 3 + 6 = 9$

(b) $^-4 + (^-5) = ^-4 - 5 = ^-9$

(c) $^-5 - (^+7) = ^-5 - 7 = ^-12$

Exercise 10
Work out:
1. ⁻3 + (⁻5)   2. ⁻5 − (⁺2)   3. 4 − (⁺3)   4. ⁻3 − (⁻4)
5. 6 − (⁻3)   6. 16 + (⁻5)   7. ⁻4 + (⁻4)   8. 20 − (⁻22)
9. ⁻6 − (⁻10)   10. 95 + (⁻80)   11. ⁻3 − (⁺4)   12. ⁻5 − (⁺4)
13. 6 + (⁻7)   14. ⁻4 + (⁻3)   15. ⁻7 − (⁻7)   16. 3 − (⁻8)
17. ⁻8 + (⁻6)   18. 7 − (⁺7)   19. 12 − (⁻5)   20. 9 − (⁺6)
21. ⁻3 − (⁻2)   22. 8 + (⁻11)   23. 10 − (⁻2)   24. ⁻7 + (⁻2)
25. 9 − (⁺6)   26. 7 + (⁻7)   27. 0 − (⁻8)   28. ⁻6 − (⁻8)

29. Copy and complete these addition squares.

(a)

+	⁻5	1	6	⁻2
3	⁻2	4		
⁻2				
6				
⁻10				

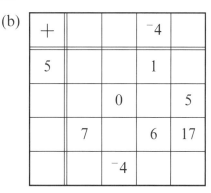

(b)

+			⁻4	
5			1	
			0	5
	7		6	17
		⁻4		

Multiplying and dividing

A directed number is a positive or negative number.

- When two directed numbers with the same sign are multiplied together, the answer is positive.

$$^+7 \times (^+3) = {}^+21$$
$$^-6 \times (^-4) = {}^+24$$

- When two directed numbers with different signs are multiplied together, the answer is negative.

$$^-8 \times (^+4) = {}^-32$$
$$^+7 \times (^-5) = {}^-35$$

- When dividing directed numbers, the rules are the same as in multiplication.

$$^-70 \div (^-2) = {}^+35$$
$$^+12 \div (^-3) = {}^-4$$
$$^-20 \div (^+4) = {}^-5$$

Exercise 1P

1. $^-3 \times (^+2)$
2. $^-4 \times (^+1)$
3. $^+5 \times (^-3)$
4. $^-3 \times (^-3)$
5. $^-4 \times (^+2)$
6. $^-5 \times (^+3)$
7. $6 \times (^-4)$
8. $3 \times (^+2)$
9. $^-3 \times (^-4)$
10. $6 \times (^-3)$
11. $^-7 \times (^+3)$
12. $^-5 \times (^-5)$
13. $6 \times (^-10)$
14. $^-3 \times (^-7)$
15. $8 \times (^+6)$
16. $^-8 \times (^+2)$
17. $^-7 \times (^+6)$
18. $^-5 \times (^-4)$
19. $^-6 \times (^+7)$
20. $11 \times (^-6)$
21. $8 \div (^-2)$
22. $^-9 \div (^+3)$
23. $^-6 \div (^-2)$
24. $10 \div (^-2)$
25. $^-12 \div (^-3)$
26. $^-16 \div (^+4)$
27. $4 \div (^-1)$
28. $8 \div (^-8)$
29. $16 \div (^-8)$
30. $^-20 \div (^-5)$
31. $^-16 \div (^+1)$
32. $18 \div (^-9)$
33. $36 \div (^-9)$
34. $^-45 \div (^-9)$
35. $^-70 \div (^+7)$
36. $^-11 \div (^-1)$
37. $^-16 \div (^-1)$
38. $1 \div (^-\frac{1}{2})$
39. $^-2 \div (^+\frac{1}{2})$
40. $50 \div (^-10)$
41. $^-8 \times (^-8)$
42. $^-9 \times (^+3)$
43. $10 \times (^-60)$
44. $^-8 \times (^-5)$
45. $^-12 \div (^-6)$
46. $^-18 \times (^-2)$
47. $^-8 \div (^+4)$
48. $^-80 \div (^+10)$

49. Copy and complete these multiplication squares.

(a)

×	4	$^-3$	0	$^-2$
$^-5$				
2				
10				
$^-1$				

(b)

×			$^-1$	
3			$^-3$	
			$^-15$	18
	$^-14$		$^-7$	$^-42$
		10		

Questions on negative numbers are more difficult when the different sorts are mixed together. The remaining questions are given in the form of four short tests.

Test 1
1. ⁻8 − 8
2. ⁻8 × (⁻8)
3. ⁻5 × 3
4. ⁻5 + 3
5. 8 − (⁻7)
6. 20 − 2
7. ⁻18 ÷ (⁻6)
8. 4 + (⁻10)
9. ⁻2 + 13
10. ⁺8 × (⁻6)
11. ⁻9 + (⁺2)
12. ⁻2 − (⁻11)
13. ⁻6 × (−1)
14. 2 − 20
15. ⁻14 − (⁻4)
16. ⁻40 ÷ (⁻5)
17. 5 − 11
18. ⁻3 × 10
19. 9 + (⁻5)
20. 7 ÷ (⁻7)

Test 2
1. ⁻2 × (⁺8)
2. ⁻2 + 8
3. ⁻7 − 6
4. ⁻7 × (⁻6)
5. ⁺36 ÷ (⁻9)
6. ⁻8 − (⁻4)
7. ⁻14 + 2
8. 5 × (⁻4)
9. 11 + (⁻5)
10. 11 − 11
11. ⁻9 × (⁻4)
12. ⁻6 + (⁻4)
13. 3 − 10
14. ⁻20 ÷ (⁻2)
15. 16 + (⁻10)
16. ⁻4 − (⁺14)
17. ⁻45 ÷ 5
18. 18 − 3
19. ⁻1 × (⁻1)
20. ⁻3 − (⁻3)

Test 3
1. ⁻10 × (⁻10)
2. ⁻10 − 10
3. ⁻8 × (⁺1)
4. ⁻8 + 1
5. 5 + (⁻9)
6. 15 − 5
7. ⁻72 ÷ (⁻8)
8. ⁻12 − (⁻2)
9. ⁻1 + 8
10. ⁻5 × (⁻7)
11. ⁻10 + (⁻10)
12. ⁻6 × (⁺4)
13. 6 − 16
14. ⁻42 ÷ (⁺6)
15. ⁻13 + (⁻6)
16. ⁻8 − (⁻7)
17. 5 × (⁻1)
18. 2 − 15
19. 21 + (⁻21)
20. ⁻16 ÷ (⁻2)

Test 4
Write down each statement and find the missing number.
1. (⁻6) × ☐ = 30
2. ☐ + (⁻2) = 0
3. ☐ ÷ (⁻2) = 10
4. ☐ − (⁻3) = 7
5. (⁻1) × ☐ = ½
6. (⁻2) − ☐ = 3
7. ☐ + (⁻10) = 2
8. 6 × ☐ = 0
9. ☐ − (⁻8) = 0
10. (⁻1)⁴ = ☐
11. 0·2 × ☐ = ⁻200
12. (⁻1)¹³ = ☐

1.7 Mixed numerical problems

Exercise 1Q

1. I have lots of 1p, 2p, 3p and 4p stamps. How many different combinations of stamps can I make which total 5p?
2. Find n if: $8 + 9 + 10 + \ldots + n = 5^3$

3. Copy and complete.
$$3^2 + 4^2 + 12^2 = 13^2$$
$$5^2 + 6^2 + 30^2 = 31^2$$
$$6^2 + 7^2 + = $$
$$x^2 + + = $$

4. You are told that 8 cakes and 6 biscuits cost 174 pence and 2 cakes and 4 biscuits cost 66 pence. Without using simultaneous equations, work out the cost of each of the following.
  (a) 4 cakes and 3 biscuits (b) 10 cakes and 10 biscuits
  (c) 3 cakes and 3 biscuits (d) 1 cake (e) 1 biscuit

5. The total mass of a jar one-quarter full of jam is 250 g. The total mass of the same jar three-quarters full of jam is 350 g. What is the mass of the empty jar?

$\frac{1}{4}$ 250g

$\frac{3}{4}$ 350g

6. On an aircraft a certain weight of luggage is carried free and a charge of £10 per kilogram is made for any excess luggage.
  (a) If the charge for 60 kg of luggage is £400, find the charge for 35 kg.
  (b) If the charge for 30 kg of luggage is £100, find the charge for 15 kg.
  (c) If a fifth of the luggage is carried free, what is the average cost per kilogram of the luggage?

7. Evaluate $\frac{1}{3} \times \frac{2}{4} \times \frac{3}{5} \times \ldots \times \frac{9}{11} \times \frac{10}{12}$.

8. Find the least positive integer n for which $(829\,642 + n)$ is exactly divisible by 7.

9. Find all the missing digits in these multiplications.

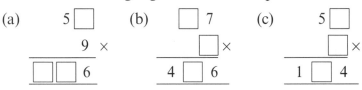

10. (a) Work out $\frac{1}{4} + \frac{1}{12}$ as a single fraction in its lowest terms.

(b) Find integers a and b such that $\dfrac{1}{a} + \dfrac{1}{b} = \dfrac{5}{8}$.

Exercise 1R

1. Copy and complete this multiplication square.

×	$\frac{2}{3}$		
$\frac{1}{2}$		$\frac{3}{8}$	
			$\frac{3}{16}$
$\frac{2}{5}$			$\frac{2}{25}$

2. Look at this number pattern.

$$7^2 = 49$$
$$67^2 = 4489$$
$$667^2 = 444\,889$$
$$6667^2 = 44\,448\,889$$

This pattern continues.
(a) Write down the next line of the pattern.
(b) Use the pattern to work out $6\,666\,667^2$.
(c) What is the square root of $4\,444\,444\,488\,888\,889$?

3. Find a pair of positive integers a and b for which:
(a) $18a + 65b = 1865$
(b) $23a + 7b = 2314$

4. If the number in the space is prime, write PRIME next to it. If it is not prime, write it as the product of its prime factors. The first two have been done for you.

47 ...PRIME...   40 ....   63 ....
26 ...2×13...   25 ....   71 ....

5. Find three consecutive square numbers whose sum is 149.

6. The diagrams show magic squares in which the sum of the numbers in any row, column or diagonal is the same. Find the value of x in each square.

(a)

	x	6
3		7
		2

(b)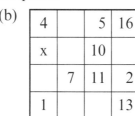

7. Work out $100 - 99 + 98 - 97 + 96 - \ldots + 4 - 3 + 2 - 1$.

8. The smallest three-digit product of a one-digit prime and a two-digit prime is:

   (a) 102   (b) 103   (c) 104   (d) 105   (e) 106

Exercise 1S

1. A group of 17 people share some money equally and each person gets £415. What is the total amount of money they have shared?

2. The 'reciprocal' of 2 is $\frac{1}{2}$. The reciprocal of 7 is $\frac{1}{7}$.

   The reciprocal of x is $\frac{1}{x}$.

   Find the square root of the reciprocal of the square root of the reciprocal of ten thousand.

3. The purpose of the set of steps given below is to work out some numbers in a number sequence.

   Step 1   Write down the number 1.
   Step 2   Add 2 to the number you have just written.
   Step 3   Write down the answer obtained in step 2.
   Step 4   If your answer in step 2 is more than 10 then stop.
   Step 5   Go to step 2.

   (a) Write down the numbers produced from the set of steps.
   (b) Change one line of the set of steps so that it could be used to produce the first six even numbers.

4. Put four different numbers in the circles so that when you add the numbers at the end of each line you always get a square number.

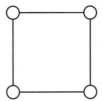

5. You are given that $41 \times 271 = 11\,111$.
   Work out the following **in your head**.
   (a) 246×271
   (b) $22\,222 \div 271$
   (c) This time you can write down a (little!) working. Work out $41^2 \times 271$.

6. Copy and complete the crossnumber puzzle.
   Across
   1. Square number
   4. (1 across) × (3 down) − 7000
   6. Next in the sequence
      1, 3, 7, 15, 31, 63, 127, __
   7. Sum of the prime numbers
      between 30 and 40

   Down
   2. South-East as a
      bearing
   3. Number of days
      in 13 weeks
   5. One-eighth of 6048
   6. Cube number

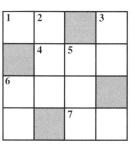

7. (a) Write down the value of $\sin A$ as a fraction.
   (b) Work out, as a fraction:
   $$\frac{5 \sin A}{(1 - \sin A)(1 + \sin A)}$$

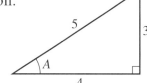

8. S_1 is the sum of all the even numbers from 2 to 1000 inclusive.
   S_2 is the sum of all the odd numbers from 1 to 999 inclusive.
   Work out $S_1 - S_2$.

Exercise 1T
1. (a) The sum of the factors of n (including 1 and n) is 7. Find n.
   (b) The sum of the factors of p (including 1 and p) is 6. Find p.

2. Each packet of washing powder carries a token and
   four tokens can be exchanged for a free packet. How
   many free packets will I receive if I buy 64 packets?

3. At birth, the mass of a kitten was 0·3 kg.
   A few weeks later its mass has increased so that the ratio:

   mass now : mass at birth = 7 : 4

   Calculate the kitten's mass now.

4. Apart from 1, 3 and 5, all odd numbers less than 100
   can be written in the form $p + 2^n$ where p is a prime
   number and n is greater than or equal to 2.
   e.g. $43 = 11 + 2^5$
   $27 = 23 + 2^2$

   For the odd numbers 7, 9, 11, ... 99, write as many as
   you can in the form $p + 2^n$.

5. The digits 1, 9, 9, 4 are used to form fractions less than 1. Each of the four digits must be used,

   e.g. $\dfrac{9}{419}$ or $\dfrac{4}{919}$

   (a) Write down, smallest first, the three smallest fractions that can be made.
   (b) Write down the largest fraction less than one that can be made.

6. (a) 411 fans went to a Capital Radio pop concert. The tickets cost £16 each.
   Calculate the total amount paid by the fans.
   (b) (i) All of these fans travelled on 52-seater coaches. What is the least number of coaches that could be used?
   (ii) On the return journey two coaches failed to turn up. How many fans had to be left behind?

Operator squares

 Each empty square contains either a number or a mathematical symbol (+, −, ×, ÷). Copy each square and fill in the details.

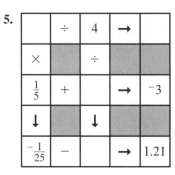

Summary

1. You can find the LCM and HCF of two numbers.

2. You can calculate with fractions.

3. You can round a number using decimal places and significant figures.

4. You can use ratios.

5. You can use negative numbers.

Check out N1

1. Find the LCM and HCF of 54 and 72.

2. Evaluate:
   (a) $\dfrac{3}{4} \times \dfrac{2}{5}$ (b) $2\dfrac{1}{4} + 3\dfrac{4}{5}$ (c) $4\dfrac{1}{5} - 1\dfrac{2}{3}$

3. (a) Round 37·3491:
       (i) to two decimal places
       (ii) to three decimal places.

   (b) Round 2493·1:
       (i) to 3 significant figures
       (ii) to 2 significant figures.

4. Divide 54 in the ratio 5:4.

5. Evaluate (a) $7 + (^-15)$
            (b) $8 \times (^-3)$ (c) $^-12 \times (^-8)$

Revision exercise N1

1. (a) Find an approximate value of $\dfrac{20 \cdot 3 \times 9 \cdot 8}{0 \cdot 497}$.

   (b) Work out: (i) $0 \cdot 07^2$ (ii) $5\tfrac{3}{4} - 2\tfrac{1}{3}$. [SEG]

2. (a) Express 36 as a product of its prime factors.
   (b) Find the Highest Common Factor (HCF) of 36 and 60. [AQA]

3. (a) **Estimate** the value of $\dfrac{908 \times 4 \cdot 92}{0 \cdot 307}$.
   (b) Evaluate $\tfrac{2}{3} - \tfrac{1}{5}$, giving your answer as a fraction. [SEG]

4. (a) Express 72 as a product of its prime factors.
       Give your answer using indices.
   (b) Given that $48 = 2^4 \times 3$, find the Lowest Common Multiple of 48 and 72. [SEG]

5. (a) You are given that $293 \times 7 \cdot 48 = 2191 \cdot 64$. Use this result to find the value of
       (i) $293 \times 0 \cdot 0748$ (ii) $293\,000 \times 0 \cdot 0748$.
   (b) Evaluate $\tfrac{3}{4} - \tfrac{2}{5}$, giving your answer as a fraction. [SEG]

N2 NUMBER 2

This unit will show you how to:

- Calculate with percentages
- Use a calculator
- Use indices
- Use standard form
- Use surds
- Substitute into formulae
- Use compound measures
- Use upper and lower bounds
- Calculate errors

Before you start:

You should know how to...	Check in N2
1. Use the BODMAS rule. For example, work out $9 + 5 \times 3$ $= 9 + 15 = 24$	1. Work out: (a) $4 + 6 \div 3$ (b) $8 \times 5 - 2 + 3 \times 7$
2. Express integers as products of their prime factors. For example, $24 = 2^3 \times 3$	2. Express as a product of their prime factors: (a) 18 (b) 108 (c) 50
3. Calculate with directed numbers. For example, $^-5 \times {}^-3 \times 8 = {}^+120$	3. Calculate: (a) $3 + {}^-7$ (b) $^-5 \div {}^-11$
4. Convert between units of metric measure.	4. Convert: (a) 7·48 m into cm (b) 743 cℓ in ℓ
5. Convert between units of imperial measure.	5. Convert: (a) 6 feet 3 inches into inches (b) 4 lb 8 oz into ounces.
6. Convert between units of metric and imperial measure.	6. Give an approximate value of: (a) 35 kg in lbs (b) 2 m in inches.
7. Do calculations involving time.	7. What is the length of time between 9.30 am and 2.45 pm?
8. Know the difference between continuous and discrete data.	8. Is (a) the height of a person (b) the number of pets in a family (c) the cost of an item a discrete or continuous variable?

MODULE 3

2.1 Percentages

Percentages, fractions and decimals

Percentages are simply a convenient way of expressing fractions or decimals.

$$25\% = \frac{25}{100} = \frac{1}{4} = 0.25$$

Example

(a) Change $\frac{3}{8}$ to a percentage. $\quad \frac{3}{8} = \left(\frac{3}{8} \times \frac{100}{1}\right)\% = 37\frac{1}{2}\%$

(b) Change $\frac{7}{8}$ to a decimal. $\quad \frac{7}{8}$ means 'divide 8 into 7'

$$8 \overline{)7 \cdot 000}^{\,0 \cdot 875}, \quad \frac{7}{8} = 0.875$$

(c) Change 0.35 to a fraction. $\quad 0.35 = \frac{35}{100} = \frac{7}{20}$

(d) Change 6% to a decimal. $\quad 6\% = \frac{6}{100} = 0.06$

Exercise 2A

1. Change these fractions to decimals.
   (a) $\frac{1}{4}$ (b) $\frac{2}{5}$ (c) $\frac{3}{8}$ (d) $\frac{5}{12}$ (e) $\frac{1}{6}$ (f) $\frac{2}{7}$

2. Change these decimals to fractions and simplify.
   (a) 0.2 (b) 0.45 (c) 0.36 (d) 0.125 (e) 1.05 (f) 0.007

3. Change to percentages.
   (a) $\frac{1}{4}$ (b) $\frac{1}{10}$ (c) 0.72 (d) 0.075 (e) 0.02 (f) $\frac{1}{3}$

4. Arrange in order of size (smallest first).
   (a) $\frac{1}{2}$; 45%; 0.6 (b) 0.38; $\frac{6}{16}$; 4%
   (c) 0.111; 11%; $\frac{1}{9}$ (d) 32%; 0.3; $\frac{1}{3}$

Change the fractions to decimals and then evaluate the following, giving the answer to 2 decimal places:

5. $\frac{1}{4} + \frac{1}{3}$ 6. $\frac{2}{3} + 0.75$ 7. $\frac{8}{9} - 0.24$
8. $\frac{7}{8} - \frac{5}{9}$ 9. $\frac{1}{3} \times 0.2$ 10. $\frac{5}{8} \times 1.1$

11. What percentage of the integers from 1 to 10 inclusive are prime numbers? [1 is **not** prime.]

12. What percentage of this shape is shaded?

13. The pass mark in a physics test was 60%. Sam got 23 out of 40. Did she pass?

14. In 2000 the prison population was 48 700 men and 1600 women. What percentage of the total prison population were men?

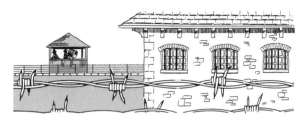

15. The table shows the results of a test for pulse rate conducted on 274 children.

	Boys	Girls	Total
High pulse	58	71	129
Low pulse	83	62	145
Total	141	133	274

(a) What percentage of the boys had a low pulse rate?
(b) What percentage of the children with a high pulse rate were girls?

16. Jasper reads a newspaper and then says, 'Great! Now prices will fall for a change.' Explain whether or not you agree with Jasper's statement.

Inflation falls from 5% to 3.5%

Working out percentages

Example
(a) Work out 7% of £3200.

Either: 7% of £3200
$= \dfrac{7}{100} \times \dfrac{3200}{1} = £224$

Or: 7% of £3200
$= 0.07 \times 3200 = £224$

(b) The price of a car costing £8400 is increased by 5%. Find the new price.

Either: 5% of £8400 $= \dfrac{5}{100} \times 8400$
$= £420$
New price $= £8400 + £420$
$= £8820$

Or: New price $= 105\%$ of £8400
$= 1.05 \times 8400$
$= £8820$

Exercise 2B

1. Calculate:
   (a) 30% of £50
   (b) 45% of 2000 kg
   (c) 4% of $70
   (d) 2·5% of 5000 people

2. In a sale, a jacket costing £40 is reduced by 20%. What is the sale price?

3. The charge for a telephone call costing 12p is increased by 10%. What is the new charge?

4. In peeling potatoes 4% of the mass of the potatoes is lost as 'peel'. How much is **left** for use from a bag containing 55 kg?

5. Write down the missing number, as a **decimal**.
   (a) 53% of 650 = ☐ × 650
   (b) 3% of 2600 = ☐ × 2600
   (c) 8·5% of 700 = ☐ × 700
   (d) 122% of 285 = ☐ × 285

6. Work out, to the nearest penny:
   (a) 6·4% of £15·95
   (b) 11·2% of £192·66
   (c) 8·6% of £25·84
   (d) 2·9% of £18·18

7. Find the total bill:
   5 golf clubs at £18·65 each
   60 golf balls at £16·50 per dozen
   1 bag at £35·80
   V.A.T. at $17\frac{1}{2}$% is added to the total cost.

8. In 1999 a club has 250 members who each pay £95 annual subscription. In 2000 the membership increases by 4% and the annual subscription is increased by 6%. What is the total income from subscriptions in 2000?

9. The cash price for a car was £7640. Mr Elder bought the car on the following hire purchase terms: 'A deposit of 20% of the cash price and 36 monthly payments of £191·60'. Calculate the total amount Mr Elder paid.

10. A motor bike costs £820. After a 12% increase the new price is 112% of £820. The 'quick' way to work this out is as follows:

$$\text{New price} = 112\% \text{ of } £820$$
$$= 1{\cdot}12 \times 820$$
$$= £918{\cdot}40$$

Use this quick method to find the new price of a lorry costing £6500 when the price is increased by 5%.

11. Find the new price of the following items

Item	Old price	Price change
Video	£190	6% increase
House	£150 000	11% increase
Boat	£2500	8% increase
Tree	£210	5% decrease
Phone	£65	15% decrease

12. In the last two weeks of a sale, prices are reduced first by 30% and then by a **further** 40% of the new price. What is the final sale price of a shirt which originally cost £15?

13. Work out the following:
    (a) 8% of 3·2 kg   (answer in grams)
    (b) 16% of £4·50   (answer in pence)
    (c) 28% of 5 cm   (answer in mm)
    (d) 5% of 10 hours   (answer in minutes)
    (e) 12% of 2 minutes   (answer in seconds)

14. A 12% increase in the value of N, followed by a 12% decrease in value can be calculated as $1{\cdot}12 \times N \times 0{\cdot}88$. Copy and complete:
    (a) An 8% increase in the value of P, followed by a 10% decrease in value can be calculated as ☐.
    (b) A ☐ increase in the value of X, followed by a further ☐ increase in value can be calculated as $1{\cdot}15 \times X \times 1{\cdot}06$.

Percentage profit/loss

In the next exercise use the formulae:

- Percentage profit = $\dfrac{\text{actual profit}}{\text{original price}} \times \dfrac{100}{1}$
- Percentage loss = $\dfrac{\text{actual loss}}{\text{original price}} \times \dfrac{100}{1}$

Example

A radio is bought for £16 and sold for £20.
What is the percentage profit?

Actual profit = £4

∴ Percentage profit = $\dfrac{4}{16} \times \dfrac{100}{1} = 25\%$

The radio is sold at a 25% profit.

Exercise 2C

1. The first figure is the cost price and the second figure is the selling price. Calculate the percentage profit or loss in each case.

   (a) £20, £25
   (b) £400, £500
   (c) £60, £54
   (d) £9000, £10 800
   (e) £460, £598
   (f) £512, £550·40
   (g) £45, £39·60
   (h) 50p, 23p

2. A car dealer buys a car for £500, gives it a clean, and then sells it for £640. What is the percentage profit?

3. A damaged carpet which cost £180 when new, is sold for £100. What is the percentage loss?

4. During the first four weeks of her life, a baby girl increases her weight from 3·2 kg to 4·7 kg. What percentage increase does this represent? (Give your answer to 3 sig. fig.)

5. When V.A.T. is added to the cost of a car tyre, its price increases from £16·50 to £18·48. What is the rate at which V.A.T. is charged?

6. In order to increase sales, the price of a Concorde airliner is reduced from £30 000 000 to £28 400 000. What percentage reduction is this?

7. A picture has the dimensions shown.
Calculate the percentage increase in the area of
the picture after both the length and width are
increased by 20%.

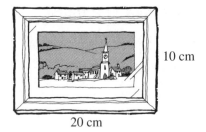

10 cm

20 cm

8. A box has a square base of side 30 cm and height 20 cm.
Calculate the percentage increase in the volume of the
box after the length and width of the base are both
increased by 10% and the height is increased by 5%.

9. The rental for a television set changed from £80 per
year to £8 per month. What is the percentage increase
in the yearly rental?

10. If an employer reduces the working week from 40 hours
to 35 hours, with no loss of weekly pay, calculate the
percentage increase in the hourly rate of pay.

11. Given that $G = ab$, find the percentage increase in G
when both a and b increase by 10%.

12. The formula connecting P, a and n is $P = an^3$.
    (a) Calculate the value of P when $a = 200$ and $n = 5$.
    (b) Calculate the percentage increase in P when a is
    increased by 2% and n is increased by 20%.

Reverse percentages

Example

(a) After an increase of 6%, the price of a motor bike is £9010.
What was the price before the increase?

A common mistake here is to work out 6% of £9010. This is
wrong because the increase is 6% of the **old** price, not 6% of
the new price.

106% of old price = £9010

∴   1% of old price = $\dfrac{9010}{106}$

∴   100% of old price = $\dfrac{9010}{106} \times 100$

Original price = £8500

(b) To increase sales, the price of a magazine is reduced by 5%. Find the original price if the new price is 190p.

This is a reduction, so the new price is 95% of the old price.

$$95\% \text{ of old price} = 190\text{p}$$
$$1\% \text{ of old price} = \frac{190}{95}$$
$$100\% \text{ of old price} = \frac{190}{95} \times 100$$

Original price = 200p

Exercise 2D

1. After an increase of 8%, the price of a car is £6696. Find the price of the car before the increase.

2. After a 12% pay rise, the salary of Mr Brown was £28 560. What was his salary before the increase?

3. Find the missing prices.

Item	Old price	New price	% change
(a) Jacket	?	£55	10% increase
(b) Dress	?	£212	6% increase
(c) CD player	?	£56·16	4% increase
(d) T.V.	?	£195	30% increase
(e) Car	?	£3960	65% increase

4. Between 1990 and 1997 the population of an island fell by 4%. The population in 1997 was 201 600. Find the population in 1990.

5. After being ill for 3 months, a man's weight went down by 12%. Find his original weight if he weighed 74·8 kg after the illness.

6. The diagram shows two rectangles. The width and height of rectangle B are both 20% greater than the width and height of rectangle A. Use the figures given to find the width and height of rectangle A.

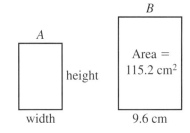

7. The label on a carton of yoghurt is shown. The figure on the right is smudged out. Work out what the figure should be.

8. A television costs £376 including $17\frac{1}{2}$% V.A.T. How much of the cost is tax?

9. During a Grand Prix car race, the tyres on a car are reduced in weight by 3%. If they weigh 388 kg at the end of the race, how much did they weigh at the start?

10. A heron has discovered Sergio's goldfish pond. Sergio lost 70% of his collection of goldfish. If he has 60 survivors, how many did he have originally?

11. The average attendance at Everton football club fell by 7% in 2000. If 2030 fewer people went to matches in 2000, how many went in 1999?

12. When heated an iron bar expands by 0·2%. If the increase in length is 1 cm, what is the original length of the bar?

13. An oven is sold for £600, thereby making a profit of 20% on the cost price. What was the cost price?

14. In 1999 Corus plc had sales of £516 000 000 which was $7\frac{1}{2}$% higher than the figure for 1998. Calculate the value of sales in 1998.

Compound interest

Example
Suppose a bank pays a fixed interest of 10% on money in deposit accounts. A man puts £500 in the bank.

After one year he has $500 + 10\%$ of $500 = £550$

After two years he has $550 + 10\%$ of $550 = £605$

   [Check that this is $1 \cdot 10^2 \times 500$].

After three years he has $605 + 10\%$ of $605 = £665 \cdot 50$

   [Check that this is $1 \cdot 10^3 \times 500$]

In general after n years the money in the bank will be £($1 \cdot 10^n \times 500$).

Multipliers

In the example, you noticed that to increase the value of the money in the deposit account by 10%, you multiply by 1·10. This is because 10% is 0·10, thus an increase of 0·10 gives a multiplier of $1 + 0·10 = 1·1$.

Thus after 8 years you multiply by $1·10^8$.

1·10 is known as a **multiplier**.

Example

Colin buys a car for £7500 and he expects its value to drop every year by 28%. What does Colin think its value will be five years later?

The value drops by 28% so the multiplier is 0·72 (i.e. $1 - 0·28$).

The value after five years will be $7500 \times 0·72^5 = £7500 \times 0·19349$
$$= £1451.19$$

Exercise 2E

1. A bank pays interest of 10% on money in deposit accounts. Mrs Wells puts £2000 in the bank. How much has she after:
   (a) one year   (b) two years   (c) three years?

2. A bank pays interest of 12%. Mr Olsen puts £5000 in the bank. How much has he after:
   (a) one year   (b) three years?

3. A computer operator is paid £10 000 a year. Assuming her pay is increased by 7% each year, what will her salary be in four years time?

4. Mrs Bergkamp's salary in 2000 is £30 000 per year. Every year her salary is increased by 5%.
   In 2001 her salary will be
   $30\,000 \times 1·05$ $= £31\,500$
   In 2002 her salary will be
   $30\,000 \times 1·05 \times 1·05$ $= £33\,075$
   In 2003 her salary will be
   $30\,000 \times 1·05 \times 1·05 \times 1·05 = £34\,728·75$
   And so on.
   (a) What will her salary be in 2004?
   (b) What will her salary be in 2006?

5. The price of a house was £90 000 in 1998. At the end of each year the price has increased by 6%.
   (a) Find the price of the house after 1 year.
   (b) Find the price of the house after 3 years.
   (c) Find the price of the house after 10 years.

6. Assuming an average inflation rate of 8%, work out the probable cost of the following items in 10 years:
   (a) car £6500   (b) T.V. £340   (c) house £50 000

7. A new car is valued at £15 000. At the end of each year its value is reduced by 15% of its value at the start of the year. What will it be worth after 3 years?

8. Twenty years ago a bus driver was paid £50 a week. He is now paid £185 a week. Assuming an average rate of inflation of 7%, has his pay kept up with inflation?

9. A bank pays interest of 11% on £6000 in a deposit account. After how many years will the money have trebled?

10. (a) Draw the graph of $y = 1 \cdot 08^x$ for values of x from 0 to 10.
    (b) Solve approximately the equation $1 \cdot 08^x = 2$.
    (c) Money is invested at 8% interest. After how many years will the money have doubled?

11. Which is the better investment over ten years:

    £20 000 at 12% compound interest
    or £30 000 at 8% compound interest?

Double percentages

In the real world, to make statements people often use percentages of quantities which are themselves percentages.

Example 1

Liz invests money in the stock market. During the first year her investment increases by 20%. During the second year her investment falls by 20%.
Is Liz's investment worth more, or less or the same as her original investment?

Using the multiplier technique:

At the end of the first year the value of Liz's investment is $1 \cdot 20 \times$ her original investment.

At the end of the second year the value of Liz's investment is $0{\cdot}80 \times$ its value at the start of the second year.
$= 0{\cdot}80 \times 1{\cdot}20 \times$ her original investment
$= 0{\cdot}96 \times$ her original investment

Thus after two years, the value of her investment is 96% of her original investment, hence it has lost 4% of its original value.

Example 2

Ben buys a car. In the first year it loses 32% of its value and during the second year it loses another 17% of its value. By what percentage has the value of the car decreased in the two years?

At the end of the first year the value of Ben's car is $0{\cdot}68$ of its original value.

At the end of the second year the value of the car is $0{\cdot}83 \times$ its value at the end of the first year.

After two years, the value of the car is $0{\cdot}83 \times 0{\cdot}68$ of its original value $= 0{\cdot}5644 = 43{\cdot}6\%$ (to 3 sf).

Therefore, Ben's car has lost 43·6% of its value in the two years.

2.2 Using a calculator

To use a calculator efficiently you sometimes have to think ahead ahead and make use of the memory, inverse $\boxed{1/x}$ and $\boxed{+/-}$ buttons.

In the example below the buttons are:

$\boxed{\sqrt{}}$	square root	$\boxed{y^x}$	raises number y to the power x
$\boxed{x^2}$	square	$\boxed{\text{Min}}$	puts number in memory [It automatically clears any number already in the memory when it puts in the new number.]
$\boxed{1/x}$	reciprocal	$\boxed{\text{MR}}$	recalls number from memory

Example

Evaluate the following to 4 significant figures:

(a) $\dfrac{2{\cdot}3}{4{\cdot}7 + 3{\cdot}61}$ Find the denominator first.

$\boxed{2{\cdot}3}$ $\boxed{\div}$ $\boxed{(}$ $\boxed{4{\cdot}7}$ $\boxed{+}$ $\boxed{3{\cdot}61}$ $\boxed{)}$ $\boxed{=}$ $\dfrac{2{\cdot}3}{4{\cdot}7 + 3{\cdot}61} = 0{\cdot}2768$ (to 4 sig. fig.)

(b) $\left(\dfrac{1}{0\cdot 084}\right)^4$

$\boxed{0\cdot 084}\ \boxed{1/x}\ \boxed{y^x}\ \boxed{4}\ \boxed{=}\quad \left(\dfrac{1}{0\cdot 084}\right)^4 = 20\,090$ (to 4 sig. fig.)

(c) $\sqrt[3]{[3\cdot 2 \times (1\cdot 7 - 1\cdot 64)]}$

$\boxed{1\cdot 7}\ \boxed{-}\ \boxed{1\cdot 64}\ \boxed{=}\ \boxed{\times}\ \boxed{3\cdot 2}\ \boxed{=}\ \boxed{y^x}\ \boxed{0\cdot 333\,333}\ \boxed{=}$

$\sqrt[3]{[3\cdot 2 \times (1\cdot 7 - 1\cdot 64)]} = 0\cdot 5769$ (to 4 sig. fig.)

Note: To find a cube root, raise to the power $\tfrac{1}{3}$.
or as a decimal $0\cdot 333\ldots$
Of course, if your calculator has a $\boxed{\sqrt[3]{\ }}$ button, use that.

Exercise 2F

Use a calculator to evaluate the following, giving the answers to 4 significant figures:

1. $\dfrac{7\cdot 351 \times 0\cdot 764}{1\cdot 847}$

2. $\dfrac{0\cdot 0741 \times 14700}{0\cdot 746}$

3. $\dfrac{0\cdot 0741 \times 9\cdot 61}{23\cdot 1}$

4. $\dfrac{417\cdot 8 \times 0\cdot 00841}{0\cdot 07324}$

5. $\dfrac{8\cdot 41}{7\cdot 601 \times 0\cdot 00847}$

6. $\dfrac{4\cdot 22}{1\cdot 701 \times 5\cdot 2}$

7. $\dfrac{9\cdot 61}{17\cdot 4 \times 1\cdot 51}$

8. $\dfrac{8\cdot 71 \times 3\cdot 62}{0\cdot 84}$

9. $\dfrac{0\cdot 76}{0\cdot 412 - 0\cdot 317}$

10. $\dfrac{81\cdot 4}{72\cdot 6 + 51\cdot 92}$

11. $\dfrac{111}{27\cdot 4 + 2960}$

12. $\dfrac{27\cdot 4 + 11\cdot 61}{5\cdot 9 - 4\cdot 763}$

13. $\dfrac{6\cdot 51 - 0\cdot 1114}{7\cdot 24 + 1\cdot 653}$

14. $\dfrac{5\cdot 71 + 6\cdot 093}{9\cdot 05 - 5\cdot 77}$

15. $\dfrac{0\cdot 943 - 0\cdot 788}{1\cdot 4 - 0\cdot 766}$

16. $\dfrac{2\cdot 6}{1\cdot 7} + \dfrac{1\cdot 9}{3\cdot 7}$

17. $\dfrac{8\cdot 06}{5\cdot 91} - \dfrac{1\cdot 594}{1\cdot 62}$

18. $\dfrac{4\cdot 7}{11\cdot 4 - 3\cdot 61} + \dfrac{1\cdot 6}{9\cdot 7}$

19. $\dfrac{3\cdot 74}{1\cdot 6 \times 2\cdot 89} - \dfrac{1}{0\cdot 741}$

20. $\dfrac{1}{7\cdot 2} - \dfrac{1}{14\cdot 6}$

21. $\dfrac{1}{0\cdot 961} \times \dfrac{1}{0\cdot 412}$

22. $\dfrac{1}{7} + \dfrac{1}{13} - \dfrac{1}{8}$

23. $4\cdot 2\left(\dfrac{1}{5\cdot 5} - \dfrac{1}{7\cdot 6}\right)$

24. $\sqrt{(9\cdot 61 + 0\cdot 1412)}$

25. $\sqrt{\left(\dfrac{8\cdot 007}{1\cdot 61}\right)}$

26. $(1\cdot 74 + 9\cdot 611)^2$

27. $\left(\dfrac{1\cdot 63}{1\cdot 7 - 0\cdot 911}\right)^2$

28. $\left(\dfrac{9\cdot 6}{2\cdot 4} - \dfrac{1\cdot 5}{0\cdot 74}\right)^2$

29. $\sqrt{\left(\dfrac{4\cdot 2 \times 1\cdot 611}{9\cdot 83 \times 1\cdot 74}\right)}$

30. $(0\cdot 741)^3$

31. $(1.562)^5$
32. $(0.32)^3 + (0.511)^4$
33. $(1.71 - 0.863)^6$
34. $\left(\dfrac{1}{0.971}\right)^4$
35. $\sqrt[3]{(4.714)}$
36. $\sqrt[3]{(0.9316)}$
37. $\sqrt[3]{\left(\dfrac{4.114}{7.93}\right)}$
38. $\sqrt[4]{(0.8145 - 0.799)}$
39. $\sqrt[5]{(8.6 \times 9.71)}$
40. $\sqrt[3]{\left(\dfrac{1.91}{4.2 - 3.766}\right)}$
41. $\left(\dfrac{1}{7.6} - \dfrac{1}{18.5}\right)^3$
42. $\dfrac{\sqrt{(4.79)} + 1.6}{9.63}$
43. $\dfrac{(0.761)^2 - \sqrt{(4.22)}}{1.96}$
44. $\sqrt[3]{\left(\dfrac{1.74 \times 0.761}{0.0896}\right)}$
45. $\left(\dfrac{8.6 \times 1.71}{0.43}\right)^3$
46. $\dfrac{9.61 - \sqrt{(9.61)}}{9.61^2}$
47. $\dfrac{9.6 \times 10^4 \times 3.75 \times 10^7}{8.88 \times 10^6}$
48. $\dfrac{8.06 \times 10^{-4}}{1.71 \times 10^{-6}}$
49. $\dfrac{3.92 \times 10^{-7}}{1.884 \times 10^{-11}}$
50. $\left(\dfrac{1.31 \times 2.71 \times 10^5}{1.91 \times 10^4}\right)^5$
51. $\left(\dfrac{1}{9.6} - \dfrac{1}{9.99}\right)^{10}$
52. $\dfrac{\sqrt[3]{(86.6)}}{\sqrt[4]{(4.71)}}$
53. $\dfrac{23.7 \times 0.0042}{12.48 - 9.7}$
54. $\dfrac{0.482 + 1.6}{0.024 \times 1.83}$
55. $\dfrac{8.52 - 1.004}{0.004 - 0.0083}$
56. $\dfrac{1.6 - 0.476}{2.398 \times 41.2}$
57. $\left(\dfrac{2.3}{0.791}\right)^7$
58. $\left(\dfrac{8.4}{28.7 - 0.47}\right)^3$
59. $\left(\dfrac{5.114}{7.332}\right)^5$
60. $\left(\dfrac{4.2}{2.3} + \dfrac{8.2}{0.52}\right)^3$

Rounding errors

A very common error occurs when numbers are rounded in the intermediate steps of a calculation.

Example

Find the area of a circle of circumference 25 cm, correct to 3 significant figures.

radius $= \dfrac{25}{2\pi} = 3.9788736\ldots$ ①

Suppose we round this number to 3 significant figures.

i.e. Take radius $= 3.98$ cm

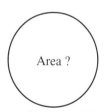

circumference = 25 cm

Then area $= \pi \times 3.98^2$
$$= 49.764084$$
$$= 49.8\,\text{cm}^2 \text{ to 3 s.f.}$$

The **correct** method is to use the radius shown in line ①.
Then area $= 49.7\,\text{cm}^2$ to 3 s.f.

Remember:
Avoid **early approximation** in calculations with several steps.

Exercise 2G

1. (a) Find the area of a circle of circumference 80 cm, giving your answer correct to 3 significant figures.
   (b) Repeat the calculation but this time round off the radius to 3 significant figures before you find the area. Compare your two answers.

2. Different makes of calculator use different logic chips which may produce slightly different answers.
   Work out $\left(\frac{1}{7}\right)^2 \times 321\,489$ and compare your answer with other people in the class (assuming there are different makes of calculator in the room!).

3. Many calculators have a $\boxed{\text{FIX}}$ mode in which the calculator will round off a number to 1 or 2 or 3 ... decimal places.
   Experiment with your calculator in $\boxed{\text{FIX}}$ mode, working out:
   (a) $7 \div 9$ (b) 11.73×6.947 (c) 0.001×0.007
   Write down what you observe.

Exercise 2H

On a calculator work out $9508^2 + 192^2 + 10^2 + 6$.
If you turn the calculator upside down and use a little imagination, you can see the word 'HEDGEHOG'.
Find the words given by the clues.

1. $19 \times 20 \times 14 - 2.66$ (Not an upstanding man)
2. $(84 + 17) \times 5$ (Dotty message)
3. $904^2 + 89\,621\,818$ (Prickly customer)
4. $(559 \times 6) + (21 \times 55)$ (What a surprise!)
5. $566 \times 711 - 23\,617$ (Bolt it down)
6. $\dfrac{9999 + 319}{8.47 + 2.53}$ (Sit up and plead)

7. $\dfrac{2601 \times 6}{4^2 + 1^2}$; $(401 - 78) \times 5^2$ (two words) (Not a great man)
8. $0.4^2 - 0.1^2$ (Little Sidney)
9. $\dfrac{(27 \times 2000 - 2)}{(0.63 \div 0.09)}$ (Not quite a mountain)
10. $(5^2 - 1^2)^4 - 14239$ (Just a name)
11. $48^4 + 102^2 - 4^2$ (Pursuits)
12. $615^2 + (7 \times 242)$ (Almost a goggle)
13. $(130 \times 135) + (23 \times 3 \times 11 \times 23)$ (Wobbly)
14. $164 \times 166^2 + 734$ (Almost big)
15. $8794^2 + 25 \times 342.28 + 120 \times 25$ (Thin skin)
16. $0.08 - (3^2 \div 10^4)$ (Ice house)
17. $235^2 - (4 \times 36.5)$ (Shiny surface)
18. $(80^2 + 60^2) \times 3 + 81^2 + 12^2 + 3013$ (Ship gunge)
19. $3 \times 17 \times (329^2 + 2 \times 173)$ (Unlimited)
20. $230 \times 230\tfrac{1}{2} + 30$ (Fit feet)
21. $33 \times 34 \times 35 + 15 \times 3$ (Beleaguer)
22. $0.32^2 + \tfrac{1}{1000}$ (Did he or didn't he?)
23. $(23 \times 24 \times 25 \times 26) + (3 \times 11 \times 10^3) - 20$ (Help)
24. $(16^2 + 16)^2 - (13^2 - 2)$ (Slander)
25. $(3 \times 661)^2 - (3^6 + 22)$ (Pester)
26. $(22^2 + 29.4) \times 10$; $(3.03^2 - 0.02^2) \times 100^2$ (Four words) (Goliath)
27. $1.25 \times 0.2^6 + 0.2^2$ (Tissue time)
28. $(710 + (1823 \times 4)) \times 4$ (Liquor)
29. $(3^3)^2 + 2^2$ (Wriggler)
30. $14 + (5 \times (83^2 + 110))$ (Bigger than a duck)
31. $2 \times 3 \times 53 \times 10^4 + 9$ (Opposite to hello, almost!)
32. $(177 \times 179 \times 182) + (85 \times 86) - 82$ (Good salesman)
33. $14^4 - 627 + 29$ (Good book, by God!)
34. $6.2 \times 0.987 \times 1\,000\,000 - 860^2 + 118$ (Flying ace)
35. $(426 \times 474) + (318 \times 487) + 22\,018$ (Close to a bubble)
36. $\dfrac{36^3}{4} - 1530$ (Foreign-sounding girl's name)

37. $(7^2 \times 100) + (7 \times 2)$ (Lofty)

38. $240^2 + 134$; $241^2 - 7^3$ (two words)   (Devil of a chime)

39. $(2 \times 2 \times 2 \times 2 \times 3)^4 + 1929$ (Unhappy ending)

40. $141\,918 + 83^3$ (Hot stuff in France)

2.3 Indices

Six basic rules

1. $a^n \times a^m = a^{n+m}$

    $7^2 \times 7^4 = (7 \times 7) \times (7 \times 7 \times 7 \times 7) = 7^6$
    so $7^2 \times 7^4 = 7^{2+4} = 7^6$

2. $a^n \div a^m = a^{n-m}$

    $3^5 \div 3^2 = \dfrac{3 \times 3 \times 3 \times \cancel{3} \times \cancel{3}}{\cancel{3} \times \cancel{3}} = 3^3$
    so $3^5 \div 3^2 = 3^{5-2} = 3^3$

3. $(a^n)^m = a^{nm}$

    $(2^2)^3 = (2 \times 2)(2 \times 2)(2 \times 2) = 2^6$
    so $(2^2)^3 = 2^{2 \times 3} = 2^6$

4. $a^{-n} = \dfrac{1}{a^n}$

    Consider this sequence
    $2^3 \quad 2^2 \quad 2^1 \quad 2^0 \quad 2^{-1} \quad 2^{-2}$
    $\downarrow \quad \downarrow \quad \downarrow \quad \downarrow \quad \downarrow \quad \downarrow$
    $8 \quad 4 \quad 2 \quad 1 \quad \tfrac{1}{2} \quad \tfrac{1}{4}$

    Check: $5^3 \div 5^5 = \dfrac{\cancel{5} \times \cancel{5} \times \cancel{5}}{\cancel{5} \times \cancel{5} \times \cancel{5} \times 5 \times 5} = \dfrac{1}{5^2} = 5^{-2}$

5. $a^{\frac{1}{n}}$ means 'the nth root of a'

    $3^{\frac{1}{2}} \times 3^{\frac{1}{2}} = \sqrt{3} \times \sqrt{3}$
    Using Rule 1:
    $3^{\frac{1}{2}} \times 3^{\frac{1}{2}} = 3^1$

6. $a^{\frac{m}{n}}$ means 'the nth root of a raised to the power m'.

    $4^{\frac{3}{2}} = (\sqrt{4})^3 = 8$

Remember:
- $x^0 = 1$ for any value of x which is not zero
- $x^{-1} = \dfrac{1}{x}$

The *reciprocal* of x is $\dfrac{1}{x}$.

Example

Simplify (a) $x^7 \times x^{13} = x^{7+13} = x^{20}$

(b) $x^3 \div x^7 = x^{3-7} = x^{-4} = \dfrac{1}{x^4}$

(c) $(x^4)^3 = x^{12}$

(d) $(3x^2)^3 = 3^3 \times (x^2)^3 = 27x^6$

(e) $3y^2 \times 4y^3 = 12y^5$

Here is the graph of $y = 2^x$.
A calculator with a button marked $\boxed{x^y}$ can be used to work out difficult powers.
e.g. for $2^{1.7}$ press $\boxed{2}$ $\boxed{x^y}$ $\boxed{1.7}$ $\boxed{=}$

$2^{1.7} = 3.249$ (4 s.f.)

Some calculators have a button marked 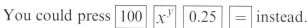 to work out roots of numbers.
e.g. the fourth root of 100 is $100^{\frac{1}{4}}$

Press $\boxed{100}$ $\boxed{x^{\frac{1}{y}}}$ $\boxed{4}$ $\boxed{=}$

$100^{\frac{1}{4}} = 3.162$ (4 s.f.)

You could press $\boxed{100}$ $\boxed{x^y}$ $\boxed{0.25}$ $\boxed{=}$ instead.

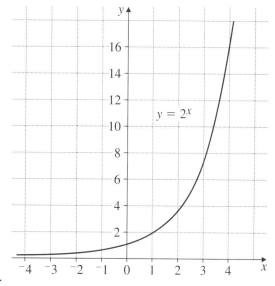

Exercise 21

Express in index form.

1. $3 \times 3 \times 3 \times 3$
2. $4 \times 4 \times 5 \times 5 \times 5$
3. $3 \times 7 \times 7 \times 7$
4. $2 \times 2 \times 2 \times 7$
5. $\dfrac{1}{10 \times 10 \times 10}$
6. $\dfrac{1}{2 \times 2 \times 3 \times 3 \times 3}$
7. $\sqrt{15}$
8. $\sqrt[3]{3}$
9. $a \times a \times a \times a \times a \times a \times a \times a$
10. $\sqrt[5]{10}$
11. $\dfrac{1}{11}$
12. $(\sqrt{5})^3$

In Questions **13** to **32** answer true or false.

13. $2^3 = 8$
14. $3^2 = 6$
15. $5^3 = 125$
16. $2^{-1} = \frac{1}{2}$
17. $10^{-2} = \frac{1}{20}$
18. $3^{-3} = \frac{1}{9}$
19. $2^2 > 2^3$
20. $2^3 < 3^2$
21. $2^{-2} > 2^{-3}$
22. $3^{-2} < 3$
23. $1^9 = 9$
24. $(^-3)^2 = {^-9}$
25. $5^{-2} = \frac{1}{10}$
26. $10^{-3} = \frac{1}{1000}$
27. $10^{-2} > 10^{-3}$
28. $5^{-1} = 0.2$
29. $10^{-1} = 0.1$
30. $2^{-2} = 0.25$
31. $5^0 = 1$
32. $16^0 = 0$

Write in a more simple form.

33. $5^2 \times 5^4$
34. $6^3 \times 6^2$
35. $2^3 \times 2^{10}$
36. $3^6 \times 3^{-2}$
37. $6^7 \div 6^2$
38. $8^5 \div 8^4$
39. $6^2 \div 6^{-2}$
40. Half of 2^6
41. Half of 2^{20}
42. $10^{100} \div 10^0$
43. $5^{-2} \times 5^{-3}$
44. $3^{-2} \times 3^{-8}$
45. $\dfrac{3^4 \times 3^5}{3^2}$
46. $\dfrac{2^8 \times 2^4}{2^5}$
47. $\dfrac{7^3 \times 7^3}{7^4}$
48. $\dfrac{5^9 \times 5^{10}}{5^{20}}$

Exercise 2J

Write in a more simple form.

1. $(3^3)^2$
2. $(5^4)^3$
3. $(7^2)^5$
4. $(x^2)^3$
5. $(2^{-1})^2$
6. $(^-1)^{102}$
7. $(7^{-1})^{-2}$
8. $(y^4)^{\frac{1}{2}}$
9. Half of 2^{22}
10. One-tenth of 10^{100}

Simplify:

11. $x^3 \times x^4$
12. $y^6 \times y^7$
13. $z^7 \div z^3$
14. $z^{50} \times z^{50}$
15. $m^3 \div m^2$
16. $e^{-3} \times e^{-2}$
17. $y^{-2} \times y^4$
18. $w^4 \div w^{-2}$
19. $y^{\frac{1}{2}} \times y^{\frac{1}{2}}$
20. $(x^2)^5$
21. $x^{-2} \div x^{-2}$
22. $w^{-3} \times w^{-2}$
23. $w^{-7} \times w^2$
24. $x^3 \div x^{-4}$
25. $(a^2)^4$
26. $(k^{\frac{1}{2}})^6$
27. $e^{-4} \times e^4$
28. $x^{-1} \times x^{30}$
29. $(y^4)^{\frac{1}{2}}$
30. $(x^{-3})^{-2}$
31. $z^2 \div z^{-2}$
32. $t^{-3} \div t$
33. $(2x^3)^2$
34. $(4y^5)^2$
35. $2x^2 \times 3x^2$
36. $5y^3 \times 2y^2$
37. $5a^3 \times 3a$
38. $(2a)^3$
39. $3x^3 \div x^3$
40. $8y^3 \div 2y$
41. $10y^2 \div 4y$
42. $8a \times 4a^3$
43. $(2x)^2 \times (3x)^3$
44. $4z^4 \times z^{-7}$
45. $6x^{-2} \div 3x^2$
46. $5y^3 \div 2y^{-2}$
47. $(x^2)^{\frac{1}{2}} \div (x^{\frac{1}{3}})^3$
48. $7w^{-2} \times 3w^{-1}$
49. $(2n)^4 \div 8n^0$
50. $4x^{\frac{3}{2}} \div 2x^{\frac{1}{2}}$

Example

Evaluate the following.

(a) $9^{\frac{1}{2}} = \sqrt{9} = 3$
(b) $5^{-1} = \frac{1}{5}$

(c) $4^{-\frac{1}{2}} = \dfrac{1}{4^{\frac{1}{2}}} = \dfrac{1}{\sqrt{4}} = \dfrac{1}{2}$ \quad (d) $25^{\frac{3}{2}} = (\sqrt{25})^3 = 5^3 = 125$

(e) $(5^{\frac{1}{2}})^3 \times 5^{\frac{1}{2}} = 5^{\frac{3}{2}} \times 5^{\frac{1}{2}} = 5^2 = 25$ \quad (f) $7^0 = 1$
$[a^0 = 1$ for any non-zero value of $a]$

Exercise 2K

Evaluate the following.

1. $3^2 \times 3$
2. 100^0
3. 3^{-2}
4. $(5^{-1})^{-2}$
5. $4^{\frac{1}{2}}$
6. $16^{\frac{1}{2}}$
7. $81^{\frac{1}{2}}$
8. $8^{\frac{1}{3}}$
9. $9^{\frac{3}{2}}$
10. $27^{\frac{1}{3}}$
11. $9^{-\frac{1}{2}}$
12. $8^{-\frac{1}{3}}$
13. $1^{\frac{5}{2}}$
14. $25^{-\frac{1}{2}}$
15. $1000^{\frac{1}{3}}$
16. $2^{-2} \times 2^5$
17. $2^4 \div 2^{-1}$
18. $8^{\frac{2}{3}}$

19. Copy and complete the crossnumber puzzle.

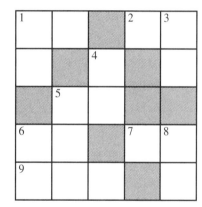

Across
1. $\sqrt{2^8}$
2. $169^{\frac{1}{2}}$
5. $7^2 \times 5^0$
6. $2^4 \times 5$
7. 3×2^5
9. 5^3

Down
1. $(1\,000\,000)^{\frac{1}{6}}$
3. $3^3 + 2^2 + 1^1$
4. $10^2 - 7^0$
5. $20^2 + 2^1$
6. $3^7 \div 27$
8. Half of 2^7

20. Find the missing index.
    (a) $\sqrt{80} = 80^{\square}$ \quad (b) $\dfrac{1}{\sqrt{3}} = 3^{\square}$ \quad (c) $(\sqrt{2})^3 = 2^{\square}$

21. Work out:
    (a) $27^{-\frac{2}{3}}$ \quad (b) $4^{-\frac{3}{2}}$ \quad (c) One per cent of 100^7

Evaluate the following.

22. $10\,000^{\frac{1}{4}}$
23. $100^{\frac{3}{2}}$
24. $(100^{\frac{1}{2}})^{-3}$
25. $(9^{\frac{1}{2}})^{-2}$
26. $(^-16 \cdot 371)^0$
27. $81^{\frac{1}{4}} \div 16^{\frac{1}{4}}$
28. $(5^{-4})^{\frac{1}{2}}$
29. $1000^{-\frac{1}{3}}$
30. $(4^{-\frac{1}{2}})^2$
31. $8^{-\frac{2}{3}}$
32. $100^{\frac{5}{2}}$
33. $1^{\frac{4}{3}}$
34. 2^{-5}
35. $(0 \cdot 01)^{\frac{1}{2}}$
36. $(0 \cdot 04)^{\frac{1}{2}}$

37. $(2 \cdot 25)^{\frac{1}{2}}$ 38. $(7 \cdot 63)^0$ 39. $3^5 \times 3^{-3}$

40. $(3\frac{3}{8})^{\frac{1}{3}}$ 41. $(11\frac{1}{9})^{-\frac{1}{2}}$ 42. $(\frac{1}{8})^{-2}$

43. $(\frac{1}{1000})^{\frac{2}{3}}$ 44. $(\frac{9}{25})^{-\frac{1}{2}}$ 45. $(10^{-6})^{\frac{1}{3}}$

46. (a) Copy and complete the sentence.
    'The inverse operation of raising a positive number to power n is raising the result of this operation to power $\square$.'
    (b) Write down two numerical examples to illustrate this result.

2.4 Standard form

When dealing with either very large or very small numbers, it is not convenient to write them out in full in the normal way. It is better to use standard form. Most calculators represent large and small numbers in this way:

The number $a \times 10^n$ is in standard form when $1 \leqslant a < 10$ and n is a positive or negative integer.

e.g.   This calculator shows $2 \cdot 3 \times 10^8$.

$$\boxed{2.3 \quad 08}$$

Example 1
Write the following numbers in standard form:

(a) $2000 = 2 \times 1000 = 2 \times 10^3$ (b) $150 = 1 \cdot 5 \times 100 = 1 \cdot 5 \times 10^2$

(c) $0 \cdot 0004 = 4 \times \dfrac{1}{10\,000} = 4 \times 10^{-4}$ (d) $450 \times 10^6 = (4 \cdot 5 \times 10^2) \times 10^6$
$\qquad\qquad\qquad\qquad\qquad\qquad\qquad\qquad = 4 \cdot 5 \times 10^8$

(e) $0 \cdot 07 \times 10^{-4} = (7 \times 10^{-2}) \times 10^{-4}$
$\qquad\qquad\quad = 7 \times 10^{-6}$

Example 2
Calculate: (a) $(4 \times 10^5) \times (8 \times 10^{-9})$ (b) $(4 \times 10^5) \div (8 \times 10^{-9})$

(a) $(4 \times 10^5) \times (8 \times 10^{-9}) = 32 \times 10^{5+(-9)}$
$\qquad\qquad\qquad\qquad\qquad = 32 \times 10^{-4}$
$\qquad\qquad\qquad\qquad\qquad = 3 \cdot 2 \times 10^{-3}$

(b) $(4 \times 10^5) \div (8 \times 10^{-9}) = \dfrac{4}{8} \times \dfrac{10^5}{10^{-9}}$

$\phantom{(b) (4 \times 10^5) \div (8 \times 10^{-9}) } = 0{\cdot}5 \times 10^{5-(-9)}$

$\phantom{(b) (4 \times 10^5) \div (8 \times 10^{-9}) } = 0{\cdot}5 \times 10^{14}$

$\phantom{(b) (4 \times 10^5) \div (8 \times 10^{-9}) } = 5 \times 10^{13}$

Exercise 2L

Write the following numbers in standard form:

1. 4000
2. 500
3. 70 000
4. 60
5. 2400
6. 380
7. 2560
8. 0·007
9. 0·0004
10. 0·0035
11. 0·421
12. 0·000 055
13. 0·01
14. 564 000
15. 19 million

16. The population of China is estimated at 1 300 000 000. Write this in standard form.

17. A hydrogen atom has a mass of 0·000 000 000 000 000 000 000 001 67 grams. Write this mass in standard form.

18. A certain virus is 0·000 000 000 25 cm in diameter. Write this in standard form.

19. Avogadro's number is 602 300 000 000 000 000 000 000. Express this in standard form.

20. A very rich oil sheikh leaves his fortune of £$3{\cdot}6 \times 10^8$ to be divided between 100 relatives. How much does each relative receive? Give the answer in standard form.

21. If $a = 512 \times 10^2$
    $b = 0{\cdot}478 \times 10^6$
    $c = 0{\cdot}0049 \times 10^7$
    arrange a, b and c in order of size (smallest first).

22. To work out $(3 \times 10^2) \times (2 \times 10^5)$, **without** a calculator:
    A Multiply: $3 \times 2 = 6$
    B Add the powers of 10: $10^2 \times 10^5 = 10^7$
    So $(3 \times 10^2) \times (2 \times 10^5) = 6 \times 10^7$

    Without a calculator, work out:
    (a) $(2 \times 10^8) \times (4 \times 10^3)$
    (b) $(2{\cdot}5 \times 10^3) \times (2 \times 10^{-10})$
    (c) $(1{\cdot}1 \times 10^{-6}) \times (5 \times 10^2)$
    (d) $(8 \times 10^{12}) \div (2 \times 10^3)$
    (e) $(9 \times 10^6) \div (3 \times 10^{-2})$
    (f) $(2 \times 10^6)^2$

23. Write out these numbers in full.
    (a) 2.3×10^4
    (b) 3×10^{-2}
    (c) 5.6×10^2
    (d) 8×10^5
    (e) 2.2×10^{-3}
    (f) 9×10^8
    (g) 6×10^{-1}
    (h) 7×10^3
    (i) 3.14×10^6

24. Without a calculator, work out the following and give your answers in standard form.
    (a) $(6 \times 10^5) + (5 \times 10^4)$
    (b) $(3 \times 10^3) - (3 \times 10^2)$
    (c) $(3.5 \times 10^6) + (4 \times 10^4)$
    (d) $(6.2 \times 10^{-3}) + (8 \times 10^{-4})$

25. Rewrite the following numbers in standard form.
    (a) 0.6×10^4
    (b) 11×10^{-2}
    (c) 450×10^6
    (d) 0.085×10^{-4}
    (e) 6 billion
    (f) $\frac{1}{2}$

Using a calculator

Example

(a) Work out $(5 \times 10^7) \times (3 \times 10^{12})$

Use the $\boxed{\text{EXP}}$ button as follows:

$\boxed{5}$ $\boxed{\text{EXP}}$ $\boxed{7}$ $\boxed{\times}$ $\boxed{3}$ $\boxed{\text{EXP}}$ $\boxed{12}$ $\boxed{=}$ Answer $= 1.5 \times 10^{20}$

Notice that you do NOT press the $\boxed{\times}$ button after the $\boxed{\text{EXP}}$ button!

(b) Work out $(3.2 \times 10^3) \div (8 \times 10^{-7})$

$\boxed{3.2}$ $\boxed{\text{EXP}}$ $\boxed{3}$ $\boxed{\div}$ $\boxed{8}$ $\boxed{\text{EXP}}$ $\boxed{7}$ $\boxed{+/-}$ $\boxed{=}$ Answer $= 4 \times 10^9$
                                                    ↑
[Press this key for negative.]

(c) $(5 \times 10^{-8})^2$

$\boxed{5}$ $\boxed{\text{EXP}}$ $\boxed{8}$ $\boxed{+/-}$ $\boxed{x^2}$ Answer $= 2.5 \times 10^{-15}$

Exercise 2M

Use a calculator to work out the following and write the answer in standard form.

1. $2000 \times 30\,000$
2. $40\,000 \times 500$
3. $25\,000 \times 600\,000$
4. $3500 \times 2\,\text{million}$
5. $600\,000 \times 1500$
6. $(40\,000)^2$
7. $18\,000 \div 400$
8. $(4 \times 10^5) \times (3 \times 10^8)$
9. $(6 \cdot 2 \times 10^4) \times (3 \times 10^6)$
10. $(5 \times 10^{-4}) \times (4 \times 10^{-7})$
11. $(3 \times 10^8) \times (2 \cdot 5 \times 10^{-20})$
12. $(5 \times 10^7) \div (2 \times 10^{-2})$
13. $(3 \times 10^5)^3$
14. $(7 \times 10^{-2}) \div (2 \times 10^4)$
15. $(9 \times 10^{-11})^2$
16. $(4 \cdot 2 \times 10^8) \times (1 \cdot 5 \times 10^5)$
17. $(3 \times 10^{-4}) \div (2 \times 10^{-20})$
18. $(4 \times 10^5) \times (2 \times 10^{-2})$
19. $(1 \cdot 4 \times 10^{-1}) \times (2 \times 10^{17})$
20. $(2 \times 10^5) \times (4 \times 10^{-8})$
21. $(8 \times 10^{-5}) \div (2 \times 10^{-3})^2$
22. $(2 \times 10^{-4})^3 \div (1 \cdot 6 \times 10^8)$
23. $10^5 \div (2 \times 10^8)$
24. $(3 \times 10^{-7}) \times 10^{-4}$

25. A patient in hospital is very ill. Between noon and midnight one day the number of viruses in her body increased from 3×10^7 to $5 \cdot 6 \times 10^9$.
    Work out the increase in the number of viruses, giving your answer in standard form.

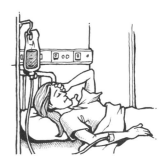

26.

Continent	Population	Area (m^2)
Europe	$6 \cdot 82 \times 10^8$	$1 \cdot 05 \times 10^{10}$
Asia	$2 \cdot 96 \times 10^9$	$4 \cdot 35 \times 10^{10}$

Population density $= \dfrac{\text{Population}}{\text{Area}}$

Which of these two continents has the larger population density?

27. Here is a statement about roads:

    At a constant 80 kilometres per hour it would take 35 years to travel over the entire road network of the world.

    Taking a year as 365 days, calculate the length of the road network of the world, giving your answer in standard form.

28. Oil flows through a pipe at a rate of $40\,\text{m}^3/\text{s}$. How long will it take to fill a tank of volume $1{\cdot}2 \times 10^5\,\text{m}^3$?

29. The mean weight of all the men, women, children and babies in the UK is $42{\cdot}1$ kg. The population of the UK is 56 million. Work out the total weight of the entire population giving your answer in kg in standard form. Give your answer to a sensible degree of accuracy.

30. A light year is the distance travelled by a beam of light in a year. Light travels at a speed of approximately 3×10^5 km/s.
    (a) Work out the length of a light year in km.
    (b) Light takes about 8 minutes to reach the Earth from the Sun. How far is the Earth from the Sun in km?

31. (a) The number 10 to the power 100 (10 000 sexdecillion) is called a 'googol'! Write 60 googols in standard form.
    If it takes $\frac{1}{5}$ second to write a zero and $\frac{1}{10}$ second to write a 'one', how long would it take to write the number 100 'googols' in full?
    (b) The number 10 to the power of a 'googol' is called a 'googolplex'. Using the same speed of writing, how long in years would it take to write 1 'googolplex' in full? You may assume that your pen has enough ink.

2.5 Rational and irrational numbers

- A rational number can always be written exactly in the form $\frac{a}{b}$ where a and b are whole numbers.
  $\frac{3}{7}$ $1\frac{1}{2} = \frac{3}{2}$ $5{\cdot}14 = \frac{257}{50}$ $0{\cdot}\dot{6} = \frac{2}{3}$ are all rational numbers.
- An irrational number cannot be written in the form $\frac{a}{b}$.
  $\sqrt{2}$, $\sqrt{5}$, π, $\sqrt[3]{2}$ are all irrational numbers.
- In general $\sqrt{n}$ is irrational unless n is a square number.

In this triangle the length of the hypotenuse is *exactly* $\sqrt{5}$. On a calculator, $\sqrt{5} = 2{\cdot}236068$. This value of $\sqrt{5}$ is *not* exact and is correct only to 6 decimal places.

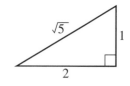

2.6 Recurring decimals

Any recurring decimal can be written as a fraction and hence is a rational number.

Example 1
Express $0.\dot{3}\dot{9}$ as a fraction.

$0.\dot{3}\dot{9}$ is the recurring decimal $0.39393939\ldots$

Let $r = 0.\dot{3}\dot{9}$

Multiplying by 100:

$100r = 39.\dot{3}\dot{9}$

(As there are **two** recurring digits, multiply by 10^2.)

Subtracting $r = 0.\dot{3}\dot{9} \Rightarrow 99r = 39$

$$r = \frac{39}{99} = \frac{13}{33}$$

Example 2
Express $0.\dot{6}3\dot{1}$ as a fraction.

Let $r = 0.631\,631\,631\ldots$

Multiplying by 1000:

$1000r = 631.631\,631\,631\ldots$

(As there are **three** recurring digits, multiply by 10^3.)

Subtract, $999r = 631$

$$r = \frac{631}{999}$$

Example 3
Express $0.4\dot{3}\dot{9}$ as a fraction.

Using $0.\dot{3}\dot{9} = \frac{13}{33}$ (from Example 1)

$$0.0\dot{3}\dot{9} = \frac{13}{330}$$

$\therefore \quad 0.4\dot{3}\dot{9} = 0.4 + 0.0\dot{3}\dot{9}$

$$= \frac{4}{10} + \frac{13}{330}$$

$$= \frac{132 + 13}{330} = \frac{145}{330} = \frac{29}{66}$$

or Let $r = 0.43\dot{9}$

$100r = 43.93\dot{9}$

(multiply by 10^2 as there are **two** recurring digits.

$99r = 43.5$

$\therefore \quad r = \dfrac{43.5}{99} = \dfrac{435}{990}$

$= \dfrac{29}{66}$

Note: You can often check using a calculator that your fraction represents the same decimal, for example: $29 \div 66$ gives $0.439\,393\,9\ldots$

Exercise 2N

1. Which of the following numbers are rational?

   $\dfrac{\pi}{2}$ $\quad \sqrt{5} \quad$ $(\sqrt{17})^2$ $\quad \sqrt{3}$

   3.14 $\quad \dfrac{\sqrt{12}}{\sqrt{3}} \quad$ π^2 $\quad 3^{-1} + 3^{-2}$

   $7^{-\frac{1}{2}}$ $\quad \dfrac{22}{7} \quad$ $\sqrt{2}+1$ $\quad \sqrt{2.25}$

2. (a) Write down any rational number between 4 and 6.
   (b) Write down any irrational number between 4 and 6.
   (c) Find a rational number between $\sqrt{2}$ and $\sqrt{3}$.
   (d) Write down any rational number between π and $\sqrt{10}$.

3. (a) For each triangle use Pythagoras' theorem to calculate the length x.
   (b) For each triangle state whether the **perimeter** is rational or irrational.
   (c) For each triangle state whether the **area** is rational or irrational.
   (d) In which triangle is $\sin \theta$ an irrational number?

Triangle A

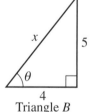
Triangle B

4. The diagram shows a circle of radius 3 cm drawn inside a square. Write down the exact value of the following and state whether the answer is rational or not:
   (a) the circumference of the circle
   (b) the diameter of the circle
   (c) the area of the square
   (d) the area of the circle
   (e) the shaded area.

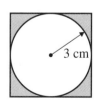

5. Write in the form $\frac{a}{b}$:
   (a) $0.\dot{2}$
   (b) $0.\dot{2}\dot{9}$
   (c) $0.\dot{5}4\dot{1}$

6. Think of two **irrational** numbers x and y such that $\frac{x}{y}$ is a **rational** number.

7. Explain the difference between a rational number and an irrational number.

8. Write as a fraction leaving your answer in its simplest form:
   (a) $0.\dot{3}\dot{5}$
   (b) $0.4\dot{5}$
   (c) $0.08\dot{2}$
   (d) $0.4\dot{3}\dot{5}$
   (e) $0.2\dot{8}\dot{9}$
   (f) $0.3\dot{1}\dot{8}$

2.7 Surds

- Numbers like $\sqrt{3}$ and $\sqrt{7}$ are called **surds**. The following rules apply:
- $\sqrt{a} \times \sqrt{b} = \sqrt{ab}$
- $\frac{\sqrt{a}}{\sqrt{b}} = \sqrt{\frac{a}{b}}$

Example
Simplify: (a) $\sqrt{2} \times \sqrt{3}$ (b) $\sqrt{80}$ (c) $\frac{\sqrt{27}}{\sqrt{3}}$

(a) $\sqrt{2} \times \sqrt{3} = \sqrt{6}$

(b) $\sqrt{80} = \sqrt{4 \times 20}$
$= \sqrt{4} \times \sqrt{20}$
$= 2 \times \sqrt{4 \times 5}$
$= 2 \times \sqrt{4} \times \sqrt{5}$
$= 2 \times 2\sqrt{5}$
$= 4 \times \sqrt{5}$

Alternatively, you could simplify $\sqrt{80}$ as
$\sqrt{80} = \sqrt{16 \times 5}$
$= \sqrt{16} \times \sqrt{5}$
$= 4\sqrt{5}$

(c) $\dfrac{\sqrt{27}}{\sqrt{3}} = \sqrt{\dfrac{27}{3}}$

$= \sqrt{9}$

$= 3$

- The fraction $\dfrac{6}{\sqrt{3}}$ can be written with a rational denominator by multiplying numerator and denominator by $\sqrt{3}$:

$$\dfrac{6}{\sqrt{3}} = \dfrac{6 \times \sqrt{3}}{\sqrt{3} \times \sqrt{3}} = \dfrac{6\sqrt{3}}{3}$$

$$= 2\sqrt{3}.$$

Example

Using $192 = 2^6 \times 3$, simplify $\dfrac{\sqrt{192}}{\sqrt{20}}$.

$$\dfrac{\sqrt{192}}{\sqrt{20}} = \dfrac{\sqrt{2^6 \times 3}}{\sqrt{4 \times 5}}$$

$$= \dfrac{\sqrt{2^6} \times \sqrt{3}}{\sqrt{4} \times \sqrt{5}}$$

$$= \dfrac{2^3 \times \sqrt{3}}{2\sqrt{5}}$$

$$= \dfrac{4\sqrt{3}}{5}$$

$$= \dfrac{4}{5}\sqrt{15}$$

Note: $\sqrt{4}$ usually means 'the positive square root of 4'. So $\sqrt{4} = 2$ only and not ± 2.

Notice that the solutions of the equation $x^2 = 4$ are $x = 2, ^-2$.
You can write $x = \sqrt[+]{4}$ or $\sqrt[-]{4}$.

Exercise 20

1. Sort these into four pairs of equal value.

   A $\sqrt{20}$ B $\sqrt{12}$ C $\sqrt{3}$ D $2\sqrt{3}$

   E 2   F $\sqrt{10} \times \sqrt{2}$ G $\dfrac{\sqrt{20}}{\sqrt{5}}$ H $\dfrac{\sqrt{15}}{\sqrt{5}}$

2. Answer 'True' or 'False'.
   (a) $\sqrt{28} = 2\sqrt{7}$
   (b) $(\sqrt{8})^2 = 8$
   (c) $\sqrt{16} = \pm 4$
   (d) $\sqrt{\frac{39}{3}} = \sqrt{13}$
   (e) $\sqrt{4} + \sqrt{4} = \sqrt{8}$
   (f) If $x^2 = 9$, $x = \pm 3$

3. Remove the brackets and simplify.
   (a) $(1 + \sqrt{2})^2$
   (b) $(2 - \sqrt{3})^2$
   (c) $(\sqrt{2} + \sqrt{8})^2$
   (d) $(3 - \sqrt{5})^2$
   (e) $(\sqrt{2} + 2)^2$
   (f) $(\sqrt{18} - \sqrt{2})^2$

4. Answer 'True' or 'False'.
   (a) $\sqrt{2} + \sqrt{2} = 2\sqrt{2}$
   (b) $\frac{\sqrt{8}}{2} = \sqrt{2}$
   (c) $\sqrt{100} = \pm 10$
   (d) $(\sqrt{2})^4 = 4$
   (e) $\sqrt{9 + 16} = \sqrt{9} + \sqrt{16}$
   (f) $(1 + \sqrt{3})^2 = 4 + 2\sqrt{3}$

5. Without using a calculator, simplify the following. Write your answers using surds where necessary.
   (a) $\sqrt{8} \times \sqrt{2}$
   (b) $\sqrt{32}$
   (c) $\sqrt{300}$
   (d) $\sqrt{18} \times 3$
   (e) $\sqrt{5} + 4\sqrt{5}$
   (f) $7\sqrt{3} - 2\sqrt{3}$
   (g) $\sqrt{20} + \sqrt{45}$
   (h) $\sqrt{75} - \sqrt{48}$
   (i) $\frac{\sqrt{8}}{\sqrt{2}}$
   (j) $\frac{\sqrt{27}}{\sqrt{12}}$
   (k) $\frac{\sqrt{125}}{\sqrt{20}}$
   (l) $\frac{\sqrt{80}}{\sqrt{45}}$

6. (a) Work out $\frac{6}{\sqrt{3}} \times \frac{\sqrt{3}}{\sqrt{3}}$.
   (b) Rationalise the denominators of these fractions.
      (i) $\frac{8}{\sqrt{2}}$
      (ii) $\frac{12}{\sqrt{3}}$
      (iii) $\frac{12}{\sqrt{8}}$
      (iv) $\frac{2}{\sqrt{3}}$

2.8 Substitution

Expressions

In general an expression contains algebraic terms and numbers but there is **no equals sign**.

Here are some expressions: $3x^2$, $2x(x - 1)$, $\frac{x + 1}{x}$, $y^3 - 3y^2$

Example

When $a = 3$, $b = ^-2$, $c = 5$, find the value of each expression.

(a) $3a + b = (3 \times 3) + (^-2)$
   $= 9 - 2$
   $= 7$

(b) $ac + b^2 = (3 \times 5) + (^-2)^2$
   $= 15 + 4$
   $= 19$

(c) $a(c - b)$
$= 3[5 - (^-2)]$
$= 3[7]$
$= 21$

Notice that working **down** the page is often easier to follow.

Exercise 2P
Evaluate the following expressions.
For questions **1** to **12** $a = 3$, $c = 2$, $e = 5$.

1. $3a - 2$
2. $4c + e$
3. $2c + 3a$
4. $5e - a$
5. $e - 2c$
6. $e - 2a$
7. $4c + 2e$
8. $7a - 5e$
9. $c - e$
10. $10a + c + e$
11. $a + c - e$
12. $a - c - e$

For questions **13** to **24** $h = 3$, $m = {}^-2$, $t = {}^-3$.

13. $2m - 3$
14. $4t + 10$
15. $3h - 12$
16. $6m + 4$
17. $9t - 3$
18. $4h + 4$
19. $2m - 6$
20. $m + 2$
21. $3h + m$
22. $t - h$
23. $4m + 2h$
24. $3t - m$

For questions **25** to **36** $x = {}^-2$, $y = {}^-1$, $k = 0$.

25. $3x + 1$
26. $2y + 5$
27. $6k + 4$
28. $3x + 2y$
29. $2k + x$
30. xy
31. xk
32. $2xy$
33. $2(x + k)$
34. $3(k + y)$
35. $5x - y$
36. $3k - 2x$

Remember:

$2x^2$ means $2(x^2)$
$(2x)^2$ means 'work out $2x$ and **then** square it'
${}^-7x$ means ${}^-7(x)$
${}^-x^2$ means ${}^-1(x^2)$

Example
When $x = {}^-2$, find the value of:

(a) $2x^2 - 5x = 2({}^-2)^2 - 5({}^-2)$
$= 2(4) + 10$
$= 18$

(b) $(3x)^2 - x^2 = (3 \times {}^-2)^2 - 1({}^-2)^2$
$= ({}^-6)^2 - 1(4)$
$= 36 - 4$
$= 32$

Exercise 2Q
If $x = {}^-3$ and $y = 2$, evaluate the following expressions.

1. x^2
2. $3x^2$
3. y^2
4. $4y^2$
5. $(2x)^2$
6. $2x^2$
7. $10 - x^2$
8. $10 - y^2$
9. $20 - 2x^2$
10. $20 - 3y^2$
11. $5 + 4x$
12. $x^2 - 2x$
13. $y^2 - 3x^2$
14. $x^2 - 3y$
15. $(2x)^2 - y^2$
16. $4x^2$
17. $(4x)^2$
18. $1 - x^2$
19. $y - x^2$
20. $x^2 + y^2$
21. $x^2 - y^2$
22. $2 - 2x^2$
23. $(3x)^2 + 3$
24. $11 - xy$
25. $12 + xy$
26. $(2x)^2 - (3y)^2$
27. $2 - 3x^2$
28. $y^2 - x^2$

Example
When $a = {}^-2$, $b = 3$, $c = {}^-3$, evaluate:

(a) $\dfrac{2a(b^2 - a)}{c}$

(b) $\sqrt{(a^2 + b^2)}$

(a) $(b^2 - a) = 9 - ({}^-2)$
$\quad = 11$
$\therefore \dfrac{2a(b^2 - a)}{c} = \dfrac{2 \times ({}^-2) \times (11)}{{}^-3}$
$\quad = 14\tfrac{2}{3}$

(b) $a^2 + b^2 = ({}^-2)^2 + (3)^2$
$\quad = 4 + 9$
$\quad = 13$
$\therefore \sqrt{(a^2 + b^2)} = \sqrt{13}$

Notice that $\sqrt{13}$ means 'the positive square root of 13' not $\pm\sqrt{13}$

Exercise 2R
Evaluate the following:
In questions **1** to **16**, $a = 4$, $b = {}^-2$, $c = {}^-3$.

1. $a(b + c)$
2. $a^2(b - c)$
3. $2c(a - c)$
4. $b^2(2a + 3c)$
5. $c^2(b - 2a)$
6. $2a^2(b + c)$
7. $2(a + b + c)$
8. $3c(a - b - c)$
9. $b^2 + 2b + a$
10. $c^2 - 3c + a$
11. $2b^2 - 3b$
12. $\sqrt{(a^2 + c^2)}$
13. $\sqrt{(ab + c^2)}$
14. $\sqrt{(c^2 - b^2)}$
15. $\dfrac{b^2}{a} + \dfrac{2c}{b}$
16. $\dfrac{c^2}{b} + \dfrac{4b}{a}$

In questions **17** to **32**, $k = {}^-3$, $m = 1$, $n = {}^-4$.

17. $k^2(2m - n)$
18. $5m\sqrt{(k^2 + n^2)}$
19. $\sqrt{(kn + 4m)}$
20. $kmn(k^2 + m^2 + n^2)$
21. $k^2m^2(m - n)$
22. $k^2 - 3k + 4$

23. $m^3 + m^2 + n^2 + n$
24. $k^3 + 3k$
25. $m(k^2 - n^2)$
26. $m\sqrt{(k-n)}$
27. $100k^2 + m$
28. $m^2(2k^2 - 3n^2)$
29. $\dfrac{2k+m}{k-n}$
30. $\dfrac{kn-k}{2m}$
31. $\dfrac{3k+2m}{2n-3k}$
32. $\dfrac{k+m+n}{k^2+m^2+n^2}$

33. Find $K = \sqrt{\left(\dfrac{a^2+b^2+c^2-2c}{a^2+b^2+4c}\right)}$ if $a = 3$, $b = {}^-2$, $c = {}^-1$.

34. Find $W = \dfrac{kmn(k+m+n)}{(k+m)(k+n)}$ if $k = \tfrac{1}{2}$, $m = {}^-\tfrac{1}{3}$, $n = \tfrac{1}{4}$.

Formulae

When a calculation is repeated many times it is often helpful to use a formula.
When a building society offers a mortgage it may use a formula like '$2\tfrac{1}{2}$ times the main salary plus the second salary'.
In a formula letters stand for defined quantities or variables.

Exercise 2S

1. The final speed v of a car is given by the formula $v = u + at$.
   [u = initial speed, a = acceleration, t = time taken]
   Find v when $u = 15$, $a = 0{\cdot}2$, $t = 30$.

2. The time period T of a simple pendulum is given by the formula
   $T = 2\pi\sqrt{\left(\dfrac{l}{g}\right)}$, where l is the length of the pendulum
   and g is the gravitational acceleration.
   Find T when $l = 0{\cdot}65$, $g = 9{\cdot}81$ and $\pi = 3{\cdot}142$.

3. The total surface area A of a cone is related to the radius r and the slant height l by the formula $A = \pi r(r+l)$.
   Find A when $r = 7$ and $l = 11$.

4. The sum S of the squares of the integers from 1 to n is given by $S = \tfrac{1}{6}n(n+1)(2n+1)$. Find S when $n = 12$.

5. The acceleration a of a train is found using the formula $a = \dfrac{v^2 - u^2}{2s}$. Find a when $v = 20$, $u = 9$ and $s = 2\cdot 5$.

6. Einstein's famous equation relating energy, mass and the speed of light is $E = mc^2$. Find E when $m = 0\cdot 0001$ and $c = 3 \times 10^8$.

7. The area A of a parallelogram with sides a and b is given by $A = ab \sin \theta$, where θ is the angle between the sides. Find A when $a = 7$, $b = 3$ and $\theta = 30°$.

8. The distance s travelled by an accelerating rocket is given by $s = ut + \frac{1}{2} at^2$. Find s when $u = 3$, $t = 100$ and $a = 0\cdot 1$.

9. The formula for the velocity of sound in air is $V = 72\sqrt{T + 273}$ where V is the velocity in km/h and T is the temperature of the air in °C.
   (a) Find the velocity of sound where the temperature is 26°C.
   (b) Find the temperature if the velocity of sound is 1200 km/h.
   (c) Find the velocity of sound where the temperature is ⁻77°C.

10. (a) Find a formula for the area, A, of the shape opposite, in terms of a, b and c.
    (b) Find the value of A when $a = 2$, $b = 7$ and $c = 10$.

11. (a) Find a formula for the area of the shaded part, S, in terms of p, q and r.
    (b) Find the value of S when $r = 8$, $p = 12\cdot 1$ and $q = 8\cdot 2$.

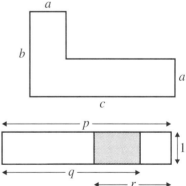

2.9 Compound measures

Speed, distance and time

Calculations involving these three quantities are simpler when the speed is **constant**. The formulae connecting the quantities are as follows:

(a) distance = speed × time (b) speed = $\dfrac{\text{distance}}{\text{time}}$ (c) time = $\dfrac{\text{distance}}{\text{speed}}$

A helpful way of remembering these formulae is to write the letters D, S and T in a triangle, thus:

to find D, cover D and we have ST.

to find S, cover S and we have $\dfrac{D}{T}$.

to find T, cover T and we have $\dfrac{D}{S}$.

Great care must be taken with the units in these questions.

Example

(a) A man is running at a speed of 8 km/h for a distance of 5200 metres. Find the time taken in minutes.

$$5200 \text{ metres} = 5.2 \text{ km}$$

$$\text{time taken in hours} = \left(\frac{D}{S}\right) = \frac{5.2}{8} = 0.65 \text{ hours}$$

time taken in minutes $= 0.65 \times 60 = 39$ minutes

(b) Change the units of a speed of 54 km/h into metres per second.

$$54 \text{ km/hour} = 54\,000 \text{ metres/hour}$$

$$= \frac{54\,000}{60} \text{ metres/minute}$$

$$= \frac{54\,000}{60 \times 60} \text{ metres/second}$$

$$= 15 \text{ m/s}$$

Exercise 2T

1. Find the time taken for the following journeys:
   (a) 100 km at a speed of 40 km/h
   (b) 250 miles at a speed of 80 miles per hour
   (c) 15 metres at a speed of 20 cm/s (answer in seconds)
   (d) 10^4 metres at a speed of 2·5 km/h.

2. A grand prix driver completed a lap of 5·2 km in exactly 2 minutes.

   What was his average speed in km/h?
   [Hint: change 2 minutes into hours.]

3. Find the speeds of the bodies which move as follows:
   (a) a distance of 600 km in 8 hours
   (b) a distance of 4×10^4 m in 10^{-2} seconds
   (c) a distance of 500 m in 10 minutes (in km/h).

4. Find the distance travelled (in metres) in the following:
   (a) at a speed of 40 km/h for $\frac{1}{4}$ hour
   (b) at a speed of 338·4 km/h for 10 minutes
   (c) at a speed of 15 m/s for 5 minutes
   (d) at a speed of 14 m/s for 1 hour.

5. A cyclist travels 25 kilometres at 20 km/h and then a further 80 kilometres at 25 km/h.

   Find:
   (a) the total time taken
   (b) the average speed for the whole journey.

6. Sebastian Coe ran two laps around a 400 m track. He completed the first lap in 50 seconds and then decreased his speed by 5% for the second lap. Find:
   (a) his speed on the first lap
   (b) his speed on the second lap
   (c) his total time for the two laps.

7. The airliner Concorde flies 2000 km at a speed of 1600 km/h and then returns due to bad weather at a speed of 1000 km/h. Find the average speed for the whole trip.

8. A train travels from A to B, a distance of 100 km, at a speed of 20 km/h. If it had gone two and a half times as fast, how much earlier would it have arrived at B?

9. Two men running towards each other at 4 m/s and 6 m/s respectively are one kilometre apart. How long will it take before they meet?

10. A car travelling at 90 km/h is 500 m behind another car travelling at 70 km/h in the same direction. How long will it take the first car to catch the second?

11. How long is a train which passes a signal in twenty seconds at a speed of 108 km/h?

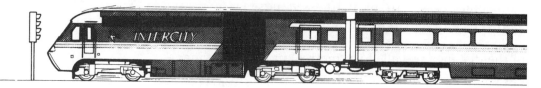

12. An earthworm of length 15 cm is crawling along at 2 cm/s. An ant overtakes the worm in 5 seconds. How fast is the ant walking?

13. A train of length 100 m is moving at a speed of 50 km/h. A horse is running alongside the train at a speed of 56 km/h. How long will it take the horse to overtake the train?

Density and other compound measures

- The density of a material is its mass per unit volume.

It is helpful to use an 'MDV' triangle similar to the one used for speed.

   $M = D \times V \qquad D = \dfrac{M}{V} \qquad V = \dfrac{M}{D}$

Exercise 2U

1. The volume of a solid object is 5 cm³ and its mass is 50 g. Calculate the density of the material.

2. A statuette is made of metal of density 8 g/cm³. Find the mass of the statue if its volume is 30 cm³.

3. Water of density 1000 kg/m³ fills a pool. The mass of the water is 35 tonnes. Calculate the volume of the pool. [1 tonne = 1000 kg]

4. The solid cuboid shown is made of metal with density $10\,g/cm^3$. Calculate the mass of the cuboid, giving your answer in kg.

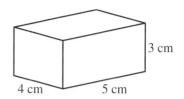

5. The population of the United Kingdom is about 60 million and the land area is about $240\,000\,km^2$. Work out the population density of the U.K. [Number of people/km^2]

6. Hong Kong has a population of about 6 million and land area $1000\,km^2$. The U.S.A. has a land area of about $9\,300\,000\,km^2$.
   (a) Work out the population density of Hong Kong.
   (b) Work out what the population of the United States would be if it was as densely populated as Hong Kong.

7. The picture shows the rate at which grass seed is sown. How much seed is needed for a field of area 2 hectares? [1 hectare = $10\,000\,m^2$]

8. A certain plastic has a density of $3\,g/cm^3$. Convert the density into kg/m^3.

9. Oil has a density of $800\,kg/m^3$. Convert this density into g/cm^3.

10. Change the units of the following speeds as indicated:
    (a) $72\,km/h$ into m/s
    (b) $30\,m/s$ into km/h
    (c) $0·012\,m/s$ into cm/s
    (d) $9000\,cm/s$ into m/s.

2.10 Measurements and bounds

Measurement is approximate

Example 1

A length of some cloth is measured for a dress. You might say the length is 145 cm to the nearest cm.

The actual length could be anything from 144·5 cm to 145·49999... cm using the normal convention which is to round up a figure of 5 or more. Clearly 145·4999... is effectively 145·5 and we say the **upper bound** is 145·5.

The **lower bound** is 144·5.

As an inequality we can write $144.5 \leqslant \text{length} < 145.5$

The upper limit often causes confusion. We use 145·5 as the upper bound simply because it is **inconvenient** to work with 145·49999...

Example 2

When measuring the length of a page in a book, you might say the length is 437 mm to the nearest mm.

In this case the actual length could be anywhere from 436·5 mm to 437·5 mm. We write 'length is between 436·5 mm and 437·5 mm'.

- In examples 1 and 2, the measurement expressed to a given unit is in **possible error** of **half** a **unit**.

Example 3

(a) Similarly if you say your weight is 57 kg to the nearest kg, you could actually weigh anything from 56·5 kg to 57·5 kg.

> The 'unit' is 1 so 'half a unit' is 0·5.

(b) If your brother was weighed on more sensitive scales and the result was 57·2 kg, his actual weight could be from 57·15 kg to 57·25 kg.

> The 'unit' is 0·1 so 'half a unit' is 0·05.

(c) The weight of a butterfly might be given as 0·032 g. The actual weight could be from 0·0315 g to 0·0325 g.

> The 'unit' is 0·001 so 'half a unit' is 0·0005.

Here are some further examples:

Measurement	Lower bound	Upper bound
The diameter of a CD is 12 cm to the nearest cm.	11·5 cm	12·5 cm
The mass of a coin is 6·2 g to the nearest 0·1 g.	6·15 g	6·25 g
The length of a fence is 330 m to the nearest 10 m.	325 m	335 m
Your age is 15 years.	15 years 0 days	15 years 364 days (unless it is a leap year!)
A television costs £84 to the nearest pound.	£83.50	£84.49

In the last example, notice that the cost of an item is a discrete variable and so the television cannot cost £84.49999...

Exercise 2V

1. In a DIY store the height of a door is given as 195 cm to the nearest cm. Write down the upper bound for the height of the door.

2. A vet weighs a sick goat at 37 kg to the nearest kg. What is the least possible weight of the goat?

3. A cook's weighing scales weigh to the nearest 0·1 kg. What is the upper bound for the weight of a chicken which she weighs at 3·2 kg?

4. A surveyor using a laser beam device can measure distances to the nearest 0·1 m. What is the least possible length of a warehouse which he measures at 95·6 m?

5. In the county sports Jill was timed at 28·6 s for the 200 m. What is the upper bound for the time she could have taken?

6. Copy and complete the table.

	Measurement	Lower bound	Upper bound
(a)	temperature in a fridge = 2°C to the nearest degree		
(b)	mass of an acorn = 2·3 g to 1 d.p.		
(c)	length of telephone cable = 64 m to nearest m		
(d)	time taken to run 100 m = 13·6 s to nearest 0·1 s		

7. The length of a telephone is measured as 193 mm, to the nearest mm. The length lies between:

    A   B   C
    192 and 194 mm   192·5 and 193·5 mm   188 and 198 mm

8. The weight of a labrador is 35 kg, to the nearest kg. The weight lies between:

    A   B   C
    30 and 40 kg   34 and 36 kg   34·5 and 35·5 kg

9. Liz and Julie each measure a different worm and they both say that their worm is 11 cm long to the nearest cm.
   (a) Does this mean that both worms are the same length?
   (b) If not, what is the maximum possible difference in the length of the two worms?

10. To the nearest cm, the length, l of a stapler is 12 cm. As an inequality we can write $11.5 \leqslant l < 12.5$.

    For parts (a) to (j) you are given a measurement. Write the possible values using an inequality as above.

    (a) mass = 17 kg   (2 s.f.)   (b) $d = 256$ km   (3 s.f.)
    (c) length = 2.4 m   (1 d.p.)   (d) $m = 0.34$ grams   (2 s.f.)
    (e) $v = 2.04$ m/s   (2 d.p.)   (f) $x = 12.0$ cm   (1 d.p.)
    (g) $T = 81.4°C$ (1 d.p.)   (h) $M = 0.3$ kg   (1 s.f.)
    (i) mass = 0.7 tonnes   (1 s.f.)   (j) $n = 52\,000$ (nearest thousand)

11. A card measuring 11.5 cm long (to the nearest 0.1 cm) is to be posted in an envelope which is 12 cm long (to the nearest cm).
    Can you guarantee that the card will fit inside the envelope? Explain your answer.

11.5 cm

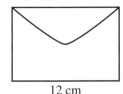

12 cm

Calculations using upper and lower bounds

Care must be taken when you are finding maximum and minimum areas.

For example, to find the maximum possible area of a rectangle you would multiply the maximum possible length by the maximum possible width.

However, given the area of a field and the length of a field, the maximum possible width of the field would be the maximum possible area divided by the **minimum** possible length.

Example 1

Here is a rectangle with sides measuring 37 cm by 19 cm to the nearest cm.

37 cm

19 cm

What are the largest and smallest possible areas of the rectangle consistent with this data?

We know the length is between 36·5 cm and 37·5 cm and the width is between 18·5 cm and 19·5 cm.

largest possible area = 37·5 × 19·5   smallest possible area = 36·5 × 18·5
$\qquad\qquad\qquad\quad$ = 731·25 cm² $\qquad\qquad\qquad\qquad\quad$ = 675·25 cm²

Example 2

The area of a field is 340 cm² and its width is 42 m. Both measurements are given to two significant figures. Find the minimum possible length of the field.

The area is between 335 and 345 m².

The width is between 41·5 and 42·5 m.

The **minimum** possible length of the field is $\dfrac{\textbf{minimum possible area}}{\textbf{maximum possible width}}$

$$= \frac{335}{42 \cdot 5} = 7 \cdot 88 \text{ m}^2$$

Example 3

You buy a VCR, three CDs and a camera. In total you pay £170; the VCR costs £90 and each CD costs £11. What is the minimum possible price of the camera?
All amounts are given to two significant figures.

The VCR costs between £89.50 and £90.49.

Each CD costs between £10.50 and £11.49.

The total cost is between £165 and £174.49

The **minimum** possible price of the camera

$\quad$ = **minimum** total price − **maximum** price of the VCR and three CDs

$\quad$ = £165 − (£90.49 + 3 × £11.49)

$\quad$ = £165 − £124.96

$\quad$ = £40.04

Errors in calculations

This section is concerned with the inaccuracy of numbers used in a calculation. We are not trying to prevent students from making mistakes!

A common problem occurs when a measured quantity is multiplied by a large number.

Example 1

A marble is weighed and found to be 12·3 grams. Find the weight of 100 identical marbles.

A simple answer is $100 \times 12\cdot3 = 1230$ g. We must realise that each marble could weigh from 12·25 g to 12·35 g.

$$100 \times 12\cdot25 = 1225 \text{ g}$$
$$100 \times 12\cdot35 = 1235 \text{ g}$$

So there is a possible error of up to 5 grams above or below our initial answer of 1230 g.

Example 2

Given that the numbers 8·6, 3·2 and 11·5 are accurate to 1 decimal place, calculate the upper and lower bounds for the calculation

$$\left(\frac{8\cdot6 - 3\cdot2}{11\cdot5}\right)$$

For the upper bound, make the top line of the fraction as large as possible and make the bottom line as small as possible.

upper bound $= \left(\dfrac{8\cdot65 - 3\cdot15}{11\cdot45}\right)$ 	lower bound $= \left(\dfrac{8\cdot55 - 3\cdot25}{11\cdot55}\right)$

$= 0\cdot480\,3493\ldots$ 	$= 0\cdot458\,8744\ldots$

Exercise 2W

1. The sides of the triangle are measured correct to the nearest cm.
   (a) Write down the upper bounds for the lengths of the three sides.
   (b) Work out the maximum possible perimeter of the triangle.

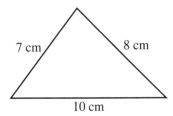

2. The dimensions of a photo are measured correct to the nearest cm. Work out the minimum possible area of the photo.

3. In this question the value of a is either exactly 4 or 5, and the value of b is either exactly 1 or 2.
   Work out:
   (a) the maximum value of $a + b$
   (b) the minimum value of $a + b$
   (c) the maximum value of ab
   (d) the maximum value of $a - b$
   (e) the minimum value of $a - b$
   (f) the maximum value of $\dfrac{a}{b}$
   (g) the minimum value of $\dfrac{a}{b}$
   (h) the maximum value of $a^2 - b^2$.

4. If $p = 7$ cm and $q = 5$ cm, both to the nearest cm, find:
   (a) the largest possible value of $p + q$
   (b) the smallest possible value of $p + q$
   (c) the largest possible value of $p - q$
   (d) the largest possible value of $\dfrac{p^2}{q}$.

5. If $a = 3\cdot1$ and $b = 7\cdot3$, correct to 1 decimal place, find the largest possible value of:
   (i) $a + b$ (ii) $b - a$

6. If $x = 5$ and $y = 7$ to 1 significant figure, find the smallest possible values of
   (i) $x + y$ (ii) $y - x$ (iii) $\dfrac{x}{y}$

7. In the diagram, $ABCD$ and $EFGH$ are rectangles with $AB = 10$ cm, $BC = 7$ cm, $EF = 7$ cm and $FG = 4$ cm, all figures accurate to the nearest cm.
   Find the largest possible value of the shaded area.

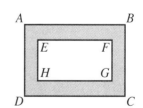

8. When a voltage V is applied to a resistance R the power consumed P is given by $P = \dfrac{V^2}{R}$.

   If you measure V as $12\cdot2$ and R as $2\cdot6$, correct to 1 d.p., calculate the smallest possible value of P.

9. While in the USA, Alan buys the equivalent of 9 British gallons of petrol.
   Petrol in the USA cost $1.30 per American gallon.
   Find the maximum amount, in pounds, which Alan could have spent on the petrol.

   The amount of petrol is given correct to 1 significant figure.
   The price per gallon is given correct to 2 significant figures.

   You should assume that:

   1 American gallon is exactly 0·8 British gallons and the exchange rate is exactly $1.46 to £1. [SEG]

10. Marion sees a video recorder for sale in summer 2000.
    In spring 2001, the price of the video recorder is reduced by $\frac{1}{3}$.
    Compared with the 2000 price, what percentage reduction did Marion obtain? [SEG]

11. A salesman estimates the number of industrial microscopes which he will sell in a year and hence estimates his expected commission.
    The number of microscopes which he actually sells is 18% above his estimate. He also finds that his commission on each sale is 20% above his estimate.

    Calculate the percentage that the actual commission was above the salesman's estimate. [SEG]

Percentage error

Suppose the answer to the calculation 4·8 × 33·4 is estimated to be 150. The exact answer to the calculation is 160·32.
The percentage error is

$$\left(\frac{160\cdot32 - 150}{160\cdot32}\right) \times \frac{100}{1} = 6\cdot4\% \quad \text{(2 s.f.)}$$

- In general, percentage error = $\left(\dfrac{\text{actual error}}{\text{exact value}}\right) \times 100$

Exercise 2X

Give the answers correct to 3 significant figures, unless told otherwise.

1. A man's salary is estimated to be £20 000. His actual salary is £21 540. Find the percentage error.

2. The speedometer of a car showed a speed of 68 m.p.h. The actual speed was 71 m.p.h. Find the percentage error.

3. The rainfall during a storm was estimated at 3·5 cm. The exact value was 3·1 cm. Find the percentage error.

4. The dimensions of a rectangle were taken as 80×30. The exact dimensions were $81·5 \times 29·5$. Find the percentage error in the area.

5. The answer to the calculation $3·91 \times 21·8$ is estimated to be 80. Find the percentage error.

6. An approximate value for π is $\frac{22}{7}$. A more accurate value is given on most calculators. Find the percentage error.

7. A wheel has a diameter of 58·2 cm. Its circumference is estimated at 3×60. Find the percentage error.

8. A speed of 60 km/h is taken to be roughly 18 m/s. Find the percentage error.

9. The fraction $\frac{19}{11}$ gives an approximate value for $\sqrt{3}$. Calculate the percentage error in using $\frac{19}{11}$ as the value of $\sqrt{3}$. Give the answer to 2 s.f.

10. The most common approximate value for π is $\frac{22}{7}$.

    Two better approximations are $\frac{355}{113}$ and $\sqrt{\left(\sqrt{\left(\frac{2143}{22}\right)}\right)}$

    Which of these two values has the smaller percentage error?

2.11 Numerical problems

Exercise 2Y

Questions **1** to **5** are about foreign currency.
Use the rates of exchange given in the table.

Country	Exchange Rate
Europe (euro)	€1·54 = £1
France (franc)	Fr 10·9 = £1
Germany (mark)	DM 3·1 = £1
Italy (lire)	lire 2900 = £1
Spain (peseta)	Ptas 260 = £1
United States (dollar)	$1·64 = £1

1. Change the amount of British money into the foreign currency stated.
   (a) £20 [French francs]
   (b) £70 [dollars]
   (c) £200 [pesetas]
   (d) £1·50 [marks]
   (e) £2·30 [lire]
   (f) 90p [euros]

2. Change the amount of foreign currency into British money.
   (a) Fr 500
   (b) $2500
   (c) DM 7·5
   (d) €900
   (e) lire 500 000
   (f) Ptas 950

3. An CD costs £12·99 in Britain and $15 in the United States. How much cheaper, in British money, is the CD when bought in the U.S.A?

4. A Renault car is sold in several countries at the prices given below.

   Britain   £15 000
   France   Fr 194 020
   USA $24 882

   Write out in order a list of the prices converted into pounds.

5. A traveller in Switzerland exchanges 1300 Swiss francs for £400. What is the exchange rate?

6. A maths teacher bought 40 calculators at £8·20 each and a number of other calculators costing £2·95 each. In all she spent £387.
   How many of the cheaper calculators did she buy?

7. At a temperature of 20°C the common amoeba reproduces by splitting in half every 24 hours. If we start with a single amoeba, how many will there be after
   (a) 8 days (b) 16 days?

8. A humming bird moves its wings 130 times per second. How many times will it move its wings in one hour? Give your answer in standard form.

9. Here are four fractions: $\frac{3}{19}$, $\frac{37}{232}$, $\frac{1}{6}$, $\frac{83}{511}$
   (a) Which of the fractions is the smallest?
   (b) Which of the fractions is the largest?

10. A wicked witch stole a newborn baby from its parents. On the baby's first birthday the witch sent the grief-stricken parents 1 penny. On the second birthday she sent 2 pence. On the third birthday she sent 4 pence and so on, doubling the amount each time. How much did the witch send the parents on the twenty-first birthday?

11. A videotape is 225 metres long. Its total playing time is three hours. Find the speed of the tape:
    (a) in metres per minute   (b) in centimetres per second.

12. In 1999 there were 21 280 000 licensed vehicles on the road. Of these, 16 486 000 were private cars. What percentage of the licensed vehicles were private cars?

13. A map is 278 mm wide and 445 mm long. When reduced on a photocopier, the copy is 360 mm long. What is the width of the copy, to the nearest millimetre?

14. Taking 1 litre to be $1\frac{3}{4}$ pints, find:
    (a) the number of pints in 20 litres
    (b) the number of litres in 9 pints, giving your answer as a decimal.

15. Measured to the nearest millimetre, the sides of a triangle are 11 mm, 14 mm and 17 mm.
    Work out the upper limit for the perimeter of the triangle.

16. (a) Fill in the empty boxes in the square so that each row, each column and each diagonal adds up to 0.
    (b) Multiply together the three numbers in the bottom row and write down your answer.

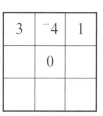

17. In 2000 Anna picked 252 kg of plums.
    This was 20% more than she picked in 1999.
    Calculate how many kilograms of plums she picked in 1999.

18. Tiger arrived at the bus station at 11:15. His bus had left half an hour before.
    (a) At what time had his bus left?
    (b) Tiger's next bus is at 12.06. How long must he wait?

19. What is the smallest number greater than 1000 that is exactly divisible by 13 and 17?

20. In 2001 Mr Giggs drove 8360 miles in his car. His car does 37 miles per gallon. Petrol costs 68 pence per litre.
    (a) By taking one gallon to be 4·55 litres, calculate, in pounds, how much Mr Giggs spent on petrol in 2001.
    (b) Show how you can use approximation to check that your answer is of the right order of magnitude.
    You **must** show all your working.

Exercise 2Z

1. A stick insect moves 3 centimetres in one minute.
   (a) Find the speed of the stick insect in kilometres per hour, giving your answer as a decimal.
   (b) Write your answer to part (a) in standard form.

2. Find the smallest value of n for which
   $$1^2 + 2^2 + 3^2 + 4^2 + 5^2 + \ldots + n^2 > 800$$

3. Pythagoras, the Greek mathematician, was also a shrewd businessman. Suppose he deposited £1 in the Bank of Athens in the year 500 B.C. at 1% compound interest. What would the investment be worth to his descendants in the year A.D. 2000?

4. Find the missing prices

Object	Old price	New price	% change
(a) Football	£9·50	?	3% increase
(b) Radio	?	£34·88	9% increase
(c) Roller blades	£52	?	15% decrease
(d) Golf club	?	£41·40	8% decrease

5. Booklets have a mass of 19 g each and they are posted in an envelope of mass 38 g. Postage charges are shown in the table below.

Mass (in grams) not more than	60	100	150	200	250	300	350	600
Postage (in pence)	24	30	37	44	51	59	67	110

   (a) A package consists of 15 booklets in an envelope. What is the total mass of the package?
   (b) The mass of a second package is 475 g. How many booklets does it contain?
   (c) What is the postage charge on a package of mass 320 g?
   (d) The postage on a third package was £1·10. What is the largest number of booklets it could contain?

6. A rabbit runs at $7\,\text{m}\,\text{s}^{-1}$ and a hedgehog at $\frac{1}{2}\,\text{m}\,\text{s}^{-1}$.
   They are 90 m apart and start to run towards each other.
   How far does the hedgehog run before they meet?

7. Mark's recipe for a cake calls for 6 fluid ounces of milk.
   Mark knows that one pint is 20 fluid ounces, that one gallon is 8 pints and that 5 litres is roughly one gallon.
   He only has a measuring jug marked in millilitres. How many millilitres of milk does he need for the cake? Give your answer to a sensible degree of accuracy.

8. 8% of 2500 + 37% of $P = 348$. Find the value of P.

9. A train leaves Paris at 15:24 and arrives in Milan at 19:44.
   A new train will cut 20% off the journey time.
   At what time will the 15:24 train now arrive in Milan?

10. In France petrol for a car costs 5·54 francs per litre.
    The exchange rate is 9·60 francs to the pound and we can assume that one gallon equals 4·5 litres.
    A family uses the car to travel a total distance of 1975 km and the owner of the car estimates the petrol consumption of the car to be 27 miles per gallon. Take 1 km equal to 0·62 miles.
    Calculate the cost of the petrol in pounds to the nearest pound.

11. A group of friends share a bill for £13·69 equally between them. How many were in the group?

12. Which is larger: 7^{77} or 77^7?

13. Write the recurring decimal 0·434 343 ... as a fraction.

Summary

1. You can calculate using percentages.

Check out N2

1. (a) (i) Find 24% of £81.
       (ii) Express 4 cm as a percentage of 6 m.
   (b) After a reduction of 30% the cost of a CD is £4.20. How much was the reduction?

2. You can use a multiplier.

2. (a) Jenny invests £4200 at 3% interest, compounded annually. How much is the value of the investment after 5 years?

   (b) A shopkeeper adds 60% for profit to the price he paid the manufacturer. In a sale he reduces all his prices by 30%. How much profit does he make when he sells an item in the sale?

3. You can use a calculator.

3. Find $\dfrac{(3\cdot 45 + \sqrt{8\cdot 12})^2}{7\cdot 12 - 1\cdot 84}$.

   Give your answer to an appropriate degree of accuracy.

4. You can calculate using indices.

4. (a) Write 54 as the product of its prime factors using index notation.

   (b) Find $10^4 \div 10^7$.

   (c) Express $\sqrt{32}$ as a power of 2.

5. You can calculate using standard form.

5. (a) Find $0\cdot 3^7$ giving your answer in standard form.

   (b) Evaluate $\dfrac{3 \times 10^5}{6 \times 10^7}$.

   (c) Evaluate $2\cdot 85 \times 10^3 + 3\cdot 91 \times 10^4$.

6. You can find the fraction equivalent of a recurring decimal.

6. Find the fraction equivalent to:

   (a) $0\cdot\dot{4}\dot{5}$ (b) $0\cdot 8\dot{2}\dot{3}$

7. You can use surds.

7. Simplify: (a) $\sqrt{108}$ (b) $\sqrt{75} + \sqrt{108}$

   (c) $\dfrac{\sqrt{1458}}{\sqrt{32} + \sqrt{50}}$ (use $1458 = 2 \times 3^6$)

8. You can use compound measures.

8. Rob travels 180 miles in 4 hours 15 minutes. What is his average speed?

9. You can:
   (a) find upper and lower bounds
   (b) calculate using upper and lower bounds.

9. The temperature in a room is 15 °C and the temperature of a fridge in the room is 4·2 °C, both measurements being given to two significant figures.

   What is (a) the maximum and (b) the minimum possible difference in temperature between the room and the fridge?

Revision exercise N2

1. Use your calculator to find the value of $\sqrt{231} \times (7 \cdot 28 + 3 \cdot 97)^2$.

   Give your answer to 2 significant figures. [SEG]

2. A plane flies form Paris to Guadeloupe, a distance of 4200 miles. The plane has an average speed of 500 miles per hour. How long does the plane take for the journey? Give your answer in hours and minutes. [SEG]

3. (a) Calculate $125^{\frac{1}{3}}$.

   (b) Calculate $49^{-\frac{1}{2}}$, giving your answer as a fraction.

   (c) Calculate $\dfrac{3^{10}}{9^6}$ giving your answer as a power of 3.

   (d) Calculate $128^{\frac{2}{7}}$. [SEG]

4. A man buys a cello for £8000. He expects the cello to increase in value by 5% in the first year and by 12% in the second year. How much does he expect the cello to be worth in 2 years' time? [SEG]

5. On holiday in Turkey, Clive buys a carpet costing 2.52×10^8 Turkish lire. The exchange rate is 4.49×10^5 Turkish lire to £1.

   (a) What is the cost of the carpet in pounds?

   Clive is given a rebate of $119.00 because he is taking the carpet out of Turkey.
   The exchange rate is $1.68 to £1.

   (b) What is the rebate as a percentage of the cost of the carpet? [SEG]

6. (a) Calculate $81^{-\frac{1}{2}}$, giving your answer as a fraction.
   (b) Calculate $\dfrac{2^3}{2^{-2}}$ leaving your answer in the form 2^p.
   [AQA]

7. Write $0.0\dot{6}\dot{3}$ as a fraction in its simplest form. [AQA]

8. Cashchem sells sun oil in special offer bottles which contain an extra 25% free.
   In a sale, Cashchem discounts all stock by 30%. What percentage is the reduction of cost of 100 ml of sun oil when bought in the special offer bottles in the sale? [AQA]

9. Jake is selling strawberries at summer fete, and expects to sell all his stock during the day. Because of the bad weather on the day of the fete, he reduces the price of the strawberries by 30%. By the end of the day, 20% of his stock is still **not** sold. Calculate the reduction in Jake's takings as a percentage of his expected takings. [SEG]

10. A train travels from Exeter to Preston in $5\frac{1}{2}$ hours. The distance from Exeter to Preston is 271 miles.

    Find the average speed of the train. Give your answer in miles per hour. [SEG]

11. George measures his bedroom floor. It is a rectangle with width 3·4 m and length 4·3 m.
    The height of the bedroom ceiling is 8 foot 3 inches.
    The metric lengths are measured to the nearest 10 cm and the imperial length is measured to the nearest 3 inches.
    Assume there are exactly 2·54 centimetres in one inch.
    There are 12 inches in one foot.
    Find the maximum possible volume of George's bedroom in cubic metres. [AQA]

12. Rebecca is using a spreader to put weed killer on her lawn. She sets the spreader to use 40 grams per square metre, to the nearest 5 grams.
    Her lawn is 300 feet in length and 20 feet in width.
    The length of the lawn is given to the nearest 5 feet, and the width is given to the nearest 6 inches.
    There are 12 inches in 1 foot. Take 1 inch to be exactly 2·54 cm.
    The cost of the weed killer is £2.10 per kg.
    Find the maximum cost of the weed killer which Rebecca uses.
    [SEG]

N3 Algebraic Techniques for Module 3

This unit will show you how to:

- Use algebraic methods to prove generalisations
- Use proportionality
- Solve equations graphically

Before you start:

You should know how to...	Check in N3
1. Factorise algebraic terms.	1. Simplify $(2m-1)+(2n-1)$.
2. Expand brackets.	2. Expand $(2m+1)(2n-1)$.
3. Substitute in formulae.	3. $y = 2x^3$ Find y when $x = 3$.
4. Change the subject of a formula.	4. $y = 2x^3$ Find x when: (a) $y = 250$ (b) $y = 17$
5. Draw the graph of a curve.	5. Draw the graph of $y = x^2 + x - 5$ for values of x from $^-3$ to $^+2$.

3.1 Conjectures

A conjecture (or hypothesis) is simply a statement which may or may not be true. In mathematics people are constantly looking for rules or formulae to make calculations easier. Suppose you carry out an investigation and after looking at the results you spot what you think might be a rule.
You could make the conjecture that:

   'x^2 is always greater than x'

or that 'the angle in a semi circle is always 90°'
or that 'the expression $x^2 + x + 41$ gives a prime number
      for all integer values of x'

There are three possibilities for any conjecture. It may be
(a) true,   (b) false,   (c) not proven.

(a) True

To show that a conjecture is true it is not enough to simply find lots of examples where it is true. We have to **prove** it for **all** values. We return to proof later in this section.

(b) False

To show that a conjecture is false it is only necessary to find one **counter-example**.

Example 1

Consider the conjecture 'x^2 is always greater than x'.

It is true that $2^2 > 2$; $3 \cdot 1^2 > 3 \cdot 1$; $(^-5)^2 > 5$

But $(\frac{1}{2})^2$ is **not** greater than $\frac{1}{2}$.

This is a counter-example so the conjecture is false.

We have **disproved** the conjecture.

Example 2

Consider the conjecture:

'the expression $x^2 + x + 41$ gives a prime number for all integer values of x'.

If we try the first few values of x, we get these results.

x	1	2	3	4	5	6
$x^2 + x + 41$	43	47	53	61	71	83

The expression does give prime numbers for these values of x and indeed, it does so for all values of x up to 39.

But when $x = 40$, $\quad x^2 + x + 41 = 1681$
$$= 41 \times 41$$

So we do not always obtain a prime number.

The conjecture is, therefore, false.

Notice that it takes only **one** counter-example to disprove a conjecture.

(c) Not proven

Suppose you have a conjecture for which you cannot find a counter-example but which you also cannot prove. In this case the conjecture is **not proven**.

Until recently, the most famous example of a conjecture not proven was 'Fermat's Last Theorem' which stated that there are no whole numbers a, b, c for which $a^3 + b^3 = c^3$ [or further, that $a^n + b^n = c^n$ $(n > 2)$].

For 358 years no mathematician could prove the conjecture but no one could find a counter-example. In 1994 Andrew Wiles, a British mathematician, finally proved the theorem after devoting himself to it for seven years. The proof is extremely long (100 pages) and it is said that only a handful of people in the world understand it.

Exercise 3A

In Questions **1** to **12**, consider the conjecture given. Some are true and some are false. Write down a counter-example where the conjecture is false.

If you cannot find a counter-example, state that the conjecture is 'not proven' as you are **not** asked to prove it.

1. $(n + 1)^2 = n^2 + 1$ for all values of n.

2. $\dfrac{1}{x}$ is always less than x.

3. $\sqrt{n + 1}$ is always smaller than $\sqrt{n} + 1$.

4. For all values of n, $2n$ is greater than $n - 2$.

5. For any set of numbers, the median is always smaller than the mean.

6. $\sqrt{x}$ is always less than x.

7. The diagonals of a parallelogram never cut at right angles.

8. The product of two irrational numbers is always another irrational number.

9. If n is even $5n^2 + n + 1$ is odd.

10. All numbers in the sequence 31, 331, 3331, 33 331, 333 331, 3 333 331, 33 333 331, 333 333 331, ... are prime.
    [Hint: Divide by 17.]

11. $n^2 + n > 0$ for all values of n.

12. The expression $a^n - a$ is always a multiple of n.

13. Here is a circle with 2 points on its circumference and 2 regions.

This circle has 3 points ($n = 3$) and 4 regions ($r = 4$).

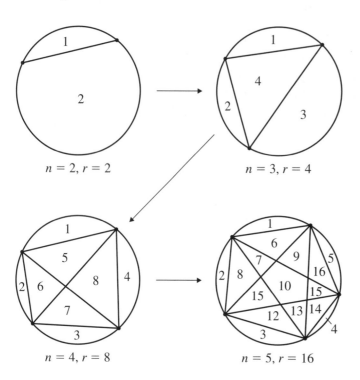

Each time, every point on the circumference is joined to all the others.

(a) Do you think a conjecture is justified at this stage?
Consider:
'The circle with 6 points on its circumference will have 32 regions'.
Or again:
'With n points on the circle the number of regions, r, is given by $r = 2^{n-1}$.

(b) Draw the next two circles in the sequence with $n = 6$ and $n = 7$.
Be careful to avoid this sort of thing:

> When you have the results for $n = 6$ and $n = 7$ you might get somewhere (although it is quite difficult) by looking at **differences** in the results.

where three lines pass through a single point.
Under a magnifying glass we might see this: and a region might be lost.
What can you say about your conjecture?

3.2 Logical argument

- Here is a statement:

    Triangle PQR is equilateral $\Rightarrow$ angle $QRP = 60°$.

    The statement can be read:

    If triangle PQR is equilateral, then angle QRP equals $60°$.

    or

    Triangle PQR is equilateral implies that angle QRP equals $60°$.

- Here is another statement.

    $a > b \Rightarrow a - b > 0$

    This can be read:

    If a is greater than b, then a less b is greater than zero

    or

    a is greater than b implies that a less b is greater than zero.

Note $x = 4 \Rightarrow x^2 = 16$ is true
but $x^2 = 16 \Rightarrow x = 4$ is not true because x could be $^-4$.

Exercise 3B

1. Write the following as sentences.
   (a) Today is Monday $\Rightarrow$ tomorrow is Tuesday.
   (b) It is raining $\Rightarrow$ there are clouds in the sky.
   (c) Abraham Lincoln was born in 1809 $\Rightarrow$ Abraham Lincoln is dead.

2. State whether the following statements are true or false.
   (a) $x - 4 = 3 \Rightarrow x = 7$
   (b) n is even $\Rightarrow n^2$ is even
   (c) $a = b \Rightarrow a^2 = b^2$
   (d) $x > 2 \Rightarrow x = 3$
   (e) $a + b$ is odd $\Rightarrow ab$ is even (a and b are whole numbers)
   (f) $x^2 = 4 \Rightarrow x = 2$
   (g) $x = 45° \Rightarrow \tan x = 1$
   (h) $5^x = 1 \Rightarrow x = 0$
   (i) $pq = 0 \Rightarrow p = 0$
   (j) a is odd $\Rightarrow a$ can be written in the form $2k + 1$, where k is an integer

3. Think of pairs of statements like those in Question **1**. Link the statements with the $\Rightarrow$ sign.

3.3 Proof

- In the field of science a theory, like Newton's theory of gravitation, can never be proved. It can only be considered highly likely using all the evidence available at the time.
- A mathematical proof is far more powerful. Once a theorem is proved mathematically it will **always** be true. A proof starts with simple facts which are accepted. The proof then argues logically to the result which is required.
- It is most important to realise that a result **cannot** be proved simply by finding thousands or even millions of results which support it.

Algebraic proof

Example 1
Prove that, if a and b are odd numbers, then $a + b$ is even.

If a is odd, there is remainder 1 when a is divided by 2.
$\therefore$ a may be written in the form $(2m + 1)$ where m is a whole number. Similarly b may be written in the form $(2n + 1)$.

$\therefore$ $a + b = 2m + 1 + 2n + 1$
$\qquad = 2(m + n + 1)$

$\therefore$ $a + b$ is even, as required.

Example 2
Prove that the answer to every line of the pattern below is 8.

$3 \times 5 - 1 \times 7$
$4 \times 6 - 2 \times 8$
$5 \times 7 - 3 \times 9$
$\vdots \qquad \vdots$

Proof
The nth line of the pattern is $(n + 2)(n + 4) - n(n + 6)$
$\qquad\qquad = n^2 + 6n + 8 - (n^2 + 6n)$
$\qquad\qquad = 8$

The result is proved.

Example 3 [Much harder but very interesting]
Prove that the product of 4 consecutive numbers is always one less than a square number.

e.g. $\quad 1 \times 2 \times 3 \times 4 = 24 \qquad$ or $\quad 3 \times 4 \times 5 \times 6 = 360$
$\qquad\qquad\qquad\quad = 25 - 1 \qquad\qquad\qquad\qquad\quad = 19^2 - 1$

Proof
Let the first number be n.
Then the next three consecutive numbers are $(n + 1)$, $(n + 2)$, $(n + 3)$.

Find the product

$$n(n + 1)(n + 2)(n + 3)$$
$$= (n^2 + n)(n^2 + 5n + 6)$$
$$= n^4 + 6n^3 + 11n^2 + 6n$$

Write the products as $(n^4 + 6n^3 + 11n^2 + 6n + 1) - 1$.
Note that this does not alter the value of the product.

The expression in the brackets factorises as

$(n^2 + 3n + 1)(n^2 + 3n + 1)$

∴ The product is $(n^2 + 3n + 1)^2 - 1$

Since $(n^2 + 3n + 1)^2$ is a square number, the result is proved.

Exercise 3C

1. If a is odd and b is even, prove that ab is even.

2. If a and b are both odd, prove that ab is odd.

3. Prove that the sum of two odd numbers is even.

4. Prove that the square of an even number is divisible by 4.

5. Prove that the answer to every line of the pattern below is 3.

   $2 \times 4 - 1 \times 5 =$
   $3 \times 5 - 2 \times 6 =$
   $4 \times 6 - 3 \times 7 =$
   $\vdots \quad \vdots$

   Hint: see the second example on page 184.

6. In any three consecutive numbers prove that the product of the first and the third number is one less than the square of the middle number.

7. Prove that the product of four consecutive numbers is divisible by 4.

8. Prove that the sum of the squares of any two consecutive integers is an odd number.

9. Prove that the sum of the squares of five consecutive numbers is divisible by 5.

   e.g. $2^2 + 3^2 + 4^2 + 5^2 + 6^2 = 90 = 5 \times 18$

   Begin by writing the middle number as n, so the other numbers are $n-2, n-1, n+1, n+2$.

10. Find the sum of the squares of three consecutive numbers and then subtract 2. Prove that the result is always 3 times a square number.

11. Here is a proof of Pythagoras' Theorem.

    Four right-angled triangles with sides a, b and c are drawn around the tilted square as shown.

    Area of large square $= (a+b)^2$ ①

    Also area of large square $=$ area of 4 triangles $+$ area of tilted square.   ②

    Equate ① and ② and hence complete the proof of Pythagoras' Theorem.

    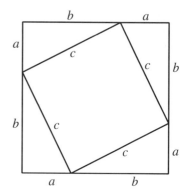

12. Here is the 'proof' that $1 = 2$.
    Let $a = b$

    $\Rightarrow ab = b^2$ [multiply by b]
    $\Rightarrow ab - a^2 = b^2 - a^2$ [subtract a^2]
    $\Rightarrow a(b-a) = (b+a)(b-a)$ [factorise]
    $\Rightarrow a = b + a$ [divide by $(b-a)$]
    $\Rightarrow a = a + a$ [from top line]
    $\Rightarrow 1 = 2$

    Which step in the argument is not allowed?

3.4 Direct and inverse proportion

Direct proportion

(a) When you buy petrol, the more you buy the more money you have to pay. So if 2·2 litres costs 176p, then 4·4 litres will cost 352p.
We say the cost of petrol is **directly proportional** to the quantity bought.
To show that quantities are proportional, we use the symbol '$\propto$'.

So in our example if the cost of petrol is c pence and the number of litres of petrol is l, we write

$$c \propto l$$

The '$\propto$' sign can always be replaced by '$= k$' where k is a constant.
So $c = kl$
From above, if $c = 176$ when $l = 2{\cdot}2$
then $176 = k \times 2{\cdot}2$

$$k = \frac{176}{2{\cdot}2} = 80$$

We can then write $c = 80l$, and this allows us to find the value of c for any value of l, and **vice versa.**

(b) If a quantity z is proportional to a quantity x, we have

$$z \propto x \quad \text{or} \quad z = kx$$

Two other expressions are sometimes used when quantities are directly proportional. We could say

'z varies as x'

or 'z varies directly as x'.

The graph connecting z and x is a straight line which passes through the origin.

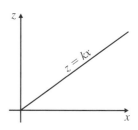

Example 1

y varies as z, and $y = 2$ when $z = 5$; find:

(a) the value of y when $z = 6$
(b) the value of z when $y = 5$.

Because $y \propto z$, then $y = kz$ where k is a constant.

$\quad\quad y = 2$ when $z = 5$
$\therefore \quad 2 = k \times 5$
$\quad\quad k = \frac{2}{5}$
So $\quad y = \frac{2}{5}z$

(a) When $z = 6$, $y = \frac{2}{5} \times 6 = 2\frac{2}{5}$.
(b) When $y = 5$, $5 = \frac{2}{5}z$; $z = \frac{25}{2} = 12\frac{1}{2}$.

Example 2

The value V of a diamond is proportional to the square of its weight W. If a diamond weighing 10 grams is worth £200, find

(a) the value of a diamond weighing 30 grams
(b) the weight of a diamond worth £5000.

or
$$V \propto W^2$$
$$V = kW^2 \text{ where } k \text{ is a constant.}$$
$$V = 200 \text{ when } W = 10$$
$$\therefore \quad 200 = k \times 10^2$$
$$k = 2$$

So $\quad V = 2W^2$

(a) When $W = 30$,
$$V = 2 \times 30^2 = 2 \times 900$$
$$V = £1800$$

So a diamond of weight 30 grams is worth £1800.

(b) When $\quad V = 5000$,
$$5000 = 2 \times W^2$$
$$W^2 = \frac{5000}{2} = 2500$$
$$W = \sqrt{2500} = 50$$

So a diamond of value £5000 weighs 50 grams.

Exercise 3D

1. Rewrite the statement connecting each pair of variables using a constant k instead of '$\propto$'.
   (a) $S \propto e$
   (b) $v \propto t$
   (c) $x \propto z^2$
   (d) $y \propto \sqrt{x}$
   (e) $T \propto \sqrt{L}$

2. y varies as t. If $y = 6$ when $t = 4$, calculate:
   (a) the value of y, when $t = 6$
   (b) the value of t, when $y = 4$.

3. z is proportional to m. If $z = 20$ when $m = 4$, calculate:
   (a) the value of z, when $m = 7$
   (b) the value of m, when $z = 55$.

4. A varies directly as r^2. If $A = 12$, when $r = 2$, calculate:
   (a) the value of A, when $r = 5$
   (b) the value of r, when $A = 48$.

5. Given that $z \propto x$, copy and complete the table.

x	1	3		$5\frac{1}{2}$
z	4		16	

6. Given that $V \propto r^3$, copy and complete the table.

r	1	2		$1\frac{1}{2}$
V	4		256	

7. The pressure of the water P at any point below the surface of the sea varies as the depth of the point below the surface d. If the pressure is 200 newtons/cm² at a depth of 3 m, calculate the pressure at a depth of 5 m.

8. The distance d through which a stone falls from rest is proportional to the square of the time taken t. If the stone falls 45 m in 3 seconds, how far will it fall in 6 seconds? How long will it take to fall 20 m?

9. The energy E stored in an elastic band is proportional to the square of the extension x. When the elastic is extended by 3 cm, the energy stored is 243 joules. What is the energy stored when the extension is 5 cm? What is the extension when the stored energy is 36 joules?

10. In an experiment, measurements of w and p were taken. Which of these laws fits the results?

    $p \propto w, \quad p \propto w^2, \quad p \propto w^3$.

w	2	5	7
p	1·6	25	68·6

11. A road research organisation recently claimed that the damage to road surfaces was proportional to the fourth power of the axle load. The axle load of a 44-ton HGV is about 15 times that of a car. Calculate the ratio of the damage to road surfaces made by a 44-ton HGV and a car.

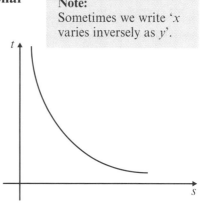

Inverse proportion

If you travel a distance of 200 m at 10 m/s, the time taken is 20 s.
If you travel the same distance at 20 m/s, the time taken is 10 s.
As you **double** the speed, you **halve** the time taken.
For a fixed journey, the time taken is **inversely proportional** to the speed at which you travel.
If t is inversely proportional to s, we write

$$s \propto \frac{1}{t} \quad \text{or} \quad s = k \times \frac{1}{t}$$

Note:
Sometimes we write 'x varies inversely as y'.

Notice that the product $s \times t$ is constant.
The graph connecting s and t is a curve.
The shape of the curve is the same as $y = \frac{1}{x}$.

Example

z is inversely proportional to t^2 and $z = 4$ when $t = 1$.
Calculate z when $t = 2$

We have $z \propto \dfrac{1}{t^2}$ or $z = k \times \dfrac{1}{t^2}$ (k is a constant)

$z = 4$ when $t = 1$,

$\therefore \quad 4 = k\left(\dfrac{1}{1^2}\right)$ so $k = 4$

$\therefore \quad z = 4 \times \dfrac{1}{t^2}$

When $t = 2$, $z = 4 \times \dfrac{1}{2^2} = 1$

Exercise 3E

1. Rewrite the statements connecting the variables using a constant of variation, k.

   (a) $x \propto \dfrac{1}{y}$ (b) $s \propto \dfrac{1}{t^2}$ (c) $t \propto \dfrac{1}{\sqrt{q}}$

   (d) m varies inversely as w
   (e) z is inversely proportional to t^2.

2. T is inversely proportional to m. If $T = 12$ when $m = 1$, find:
   (a) T when $m = 2$ (b) T when $m = 24$.

3. L is inversely proportional to x. If $L = 24$ when $x = 2$, find:
   (a) L when $x = 8$ (b) L when $x = 32$.

4. b varies inversely as e. If $b = 6$ when $e = 2$, calculate:
   (a) the value of b when $e = 12$
   (b) the value of e when $b = 3$.

5. x is inversely proportional to y^2. If $x = 4$ when $y = 3$, calculate:
   (a) the value of x when $y = 1$
   (b) the value of y when $x = 2\tfrac{1}{4}$.

6. p is inversely proportional to $\sqrt{y}$. If $p = 1{\cdot}2$ when $y = 100$, calculate:
   (a) the value of p when $y = 4$
   (b) the value of y when $p = 3$.

7. Given that $z \propto \dfrac{1}{y}$, copy and complete the table:

y	2	4		$\tfrac{1}{4}$
z	8		16	

8. Given that $v \propto \dfrac{1}{t^2}$, copy and complete the table:

t	2	5		10
v	25		$\frac{1}{4}$	

9. e varies inversely as $(y - 2)$. If $e = 12$ when $y = 4$, find:
   (a) e when $y = 6$
   (b) y when $e = \frac{1}{2}$.

10. The volume V of a given mass of gas varies inversely as the pressure P. When $V = 2\,\text{m}^3$, $P = 500\,\text{N/m}^2$. Find the volume when the pressure is $400\,\text{N/m}^2$. Find the pressure when the volume is $5\,\text{m}^3$.

11. The number of hours N required to dig a certain hole is inversely proportional to the number of men available x. When 6 men are digging, the hole takes 4 hours.
Find the time taken when 8 men are available. If it takes $\frac{1}{2}$ hour to dig the hole, how many men are there?

12. The force of attraction F between two magnets varies inversely as the square of the distance d between them. When the magnets are 2 cm apart, the force of attraction is 18 newtons. How far apart are they if the attractive force is 2 newtons?

13. The number of tiles, n, that can be pasted using one tin of tile paste is inversely proportional to the square of the side d, of the tile. One tin is enough for 180 tiles of side 10 cm. How many tiles of side 15 cm can be pasted using one tin?

14. When cooking snacks in a microwave oven, a French chef assumes that the cooking time is inversely proportional to the power used. The five levels on his microwave have the powers shown in the table.

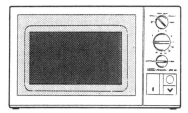

Level	Power used
Full	600 W
Roast	400 W
Simmer	200 W
Defrost	100 W
Warm	50 W

(a) Escargots de Bourgogne take 5 minutes on 'Simmer'. How long will they take on 'Warm'?
(b) Escargots à la Provençale are normally cooked on 'Roast' for 3 minutes. How long will they take on 'Full'?

5.5 Graphical solution of equations

Accurately drawn graphs enable approximate solutions to be found for a wide range of equations, many of which are impossible to solve exactly by other methods.

Example 1

Draw the graph of the function $y = 2x^2 - x - 3$ for $-2 \leqslant x \leqslant 3$.
Use the graph to find approximate solutions to the following equations

(a) $2x^2 - x - 3 = 6$

(b) $2x^2 - x = x + 5$

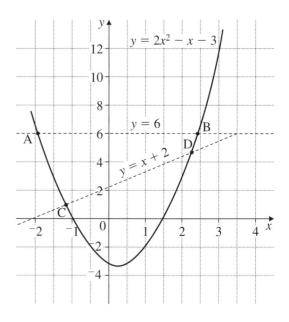

(a) To solve the equation $2x^2 - x - 3 = 6$, the line $y = 6$ is drawn.
At the points of intersection (A and B), y simultaneously equals both 6 and $(2x^2 - x - 3)$.
So we may write
$2x^2 - x - 3 = 6$

The solutions are the x-values of the points A and B.
i.e. $x = -1.9$ and $x = 2.4$ approx.

(b) To solve the equation $2x^2 - x = x + 5$, we rearrange the equation to obtain the function $(2x^2 - x - 3)$ on the left-hand side. In this case, subtract 3 from both sides.
$2x^2 - x - 3 = x + 5 - 3$
$2x^2 - x - 3 = x + 2$

If we now draw the line $y = x + 2$, the solutions of the equation are given by the x-values of C and D, the points of intersection.
i.e. $x = -1.2$ and $x = 2.2$ approx.

Example 2

Assuming that the graph of $y = x^2 - 3x + 1$ has been drawn, find the equation of the line which should be drawn to solve the equation $x^2 - 4x + 3 = 0$.

Rearrange $x^2 - 4x + 3 = 0$ in order to obtain $(x^2 - 3x + 1)$ on the left-hand side.

$$x^2 - 4x + 3 = 0$$
add x $\quad x^2 - 3x + 3 = x$
subtract 2 $\quad x^2 - 3x + 1 = x - 2$

Therefore draw the line $y = x - 2$ to solve the equation.

Exercise 3F

1. In the diagram shown, the graphs of $y = x^2 - 2x - 3$, $y = {}^-2$ and $y = x$ have been drawn.

   Use the graphs to find approximate solutions to the following equations.
   (a) $x^2 - 2x - 3 = {}^-2$
   (b) $x^2 - 2x - 3 = x$
   (c) $x^2 - 2x - 3 = 0$
   (d) $x^2 - 2x - 1 = 0$
   [Reminder: Only the x values are required.]

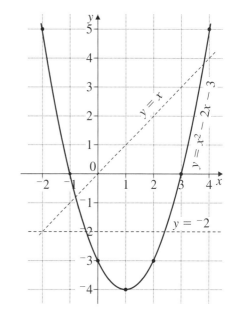

2. The graphs of $y = x^2 - 2$, $y = 2x$ and $y = 2 - x$ are shown.

   Use the graphs to solve:
   (a) $x^2 - 2 = 2 - x$
   (b) $x^2 - 2 = 2x$
   (c) $x^2 - 2 = 2$
   (d) $2 - x = 2x$

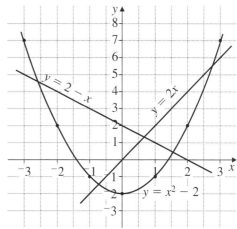

In Questions **3** and **4**, use a scale of 2 cm to 1 unit for x and 1 cm to 1 unit for y.

3. Draw the graphs of the functions $y = x^2 - 2x$ and $y = x + 1$ for ${}^-1 \leqslant x \leqslant 4$. Hence find approximate solutions of the equation $x^2 - 2x = x + 1$.

4. Draw the graphs of the functions $y = 6x - x^2$ and $y = 2x + 1$ for $0 \leq x \leq 5$. Hence find approximate solutions of the equation $6x - x^2 = 2x + 1$.

5. (a) Complete the table and then draw the graph of $y = x^2 - 4x + 1$.

x	-1	0	1	2	3	4
y	6		-2		-2	1

   (b) On the same axes, draw the graph of $y = x - 3$.
   (c) Find solutions of the equations:
      (i) $x^2 - 4x + 1 = x - 3$
      (ii) $x^2 - 4x + 1 = 0$ [answers to 1 d.p.]

In Questions **6** and **7**, do **not** draw any graphs.

6. Assuming the graph of $y = x^2 - 5x$ has been drawn, find the equation of the line which should be drawn to solve the equations:
   (a) $x^2 - 5x = 3$ (b) $x^2 - 5x = -2$
   (c) $x^2 - 5x = x + 4$ (d) $x^2 - 6x = 0$
   (e) $x^2 - 5x - 6 = 0$

7. Assuming the graph of $y = x^2 + x + 1$ has been drawn, find the equation of the line which should be drawn to solve the equations:
   (a) $x^2 + x + 1 = 6$ (b) $x^2 + x + 1 = 0$
   (c) $x^2 + x - 3 = 0$ (d) $x^2 - x + 1 = 0$
   (e) $x^2 - x - 3 = 0$

For Questions **8** to **10**, use scales of 2 cm to 1 unit for x and 1 cm to 1 unit for y.

8. (a) Complete the table and then draw the graph of $y = x^2 + 3x - 1$.

x	-5	-4	-3	-2	-1	0	1	2
y	9	3		-3	-3			9

   (b) By drawing other graphs, solve the equations:
      (i) $x^2 + 3x - 1 = 0$
      (ii) $x^2 + 3x = 7$
      (iii) $x^2 + 3x - 3 = x$

9. Draw the graph of $y = x^2 - 2x + 2$ for $-2 \leqslant x \leqslant 4$. By drawing other graphs, solve the equations:
   (a) $x^2 - 2x + 2 = 8$
   (b) $x^2 - 2x + 2 = 5 - x$
   (c) $x^2 - 2x - 5 = 0$

10. Draw the graph of $y = x^2 - 7x$ for $0 \leqslant x \leqslant 7$.
    Draw suitable straight lines to solve the equations:
    (a) $x^2 - 7x + 9 = 0$
    (b) $x^2 - 5x + 1 = 0$

11. Draw the graph of $y = \dfrac{18}{x}$ for $1 \leqslant x \leqslant 10$, using scales of 1 cm to one unit on both axes. Use the graph to solve approximately:
    (a) $\dfrac{18}{x} = x + 2$
    (b) $\dfrac{18}{x} + x = 10$
    (c) $x^2 = 18$

12. Draw the graph of $y = \tfrac{1}{2}x^2 - 6$ for $-4 \leqslant x \leqslant 4$, taking 2 cm to 1 unit on each axis.
    (a) Use your graph to solve approximately the equation $\tfrac{1}{2}x^2 - 6 = 1$.
    (b) Using tables or a calculator confirm that your solutions are approximately $\pm\sqrt{14}$ and explain why this is so.
    (c) Use your graph to find the square roots of 8.

13. Here are five sketch graphs.

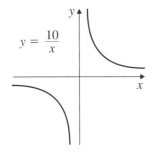
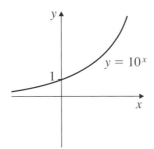
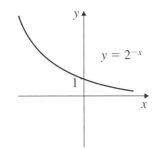

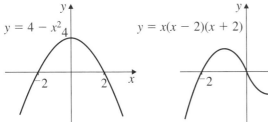

Make your own sketch graphs to find the **number of solutions** of the following equations:

(a) $\dfrac{10}{x} = 10^x$
(b) $4 - x^2 = 10^x$
(c) $x(x-2)(x+2) = \dfrac{10}{x}$

(d) $2^{-x} = 10^x$
(e) $4 - x^2 = 2^{-x}$
(f) $x(x-2)(x+2) = 0$

Summary

1. You can expand brackets.

2. You can use algebra to generalise numerical proofs.

3. You can calculate using proportionality.

4. You can use a graph to solve an equation.

Check out N3

1. Simplify $(\sqrt{6} + \sqrt{3})^2$.

2. (a) Prove that the sum of two odd numbers is an even number.
   (b) Prove that the product of two odd numbers is an odd number.

3. y varies as the square of x. When $x = 2$, $y = 7$. Find:
   (a) an equation connecting x and y
   (b) y when $x = 5$
   (c) x when $y = 11$.

4. (a) Draw the graph of $y = x^2 - 3x + 6$ for values of x from $^-2$ to $^+5$.
   (b) Hence solve the equation $x^2 - 3x + 5 = 0$.
   (c) Draw the graph of an appropriate straight line to solve the equation $x^2 - 4x - 2 = 0$.

Revision exercise N3

1. (a) If $p = \sqrt{a}$, where a is an integer, prove that it is always possible to find a number q, $p \neq q$, so that pq is also an integer.
   (b) Write $(\sqrt{108} - \sqrt{2})^2$ in the form $p - q\sqrt{6}$ where p and q are integers. [AQA]

2. Zoe states that the product of two consecutive integers is never divisible by 50. By means of an example, show that Zoe is **not** correct. [AQA]

3. The volume of olive oil, v litres, in a container, is proportional to the cube of its height, h centimetres. When the height is 10 cm, the volume is 0.3 litres.
   (a) Find the equation connecting v and h.
   (b) Find the height of a container which contains 2 litres of olive oil. [SEG]

4. A gravitational force of F newtons exists between two bodies. The force is inversely proportional to the square of the distance, d metres, between the bodies. The force is 64 newtons when the distance between the bodies is 100 metres.

(a) Find the equation connecting F and d.
(b) Find the force when the distance is 400 metres.
(c) Find the distance when the force is 16 newtons. [SEG]

5. The number of fish, F, needed to stock a lake is proportional to the cube of the width, W metres, of the lake. The number of fish needed to stock a lake of width 100 m is 2700.

(a) Find the equation connecting F and W.
(b) Find the number of fish when the width of the lake is 200 metres.
(c) Find the width of the lake when 4800 fish are needed. [SEG]

6. In an experiment, weights are attached to springs of equal length. The extension of the spring, E metres, is found to be inversely proportional to the stiffness, S, of the spring. The extension is 3 metres when the stiffness is 8.

(a) Find the equation connecting E and S.
(b) Find the extension for a spring with stiffness 12.
(c) Find the stiffness when the extension is 4 metres. [SEG]

7. (a) Copy and complete the table of values for $y = x^2 - 3x - 4$.

x	$^-3$	$^-2$	$^-1$	0	1	2	3	4	5	6
y	14		0	$^-4$	$^-6$	$^-6$		0	6	14

(b) On a grid of $4 \leqslant x \leqslant 6$ and $^-8 \leqslant y \leqslant 14$, draw the graph $y = x^2 - 3x - 4$ for values of x between $^-3$ and $^+6$.
(c) Write down the solutions of $x^2 - 3x - 4 = 0$.
(d) By drawing an appropriate linear graph, write down the solutions of:

$$x^2 - 4x - 1 = 0$$ [AQA]

8. The graph of $y = x^2 + 2x - 7$ is drawn below for values of x between $^-5$ and $^+5$.

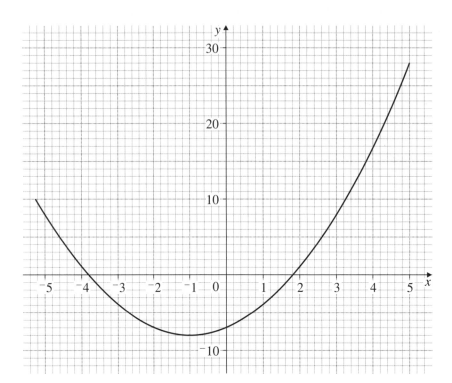

(a) Using the graph, find the solutions of $x^2 + 2x - 7 = 0$, giving your answers to 1 decimal place.

(b) By drawing an appropriate linear graph, write down the solutions of:
$$x^2 - x - 11 = 0.$$
[AQA]

9. The area, A, of an envelope is proportional to the square of its width, W.
When its width is 10 cm, the area is 75 cm^2.

(a) Find the equation connecting A and W.
(b) Find the area of an envelope with a width of 4 cm. [SEG]

Module 3 Practice Calculator Test

1. Find the value of $\dfrac{121.32^2}{19.31 - 7.84}$.

   Give your answer to an appropriate degree of accuracy.

   (3 marks)

2. In 1978 a scientific calculator cost £53.50.
   In 1999 a similar scientific calculator cost £5.95.
   What was the percentage reduction in cost between 1978 and 1999? **(3 marks)** [SEG]

3. In a sale, everything is reduced by 17%. Dave pays £84.66 for a DVD player in the sale.
   How much money does Dave save by buying the DVD player in the sale? **(3 marks)**

4. Harry invests £3000 in an investment account which pays him 4.2% annual compound interest.
   Calculate the amount of his investment after 5 years.

   (3 marks)

5. The tale below shows values for $y = x^2 - 5x + 1$.

x	⁻1	0	1	2	3	4	5	6
y	7	1	⁻3	⁻5	⁻5	⁻3	1	7

   (a) On a grid, draw the graph of $y = x^2 - 5x + 1$ for values of x between ⁻1 and ⁺6.

   (b) Write down the solutions of $x^2 - 5x + 1 = 0$.

   (c) By drawing an appropriate linear graph, write down the solutions of:
   $$x^2 - 4x - 3 = 0.$$ **(6 marks)**

6. Liz and Ken run a hotel and estimate that next month 70% of their rooms will be used. At the end of the month they find that 84% of their rooms have been used and the average amount paid by guests per room has been 15% above their estimate.
   Find the increase in income as a percentage of their estimate.
   (4 marks) [SEG]

7. (a) Write $0.5\dot{4}$ as a fraction in its simplest form.
   (b) Write $0.7\dot{5}\dot{4}$ as a fraction in its simplest form.

   (5 marks)

8. Jim is training for a 400 metre race.
   He states that he runs 400 metres in 46 seconds.

   Both of these figures are given to two significant figures.

   Find his maximum speed in miles per hour.
   Give your answer to three significant figures.

   You may use the exact conversions:

   2.54 cm = 1 inch
   1 mile = 5280 feet

   You **must** show all of your working.   **(5 marks)**   [AQA]

Module 3   Practice Non-Calculator Test

1. Steff spends £77 in total on buying a personal CD player and a CD.
   The ratio of the cost of the CD player to the cost of the CD is in the ratio of 5 : 2.
   Find how much Steff pays for the CD player.   **(2 marks)**

2. A greengrocer pays £609 for 2130 melons.
   By rounding each number to one significant figure, estimate the cost of one melon.   **(3 marks)**

3. (a) Evaluate $\dfrac{2}{5} + \dfrac{3}{7}$.

   (b) Evaluate $4\dfrac{2}{5} - 1\dfrac{2}{3}$.

   (c) Work out $\dfrac{3 \times 10^5}{6 \times 10^{-7}}$.

   Give your answer in standard form.   **(8 marks)**

4. Chloe states that the division of two integers is never divisible by 20.
   By means of an example, show that Chloe is not correct.

   (2 marks)

5. In a large house, the area of each mirror, A, in square feet is proportional to the square of its width, W, in feet.
   A mirror with an area of 15 square feet is 3 feet wide.
   (a) Find an equation connecting A and W.
   (b) Find the area of a mirror with a width of 6 feet.
   (c) A third mirror has an area of 2.4 square feet.
       Find its width. **(7 marks)** [SEG]

6. (a) Express in its simplest form:
       (i) 0.25^{-1}
       (ii) $81^{\frac{3}{4}}$
   (b) Express $\dfrac{1}{32^3}$ as a power of 2. **(5 marks)** [SEG]

7. (a) Simplify fully the following expression, leaving your answer in surd form:
       $$\sqrt{50} - \sqrt{8}$$
   (b) Given that $486 = 3^5 \times 2$, simplify, the expression
       $\dfrac{\sqrt{486}}{\sqrt{50} - \sqrt{8}}$. Give your answer in surd form.
       (5 marks) [SEG]

AS1 Algebra 1

This unit will show you how to:

- Expand and factorise expressions
- Solve linear equations
- Use trial and improvement
- Find the nth term in a sequence
- Draw, describe and use straight line graphs
- Solve simultaneous equations

Before you start:

You should know how to...	Check in AS1
1. Calculate with indices. For example, $3^5 \times 3^4 = 3^{5+4} = 3^9$ $5^2 \div 5^6 = 5^{2-6} = 5^{-4}$	1. Find the value of: (a) $2^7 \times 2^8$. (b) $3^4 \div 3^{-6}$
2. Use directed numbers. For example, $7 + {}^-3 = 7 - 3 = 4$ ${}^-11 + {}^-4 = {}^-15$ ${}^-5 \times {}^+4 = {}^-20$	2. Find the value of: (a) $5 + {}^-7$ (b) $8 - {}^-5$ (c) $7 \times {}^-3$ (d) ${}^-8 \times {}^-5$
3. Substitute in formulae. For example, find the value of y when $y = 7x - 5$ and $x = 3$. $y = 7 \times 3 - 5 = 21 - 5 = 16$	3. Find the value of y when (a) $y = 2x + 11$, $x = 5$ (b) $y = 4 - 3x$, $x = 8$.

1.1 Basic algebra

Using letters for numbers

Exercise 1A

1. Find what number I am left with if:
    (a) I start with x, double it and then subtract 6
    (b) I start with x, add 4 and then square the result
    (c) I start with x, take away 5, double the result and then divide by 3
    (d) I start with w, subtract x and then square the result
    (e) I start with n, add p, cube the result and then divide by a
    (f) I start with t, subtract 1, treble the result and then square the new result?

2. A brick weighs w kg. How much do n bricks weigh?

3. A man shares a sum of x pence equally between n children. How much does each child receive?

4. A cake weighing y kg is cut into n equal pieces. How much does each piece weigh?

5. A sum of £p is shared equally between you and four others. How much does each person receive?

6. Gary is n years older than Mark who is y years old. How old will Gary be in six years time?

7. Sangita used to earn £n per week. She then had a rise of £r per week. How much will she earn in w weeks?

8. The height of a balloon increases at a steady rate of x metres in t hours. How far will the balloon rise in n hours?

9. How many drinks costing x pence each can be bought for £n?

10. A prize of £n is shared equally between you and x other people. How much does each person receive in **pence**?

Collecting terms

- '$3x + 4$' is an **expression**. It is **not** an equation because there is no equals sign. $3x$ and 4 are two **terms** in the expression.
- The expression $7x + 2x$ consists of two **like terms**, $7x$ and $2x$. Like terms can be added or subtracted. So $7x + 2x = 9x$.
- The expression $5x - 3y$ consists of two **unlike** terms, $5x$ and $3y$. Unlike terms cannot be added or subtracted.

Example

(a) $8a + 7a - 3a = 12a$
(b) $3x + y + 2x = 5x + y$
(c) $x^2 + 3x^2 + 2x = 4x^2 + 2x$
(d) $5 + 3y + 6 + y^2 = 11 + 3y + y^2$

Exercise 1B

In Questions **1** to **20** collect like terms together.

1. $2x + 3 + 3x + 5$
2. $4x + 8 + 5x - 3$
3. $5x - 3 + 2x + 7$
4. $6x + 1 + x + 3$
5. $4x - 3 + 2x + 10 + x$
6. $5x + 8 + x + 4 + 2x$
7. $7x - 9 + 2x + 3 + 3x$
8. $5x + 7 - 3x - 2$
9. $4x - 6 - 2x + 1$
10. $10x + 5 - 9x - 10 + x$
11. $4a + 6b + 3 + 9a - 3b - 4$
12. $8m - 3n + 1 + 6n + 2m + 7$

13. $6p - 4 + 5q - 3p - 4 - 7q$
14. $12s - 3t + 2 - 10s - 4t + 12$
15. $a - 2b - 7 + a + 2b + 8$
16. $3x + 2y + 5z - 2x - y + 2z$
17. $6x - 5y + 3z - x + y + z$
18. $2k - 3m + n + 3k - m - n$
19. $12a - 3 + 2b - 6 - 8a + 3b$
20. $3a + x + e - 2a - 5x - 6e$

21. Simplify where possible.
    (a) $x^2 - 3x + 1 + 2x^2 + 3x$ (b) $5a + ab - 3a + 4ab$
    (c) $x^3 - 7x + 4x^2$ (d) $ab + 3a^2 - 7a - ab + a^2$

22. Which of these expressions is equivalent to $x - 2$?

    A $x^2 - 7x - 1 - x^2 + 8x + 3$

    B $x^2 + 7x + 2x - 8x - x^2 - 2$

    C $5 + 7x - x - 4 - 6x + x^2$

    D $5x - 7 + 4 - x + 1 - 3x$

Collect like terms together.

23. $x^2 + 5x + 2 - 2x + 1$
24. $x^2 + 2x + 2x^2 + 4x + 5$
25. $x^2 + 5x + x^2 + x - 7$
26. $2x^2 - 3x + 8 + x^2 + 4x + 4$
27. $3x^2 + 4x + 6 - x^2 - 3x - 3$
28. $5x^2 - 3x + 2 - 3x^2 + 2x - 2$
29. $2x^2 - 2x + 3 - x^2 - 2x - 5$
30. $6x^2 - 7x + 8 - 3x^2 + 5x - 10$
31. $3y^2 - 6x + y^2 + x^2 + 7x + 4x^2$
32. $8 - 5x - 2x^2 + 4 + 6x + 2x^2$
33. $5 + 2y + 3y^2 - 8y - 6 + 2y^2 + 3$
34. $ab + a^2 - 3b + 2ab - a^2$
35. $3c^2 - d^2 + 2cd - 3c^2 - d^2$
36. $ab + 2a^2 + 3ab - 4a^2 + 2a$
37. $x^3 + 2x^2 - x + 3x^2 + x^3 + x$
38. $5 - x^2 - 2x^3 + 6 + 2x^2 + 3x^3$
39. $xy + ab - cd + 2xy - ab + dc$
40. $pq - 3qp + p^2 + 2qp - q^2$

Simplifying terms and brackets

Example

(a) $3 \times 4x = 12x$ $5x \times 4x = 20x^2$ $-3 \times 2x = -6x$
    $5(2a \times 3a) = 30a^2$ $3m \times 2n = 6mn$ $5b \times 2bc = 10b^2c$

(b) $3(2x - 1) = 6x - 3$ $x(3x + 4) = 3x^2 + 4x$
    $a(a + 2b) = a^2 + 2ab$ $t^2(3t - 2) = 3t^3 - 2t^2$

(c) $2(x + 3) + 5(x - 1)$ $5(a - b) - 2(2a - b)$
    $= 2x + 6 + 5x - 5$ $= 5a - 5b - 4a + 2b$
    $= 7x + 1$ $= a - 3b$

Exercise 1C

1. Write in a more simple form.
   (a) $2 \times 3x$ (b) $3a \times 5$ (c) $5t \times (^-2)$ (d) $100 \times 10a$
   (e) $x \times 2x$ (f) $x \times 4x$ (g) $x \times 2x^2$ (h) $4t \times 5t$

2. Remove the brackets.
   (a) $2(2x+1)$
   (b) $5(2x+3)$
   (c) $a(a-3)$
   (d) $2n(n+1)$
   (e) $^-2(2x+3)$
   (f) $2x(x+y)$
   (g) $^-3(a-2)$
   (h) $p(q-p)$
   (i) $a(b+a)$
   (j) $2b(b-1)$
   (k) $2n(n+2)$
   (l) $3x(2x+3)$

3. Find four pairs of equivalent expressions or terms.
   A $2a \times 3a^2$ B $6 \times a \times a$ C $4a^2$ D $6a^2$
   E $(2a)^2$ F $3a \times 3a$ G $6a^3$ H $9a^2$

4. Multiply these terms.
   (a) $2a \times 3b$
   (b) $x \times 3x^3$
   (c) $y \times 2x$
   (d) $2p \times 5q$
   (e) $3x \times 5y$
   (f) $6x \times 3x^2$
   (g) $3a \times 8a^3$
   (h) $ab \times 2a$
   (i) $xy \times 3y$
   (j) $cd \times 5c$
   (k) $ab \times ab$
   (l) $2xy \times xy$
   (m) $3d \times d \times d$
   (n) $5(2x \times xy)$
   (o) $3a \times (^-2a)$
   (p) $^-a \times (^-2ab)$

5. A solid rectangular block measures x cm by x cm by $(x+3)$ cm. Find a simplified expression for its surface area in cm^2.

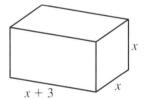

6. Remove the brackets and simplify.
   (a) $3(x+2) + 4(x+1)$
   (b) $5(x-2) + 3(x+4)$
   (c) $2(a-3) + 3(a+1)$
   (d) $5(a+1) + 6(a+2)$
   (e) $3(x+2) - 2(x+1)$
   (f) $4(x+3) - 3(x+2)$

7. Three rods A, B and C have lengths of x, $(x+1)$ and $(x-2)$ cm respectively, as shown.

A	B	C
x cm	$(x+1)$ cm	$(x-2)$ cm

   In the diagrams below express the length l in terms of x. Give your answers in their simplest form.

(a)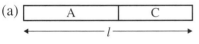

(b)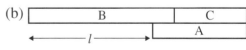

(c)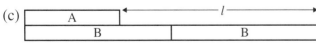

(d)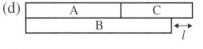

(e)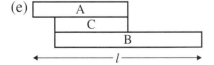

(f)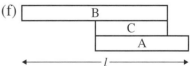

8. Simplify the following expressions.
   (a) $x(2x+1) + 3(x+2)$
   (b) $x(2x-3) + 5(x+1)$
   (c) $x(2x+1) + x(3x+1)$
   (d) $a(2a+3) - a(a+1)$

Expanding brackets

To work out the expression $(x+3)(x+2)$, the area of a rectangle can be a help.
From the diagram, the total area of the rectangle

$$= (x+3)(x+2)$$
$$= x^2 + 2x + 3x + 6$$
$$= x^2 + 5x + 6$$

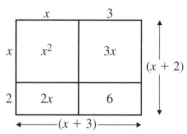

After a little practice, it is possible to do without the diagram.

Example

(a) Remove the brackets and simplify $(3x-2)(2x-1)$.

$$(3x-2)(2x-1) = 3x(2x-1) - 2(2x-1)$$
$$= 6x^2 - 3x - 4x + 2$$
$$= 6x^2 - 7x + 2$$

(b) Remove the brackets and simplify $(x-3)^2$.

$$(x-3)^2 = (x-3)(x-3)$$
$$= x(x-3) - 3(x-3)$$
$$= x^2 - 3x - 3x + 9$$
$$= x^2 - 6x + 9$$

> Be careful with this expression. It is not $x^2 - 9$, nor even $x^2 + 9$.

Exercise 1D

Remove the brackets and simplify.

1. $(x+1)(x+3)$
2. $(x+3)(x+2)$
3. $(y+4)(y+5)$
4. $(x-3)(x+4)$
5. $(x+5)(x-2)$
6. $(x-3)(x-2)$
7. $(a-7)(a+5)$
8. $(z+9)(z-2)$
9. $(x-3)(x+3)$
10. $(k-11)(k+11)$
11. $(2x+1)(x-3)$
12. $(3x+4)(x-2)$
13. $(2y-3)(y+1)$
14. $(7y-1)(7y+1)$
15. $(x+4)^2$
16. $(x+2)^2$
17. $(x-2)^2$
18. $(2x+1)^2$
19. $(x+1)^2 + (x+2)^2$
20. $(x-2)^2 + (x+3)^2$
21. $(x+2)^2 + (2x+1)^2$
22. $(y-3)^2 + (y-4)^2$
23. $(x+2)^2 - (x-3)^2$
24. $(x-3)^2 - (x+1)^2$

25. Four identical tiles surround the shaded square.
    (a) Find an expression for the sum of the areas of the four tiles.
    (b) Find the area of square $ABCD$.
    (c) Hence find the area of the shaded square.

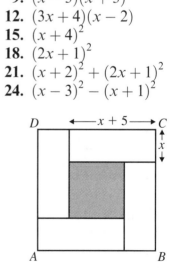

26. The expression $(x+1)(x+2)(x+3)$ is expanded in stages. Complete the working

$(x+1)[(x+2)(x+3)]$
$= (x+1)(x^2 + 5x + 6)$
$= \ldots$

27. Remove the brackets and simplify.
   (a) $(x+1)(x+3)(x+4)$
   (b) $(x+1)^3$
   (c) $(x-1)(x+2)^2$

28. (a) If n is an integer, explain why $(2n-1)$ is an odd number.
   (b) Write down the next odd number after $(2n-1)$ and show that the sum of these two numbers is a multiple of 4.
   (c) Write down the next odd number after these two and show that the sum of the three odd numbers is a multiple of 3.
   (d) Show that the sum of the squares of these three odd numbers is $12n^2 + 12n + 11$.
   Is it possible for this expression to be a multiple of 11? If so give two possible values of n.

Factorising expressions

Earlier we expanded expressions such as $x(3x-1)$ to give $3x^2 - x$. The reverse of this process is called factorising.

Example
Factorise: (a) $x^2 + 7x$ (b) $3y^2 - 12y$ (c) $6a^2b - 10ab^2$

(a) x is common to x^2 and $7x$. $\therefore$ $x^2 + 7x = x(x+7)$.
The factors are x and $(x+7)$.
(b) $3y$ is common $\therefore$ $3y^2 - 12y = 3y(y-4)$
(c) $2ab$ is common $\therefore$ $6a^2b - 10ab^2 = 2ab(3a - 5b)$

Exercise 1E
Factorise the following expressions completely.

1. $x^2 + 5x$
2. $x^2 - 6x$
3. $7x - x^2$
4. $y^2 + 8y$
5. $2y^2 + 3y$
6. $6y^2 - 4y$
7. $3x^2 - 21x$
8. $16a - 2a^2$
9. $6c^2 - 21c$
10. $15x - 9x^2$
11. $56y - 21y^2$
12. $ax + bx + 2cx$
13. $x^2 + xy + 3xz$
14. $x^2y + y^3 + z^2y$
15. $3a^2b + 2ab^2$
16. $x^2y + xy^2$
17. $6a^2 + 4ab + 2ac$
18. $ma + 2bm + m^2$
19. $2kx + 6ky + 4kz$
20. $ax^2 + ay + 2ab$
21. $7x^2 + x$
22. $4y^2 - 4y$
23. $p^2 - 2p$
24. $6a^2 + 2a$

25. $4 - 8x^2$
26. $5x - 10x^3$
27. $4\pi r + \pi h$
28. $\pi r^2 + 2\pi r$
29. $3\pi r^2 + \pi r h$
30. $3xy + 2x$
31. $x^2 k + xk^2$
32. $a^3 b + 2ab^2$
33. $abc - 3b^2 c$
34. $2a^2 e - 5ae^2$
35. $a^3 b + ab^3$
36. $x^3 y + x^2 y^2$
37. $6xy^2 - 4x^2 y$
38. $3ab^3 - 3a^3 b$
39. $2a^3 b + 5a^2 b^2$
40. $ax^2 y - 2ax^2 z$
41. $2abx + 2ab^2 + 2a^2 b$
42. $ayx + yx^3 - 2y^2 x^2$

43. In these two rectangles you are given one side and the area. Find the perimeter of each rectangle.

(a) area = $2xy + 4y^2$, side = $2y$

(b) area = $6a^2 - 2ax$, side = $2a$

Mathematical language

In algebra, letter symbols are used to represent unknowns in a variety of situations.

- In **equations** the letters stand for particular numbers (the solutions of the equation), for example, $3x - 1 = 2(x + 1)$.
- In an **expression** there is no equals sign, for example, $5x^2 - 3x + 1$.
- In **formulae** letters stand for defined quantities or variables, for example, $F = ma$.
- In an **identity** there is an equals sign (sometimes written $\equiv$) but the equality holds for **all values** of the unknown, for example, $(x^2 - 1) = (x - 1)(x + 1)$.
- In the definition of **functions**, for example, $y = x^2 + 7x$, $f(x) = 3x^2$.

Exercise 1F

Copy and complete the table by putting the contents of each box in the correct column.

$x(x + 1) = x^2 + x$

$7y + 10$

$V = IR$

$x^2 - 7x = 0$

$7x + 11 = x - 9$

$x^2 - 3x + 10$

$(x + 1)^2 = x^2 + 2x + 1$

$A = \pi r^2$

equation	expression	identity	formula

1.2 Linear equations

Many questions in mathematics are easier to answer if algebra is used. It is often best to let the unknown quantity be x and to try to translate the question into the form of an equation.

Basic rules

Rule 1. Treat both sides the same.

(a) We can add the same to both sides, for example:
$$3x - 1 = 4$$
$$3x - 1 + 1 = 4 + 1 \qquad \text{[add 1]}$$

(b) We can subtract the same from both sides, for example:
$$5x + 7 = 10$$
$$5x + 7 - 7 = 10 - 7 \qquad \text{[take away 7]}$$

(c) We can multiply both sides by the same factor, for example:
$$\frac{x}{4} = 3$$
$$4\left(\frac{x}{4}\right) = 4 \times 3 \qquad \text{[multiply by 4]}$$

(d) We can divide both sides by the same factor.
$$3x = 8$$
$$\frac{3x}{3} = \frac{8}{3} \qquad \text{[divide by 3]}$$

Rule 2. If the x term is negative, take it to the other side where it becomes positive.

Rule 3. If there are x terms on both sides, collect them on one side.

Example 1

(a) $\quad 4 - 3x = 2$
$\quad\quad\; 4 = 2 + 3x$
$\quad\quad\; 2 = 3x$
$\quad\quad\; \dfrac{2}{3} = x$

(b) $\quad 2x - 7 = 5 - 3x$
$\quad\quad\; 2x + 3x = 5 + 7$
$\quad\quad\; 5x = 12$
$\quad\quad\; x = \dfrac{12}{5} = 2\tfrac{2}{5}$

Rule 4. If there is a fraction in the x term, multiply out to simplify the equation.

Example 2

(a) $\dfrac{2x}{3} = 10$

$2x = 30$

$x = \dfrac{30}{2} = 15$

(b) $3 = \dfrac{x}{4} - 4$

$7 = \dfrac{x}{4}$

$28 = x$

(c) $\dfrac{a}{3} + 7 = 5$

$\dfrac{a}{3} = {}^-2$

$a = {}^-6$

Exercise 1G

Solve the following equations:

1. $2x - 5 = 11$
2. $3x - 7 = 20$
3. $2x + 6 = 20$
4. $5x + 10 = 60$
5. $8 = 7 + 3x$
6. $12 = 2x - 8$
7. ${}^-7 = 2x - 10$
8. $3x - 7 = {}^-10$
9. $12 = 15 + 2x$
10. $5 + 6x = 7$
11. $100x - 1 = 98$
12. $7 = 7 + 7x$
13. $\dfrac{x}{100} + 10 = 20$
14. $1000x - 5 = -6$
15. ${}^-4 = {}^-7 + 3x$
16. $2x + 4 = x - 3$
17. $x - 3 = 3x + 7$
18. $5x - 4 = 3 - x$
19. $3 - 2x = x + 12$
20. $6 + 2a = 3$
21. $a - 3 = 3a - 7$
22. $2y - 1 = 4 - 3y$
23. $7 - 2x = 2x - 7$
24. $7 - 3x = 5 - 2x$
25. $8 - 2y = 5 - 5y$
26. $x - 16 = 16 - 2x$
27. $x + 2 = 3 \cdot 1$
28. ${}^-x - 4 = {}^-3$
29. ${}^-3 - x = {}^-5$
30. $\dfrac{x}{5} = 7$
31. $\dfrac{x}{10} = 13$
32. $7 = \dfrac{x}{2}$
33. $\dfrac{x}{2} = \dfrac{1}{3}$
34. $\dfrac{3}{4} = \dfrac{2x}{3}$
35. $\dfrac{5x}{6} = \dfrac{1}{4}$
36. $-\dfrac{3}{4} = \dfrac{3x}{5}$
37. $\dfrac{x}{2} + 7 = 12$
38. $\dfrac{x}{3} - 7 = 2$
39. $\dfrac{x}{5} - 6 = {}^-2$
40. $-\dfrac{x}{2} + 1 = -\dfrac{1}{4}$
41. $-\dfrac{3}{5} + \dfrac{x}{10} = -\dfrac{1}{5} - \dfrac{x}{5}$

Rule 5. When an equation has brackets, multiply out the brackets first.

Example

(a) $x - 2(x - 1) = 1 - 4(x + 1)$

$x - 2x + 2 = 1 - 4x - 4$

$x - 2x + 4x = 1 - 4 - 2$

$3x = {}^-5$

$x = -\dfrac{5}{3}$

(b) $\dfrac{5}{x} = 2$

$5 = 2x$

$\dfrac{5}{2} = x$

Exercise 1H

Solve the following equations.

1. $x + 3(x + 1) = 2x$
2. $1 + 3(x - 1) = 4$
3. $2x - 2(x + 1) = 5x$
4. $2(3x - 1) = 3(x - 1)$
5. $4(x - 1) = 2(3 - x)$
6. $4(x - 1) - 2 = 3x$
7. $4(1 - 2x) = 3(2 - x)$
8. $3 - 2(2x + 1) = x + 17$
9. $4x = x - (x - 2)$
10. $7x = 3x - (x + 20)$
11. $5x - 3(x - 1) = 39$
12. $3x + 2(x - 5) = 15$
13. $7 - (x + 1) = 9 - (2x - 1)$
14. $10x - (2x + 3) = 21$
15. $3(2x + 1) + 2(x - 1) = 23$
16. $5(1 - 2x) - 3(4 + 4x) = 0$
17. $7x - (2 - x) = 0$
18. $3(x + 1) = 4 - (x - 3)$
19. $3y + 7 + 3(y - 1) = 2(2y + 6)$
20. $4(y - 1) + 3(y + 2) = 5(y - 4)$
21. $\dfrac{7}{x} = 21$
22. $30 = \dfrac{6}{x}$
23. $\dfrac{5}{x} = 3$
24. $\dfrac{9}{x} = {}^-3$
25. $11 = \dfrac{5}{x}$
26. ${}^-2 = \dfrac{4}{x}$
27. $\dfrac{x + 1}{3} = \dfrac{x - 1}{4}$
28. $\dfrac{x + 3}{2} = \dfrac{x - 4}{5}$
29. $\dfrac{2x - 1}{3} = \dfrac{x}{2}$
30. $\dfrac{3x + 1}{5} = \dfrac{2x}{3}$
31. $\dfrac{5}{x - 1} = \dfrac{10}{x}$
32. $\dfrac{12}{2x - 3} = 4$
33. $\dfrac{6}{x} - 3 = 7$
34. $\dfrac{9}{x} - 7 = 1$
35. ${}^-2 = 1 + \dfrac{3}{x}$
36. $4 - \dfrac{4}{x} = 0$
37. $5 - \dfrac{6}{x} = {}^-1$
38. $\dfrac{x}{3} + \dfrac{x}{4} = 1$

Solving problems using linear equations

So far we have concentrated on solving given equations. Making up our own equations helps to solve problems which are difficult to solve.

There are four steps.

(a) Let the unknown quantity be x (or any other suitable letter) and state the units where appropriate.
(b) Write the problem in the form of an equation.
(c) Solve the equation and give the answer in words.
(d) Check your solution using the problem and **not** your equation.

Example

Find three consecutive even numbers which add up to 792.

(a) Let the smallest number be x.
Then the other numbers are $(x+2)$ and $(x+4)$ because they are consecutive **even** numbers.

(b) Form an equation: $x + (x+2) + (x+4) = 792$

(c) Solve: $3x + 6 = 792$
$$3x = 786$$
$$x = 262$$
The three numbers are 262, 264 and 266.

(d) Check: $262 + 264 + 266 = 792$ ✓

Exercise 1I

Solve each problem by forming an equation.

1. The length of a rectangle is twice the width. If the perimeter is 20 cm, find the width.

2. The width of a rectangle is one third of the length. If the perimeter is 96 cm, find the width.

3. The sum of three consecutive numbers is 276. Find the numbers. Let the first number be x.

4. The sum of four consecutive numbers is 90. Find the numbers.

5. The sum of three consecutive odd numbers is 177. Find the numbers.

6. Find three consecutive even numbers which add up to 1524.

7. When a number is doubled and then added to 13, the result is 38. Find the number.

8. If AB is a straight line, find x.

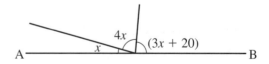

9. The difference between two numbers is 9. Find the numbers, if their sum is 46.

10. The three angles in a triangle are in the ratio $1:3:5$. Find the angles.

11. The sum of three numbers is 28. The second number is three times the first and the third is 7 less than the second. What are the numbers?

12. If the perimeter of the triangle is 22 cm, find the length of the shortest side.

    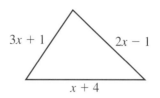

13. David weighs 5 kg less than John, who in turn is 8 kg lighter than Paul. If their total weight is 197 kg, how heavy is each person?

14. If the perimeter of the rectangle is 34 cm, find x.

    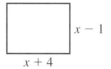

15. The diagram shows a rectangular lawn surrounded by a footpath x m wide.
    (a) Show that the area of the path is $4x^2 + 14x$.
    (b) Find an expression, in terms of x for the distance around the outside edge of the path.
    (c) Find the value of x when this perimeter is 20 m.

    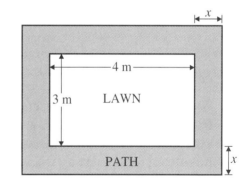

16. Find the value of x so that the areas of the shaded rectangles are equal.

    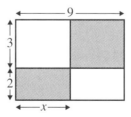

17. Two angles of an isosceles triangle are $a°$ and $(a + 10)°$. Find two possible values of a.

18. A man is 32 years older than his son. Ten years ago he was three times as old as his son was then. Find the present age of each.

19. A car completes a journey in 10 minutes. For the first half of the distance the speed was 60 km/h and for the second half the speed was 40 km/h. How far is the journey?

20. A lemming runs from a point A to a cliff at 4 m/s, jumps over the edge at B and falls to C at an average speed of 25 m/s.
If the total distance from A to C is 500 m and the time taken for the journey is 41 seconds, find the height BC of the cliff.

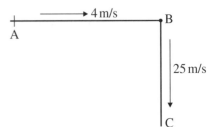

21. A bus is travelling with 52 passengers. When it arrives at a stop, y passengers get off and 4 get on. At the next stop one third of the passengers get off and 3 get on. There are now 25 passengers. Find y.

22. Mr Lee left his fortune to his 3 sons, 4 daughters and his wife. Each son received twice as much as each daughter and his wife received £6000, which was a quarter of the money. How much did each son receive?

23. In a regular polygon with n sides each interior angle is $180 - \dfrac{360}{n}$ degrees. How many sides does a polygon have if each angle is 156°?

24. A sparrow flies to see a friend at a speed of 4 km/h. His friend is out, so the sparrow immediately returns home at a speed of 5 km/h. The complete journey took 54 minutes. How far away does his friend live?

25. Consider the equation $an^2 = 182$ where a is any number between 2 and 5 and n is a positive integer. What are the possible values of n?

26.

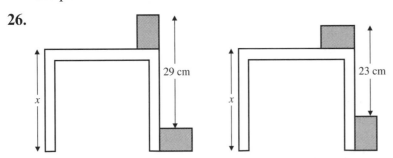

The diagrams show a table with two identical wooden blocks. Calculate the height of the table, x.

Solving equations involving brackets

Example

Solve the equation $(x+3)^2 = (x+2)^2 + 3^2$

$$(x+3)^2 = (x+2)^2 + 3^2$$
$$(x+3)(x+3) = (x+2)(x+2) + 9$$
$$x^2 + 6x + 9 = x^2 + 4x + 4 + 9$$
$$6x + 9 = 4x + 13$$
$$2x = 4$$
$$x = 2$$

Exercise 1J

Solve the following equations:

1. $x^2 + 4 = (x+1)(x+3)$
2. $x^2 + 3x = (x+3)(x+1)$
3. $(x+3)(x-1) = x^2 + 5$
4. $(x+1)(x+4) = (x-7)(x+6)$
5. $(x-2)(x+3) = (x-7)(x+7)$
6. $(x-5)(x+4) = (x+7)(x-6)$
7. $2x^2 + 3x = (2x-1)(x+1)$
8. $(2x-1)(x-3) = (2x-3)(x-1)$
9. $x^2 + (x+1)^2 = (2x-1)(x+4)$
10. $x(2x+6) = 2(x^2 - 5)$

In Questions **11** and **12**, form an equation in x by means of Pythagoras' Theorem, and hence find the length of each side of the triangle.
(All the lengths are in cm.)

11.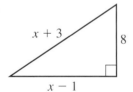

12.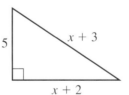

13. The area of the rectangle shown exceeds the area of the square by $2\,\text{cm}^2$. Find x.

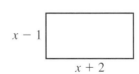

14. The area of the square exceeds the area of the rectangle by $13\,\text{m}^2$. Find y.

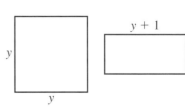

1.3 Trial and improvement

Some problems cannot be solved using linear equations. In such cases, the method of 'trial and improvement' is often a help.

Exercise 1K

1. Think of a rectangle having an area of 72 cm² whose base is twice its height.

   Write down the length of the base.

   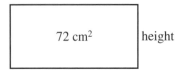

2. Find a rectangle of area 75 cm² so that its base is three times its height.

3. In each of the rectangles below, the base is twice the height. The area is shown inside the rectangle. Find the base and the height.

   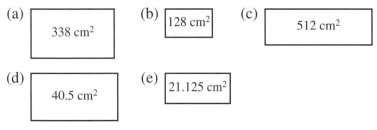

4. In each of the rectangles below, the base is 1 cm more than the height. Find each base and height.

   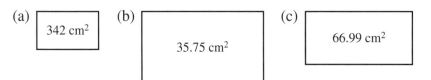

Inexact answers

In some questions it is not possible ever to find an answer which is precisely correct. However, we can find answers which are nearer and nearer to the exact one, perhaps to the nearest 0·1 cm or even to the nearest 0·01 cm.

In this rectangle, the base is 1 cm more than the height h cm. The area is 80 cm². Find the height h, correct to 1 d.p. Here we are solving the equation $h(h+1) = 80$.

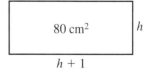

(a) Try different values for h. **Be systematic**.

$h = 8 \ : 8(8+1) \quad = 72 \qquad$ Too small
$h = 9 \ : 9(9+1) \quad = 90 \qquad$ Too large
$h = 8.5 : (8.5+1) \quad = 80.75 \quad$ Too large
$h = 8.4 : 8.4(8.4+1) = 78.96 \quad$ Too small

(b) We now see that the answer is between 8·4 and 8·5. We also see that the value of 8·5 for h gave the value closest to 80. We will take the answer as $h = 8.5$, correct to 1 d.p.

Note: Strictly speaking, to ensure that our answer **is** correct to 1 decimal place, we should try $h = 8.45$. This degree of complexity is not usually necessary in most situations.

Example

Solve the equation $x^3 + 2x = 90$, correct to 1 d.p.

Try $x = 4 \quad : \quad 4^3 + (2 \times 4) \quad = 72 \qquad$ Too small
$x = 5 \quad : \quad 5^3 + (2 \times 5) \quad = 135 \qquad$ Too large
$x = 4.5 \ : \ 4.5^3 + (2 \times 4.5) \ = 100.125 \ $ Too large
$x = 4.4 \ : \ 4.4^3 + (2 \times 4.4) \ = 93.984 \ $ Too large
$x = 4.3 \ : \ 4.3^3 + (2 \times 4.3) \ = 88.107 \ $ Too small
$x = 4.35 : 4.35^3 + (2 \times 4.35) = 91.013 \ $ Too large

∴ The solution is $x = 4.3$, correct to 1 decimal place

Exercise 1L

1. In these two rectangles, the base is 1 cm more than the height. Find the value of h for each one, correct to 1 decimal place.

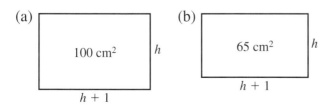

2. Find solutions to the following equations, giving the answer correct to 1 decimal place.
(a) $x(x - 3) = 11$ (b) $x(x - 2) = 7$
(c) $x^3 = 300$ (d) $x^2(x + 1) = 50$

3. Here you need a calculator with a $\boxed{x^y}$ button.

Find x, correct to 1 decimal place.
(a) $x^5 = 313$ (b) $5^x = 77$ (c) $x^x = 100$

4. The number n has two digits after the decimal point and n to the power 10 is 2110 to the nearest whole number. Find n.

5. An engineer wants to make a solid metal cube of volume 526 cm^3.
Call the edge of the cube x and write down an equation.
Find x giving your answer correct to 1 d.p.

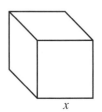

6. In this rectangle, the base is 2 cm more than the height. If the diagonal is 15 cm, find h correct to 2 d.p.

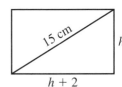

7. A designer for a supermarket chain wants to make a cardboard box of depth 6 cm. He has to make the length of the box 10 cm more than the width.
(a) What is the length of the box in terms of x?
The box is designed so that its volume is to be 9000 cm^3.
(b) Form an equation involving x.
(c) Solve the equation and hence give the dimensions of the box to the nearest cm.

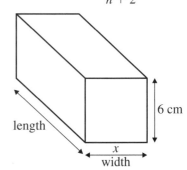

8. The diagram represents a rectangular piece of paper $ABCD$ which has been folded along EF so that C has moved to G.
(a) Calculate the area of $\triangle ECF$.
(b) Find an expression for the shaded area $ABFGED$ in terms of x.

Given that the shaded area is 20 cm^2, show that $x(x + 3) = 26$.

Solve this equation, giving your answer correct to one decimal place.

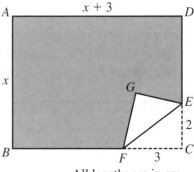

All lengths are in cm.

9. In the rectangle $PQRS$, $PQ = x$ cm and $QR = 1$ cm. The line LM is drawn so that $PLMS$ is a square.
   (a) Write down, in terms of x, the length LQ.
   (b) If $\dfrac{PQ}{QR} = \dfrac{QR}{LQ}$, obtain an equation in x.

   Hence find x correct to 2 decimal places.

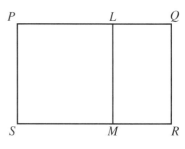

10. Find the number whose cube is 100 times as large as its square root. Give your answer correct to 1 decimal place.

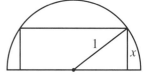

11. In triangle ABC, angle $ABC = x°$, $BC = x$ cm and $AB = (x + 10)$ cm.

    Find the value of x, correct to the nearest 0.5.

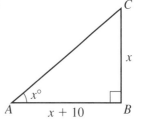

12. A rectangle is drawn inside a semicircle of radius 1 unit.
    (a) Find an expression, in terms of x, for the area of the rectangle.
    (b) Use trial and improvement, possibly using a spreadsheet program, to find the value of x, correct to 2 d.p., which gives the largest value for the area of the rectangle.

13. The net for an open box is made by cutting corners from a square card as shown.
    (a) Show that the volume of the box is given by the formula $V = x(9 - 2x)(9 - 2x)$.
    (b) Find the value of x which gives the maximum volume of the box. Again you may find it helpful to use a spreadsheet.

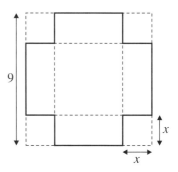

1.4 Finding a rule

Here is a sequence of 'houses' made from matches.

The table on the right records the number of houses *h* and the number of matches *m*.

If the number in the *h* column goes up one at a time, look at the number in the *m* column. If it goes up (or down) by the same number each time, the function connecting *m* and *h* is linear. This means that there are no terms in h^2 or anything more complicated.

h		m
1		5
2		9
3		13
4		17

In this case, the numbers in the *m* column go up by 4 each time. This suggests that a column for 4*h* might help.

Now it is fairly clear that *m* is one more than 4*h*.
So the formula linking *m* and *h* is: $m = 4h + 1$

h	4h	m
1	4	5
2	8	9
3	12	13
4	16	17

Example
The table shows how *r* changes with *n*. What is the formula linking *r* with *n*?

n		r
2		3
3		8
4		13
5		18

Because *r* goes up by 5 each time, write another column for 5*n*. The table shows that *r* is always 7 less than 5*n*, so the formula linking *r* with *n* is $r = 5n - 7$

The method only works when the first set of numbers goes up by one each time.

n	5n	r
2	10	3
3	15	8
4	20	13
5	25	18

Exercise 1M

1. Below is a sequence of diagrams showing black tiles *b* and white tiles *w* with the related table.

 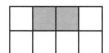 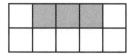

b		w
1		5
2		6
3		7
4		8

What is the formula for *w* in terms of *b*? Write it as $w = \ldots$

2. This is a different sequence with black tiles *b* and white tiles *w* and the related table.

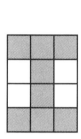

b	w
2	10
3	12
4	14
5	16

What is the formula? Write it as $w = \ldots$

3. Here is a sequence of I-shapes.

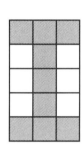

Make your own table for black tiles *b* and white tiles *w*.
What is the formula for *w* in terms of *b*?

4. This sequence shows matches *m* arranged in triangles *t*.

Make a table for *t* and *m* starting like this:
Continue the table and find a formula for *m* in terms of *t*. Write it as $m = \ldots$

t	m
1	3
2	5
⋮	⋮

5. Here is a different sequence of matches and triangles.

Make a table and find a formula connecting *m* and *t*.

6. In this sequence, there are triangles t and squares s around the outside.

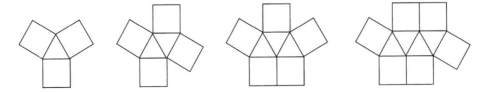

What is the formula connecting t and s?

7. Look at the tables below. In each case, find a formula connecting the two letters.

(a)
n	p
1	3
2	8
3	13
4	18

(b)
n	k
2	17
3	24
4	31
5	38

(c)
n	w
3	17
4	19
5	21
6	23

8. This is one member of a sequence of cubes c made from matches m.

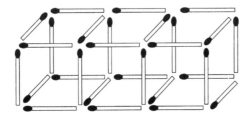

Find a formula connecting m and c.

9. In these tables, the numbers on the left do not go up by one each time. Try to find a formula in each case.

(a)
n	y
1	4
3	10
7	22
8	25

(b)
n	h
2	5
3	9
6	21
10	37

(c)
n	k
3	14
7	26
9	32
12	41

10. Some attractive loops can be made by fitting pentagon tiles together.

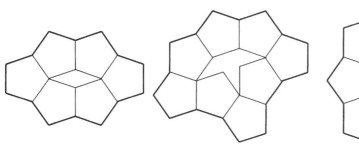
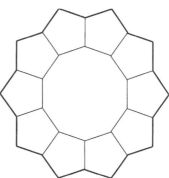

Diagram $n = 1$
6 tiles t
14 external edges e

Diagram $n = 2$
8 tiles t
17 external edges e

Diagram $n = 3$
10 tiles t
20 external edges e

Make your own tables involving n, t and e.
(a) Find a formula connecting n and t.
(b) Find a formula connecting n and e.
(c) Find a formula connecting t and e.

11. In each diagram there are black squares along one diagonal.
Find a formula for the number of white squares in the $n \times n$ square.

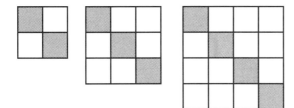

The nth term in a sequence

For the sequence 3, 6, 9, 12, 15 ... the rule is 'add 3'.

Here is the **mapping diagram** for the sequence.

The terms are found by multiplying the term number by 3.
So the 10th term is 30, the 33rd term is 99.

A **general** term in the sequence is the nth term, where n stands for any number.
The nth term of this sequence is $3n$.

Term number (n)		Term
1	→	3
2	→	6
3	→	9
⋮		⋮
10	→	30
⋮		
n	→	$3n$

Here is a more difficult sequence:
3, 7, 11, 15, ...

The rule is 'add 4' so, in the mapping diagram below, we have written a column for 4 times the term number [i.e. $4n$].

Term number (n)		$4n$		Term
1	→	4	→	3
2	→	8	→	7
3	→	12	→	11
4	→	16	→	15

We see that each term is 1 less than $4n$. So, the 10th term is $(4 \times 10) - 1 = 39$ and the nth term is $(4 \times n) - 1 = 4n - 1$.

Exercise 1N

1. Write down each sequence and select the correct expression for the nth term from the list given.
   (a) 10, 20, 30, 40, ...
   (b) 5, 10, 15, 20, ...
   (c) 3, 4, 5, 6, ...
   (d) 3, 5, 7, 9, 11, ...
   (e) 30, 60, 90, 120, ...
   (f) 5, 11, 17, 23, ...
   (g) 1, 4, 7, 10, 13, ...

   $n + 2$ $5n$ $6n - 1$ $2n + 1$ $10n$ $30n$ $3n - 2$

2. Copy and complete the mapping diagrams. Notice that an extra column has been written.

   (a)
Term number (n)		$2n$		Term
1	→	2	→	5
2	→	4	→	7
3	→	6	→	9
4	→	8	→	11
⋮				
10	→	☐	→	☐
⋮				
n	→	☐	→	☐

   (b)
Term number (n)		$3n$		Term
1	→	3	→	4
2	→	6	→	7
3	→	9	→	10
⋮				
20	→	☐	→	☐
⋮				
n	→	☐	→	☐

3. Here are the first five terms of a sequence:
   6, 11, 16, 21, 26, ...
   (a) Draw a mapping diagram similar to those above.
   (b) Write down: (i) the 10th term
   (ii) the nth term.

   Hint:
   The rule for the sequence is 'add 5' so write a column for '$5n$' in the diagram.

4. Here are three sequences:
   A: 1, 4, 7, 10, 13, …
   B: 6, 10, 14, 18, 22, …
   C: 5, 12, 19, 26, 33, …

   For each sequence: (a) draw a mapping diagram
                      (b) find the nth term.

5. Look at the sequences A1, A2, A3, A4 … and B1, B2, B3, B4 …
   Write down
   (a) A10 (b) B10
   (c) the nth term in the 'A' sequence.
   (d) the nth term in the 'B' sequence.

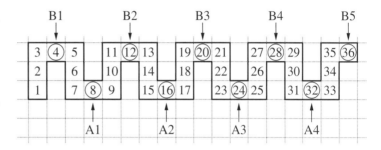

6. Here are two more difficult sequences. Copy and complete the mapping diagrams.

   (a)
Term number (n)		Term
1	→ 1 × 2 →	2
2	→ 2 × 3 →	6
3	→ 3 × 4 →	12
4	→ 4 × 5 →	20
⋮		
10	→ □ →	□
⋮		
n	→ □ →	□

   (b)
Term number (n)		Term
1	→ $1^2 + 1$ →	2
2	→ $2^2 + 1$ →	5
3	→ $3^2 + 1$ →	10
4	→ $4^2 + 1$ →	17
⋮		
10	→ □ →	□
⋮		
n	→ □ →	□

7. Look at the sequence: 5, 8, 13, 20, …

   Decide which of the following is the correct expression for the nth term of the sequence.

   $4n + 1$ $3n + 2$ $n^2 + 4$

In Questions **8** to **12** look carefully at how each sequence is formed. Write down:
(a) the 10th term
(b) the nth term.

8. $1^2, 2^2, 3^2, 4^2, \ldots$

9. $(1 \times 2), (2 \times 3), (3 \times 4), (4 \times 5), \ldots$

10. $\frac{1}{2}, \frac{2}{3}, \frac{3}{4}, \frac{4}{5}, \ldots$

11. $\frac{5}{1^2}, \frac{5}{2^2}, \frac{5}{3^2}, \frac{5}{4^2}, \ldots$

12. $(1 \times 3), (2 \times 4), (3 \times 5), (4 \times 6), \ldots$

13. The nth term of a sequence is t_n.

   If $t_n = n^3$: the first term, $t_1 = 1^3$
   the second term, $t_2 = 2^3$
   the third term, $t_3 = 3^3$ etc.

   Write down the first 3 terms of the sequences where the nth term is given.

   (a) $t_n = 2n + 1$
   (b) $t_n = 5n - 3$
   (c) $t_n = 100 - 2n$
   (d) $t_n = \dfrac{n}{n + 2}$
   (e) $t_n = 10^n$
   (f) $t_n = n(n + 1)$

14. Two pupils, Anna and Dipika, did the same investigation and got the same results. The patterns they noticed were different.

   Anna's table

n	t	pattern
1	3	1×3
2	8	2×4
3	15	3×5
4	24	4×6

   Dipika's table

n	t	pattern
1	3	$1^2 + 2 \times 1$
2	8	$2^2 + 2 \times 2$
3	15	$3^2 + 2 \times 3$
4	24	$4^2 + 2 \times 4$

   (a) What formula do you think Anna got ($t = \ldots$)?
   (b) What formula do you think Dipika got ($t = \ldots$)?
   (c) Are they in fact the same formula, written differently? Explain your answer.

15. Find an expression for the nth term of each sequence.

   (a) $1, \frac{1}{4}, \frac{1}{9}, \frac{1}{16}, \ldots$
   (b) $2, 4, 8, 16, 32, \ldots$
   (c) $(3 \times 2), (3 \times 4), (3 \times 8), (3 \times 16), \ldots$
   (d) $9, 99, 999, 9999, \ldots$
   (e) $(2 \times 2), (4 \times 3), (6 \times 4), (8 \times 5), \ldots$
   (f) $\frac{1}{3}, \frac{4}{9}, \frac{9}{27}, \frac{16}{81}, \ldots$

16. Diagonals are drawn in a sequence of polygons.
   Draw polygons with 6, 7, 8 ... sides and use your results to find an expression for the number of diagonals in a polygon with n sides.

17. (a) The nth term of the sequence 6, 14, 24, 36, ... is $n^2 + kn$. Find k.
    (b) The nth term of the sequence 4, 5, 8, 13, ... is $n^2 + an + b$. Find a and b.
    (c) The nth term of the sequence 8, 17, 32, 53, ... is $cn^2 + d$. Find c and d.

1.5 Linear graphs, $y = mx + c$

A **linear** graph is a graph which is a straight line. The graphs of $y = 3x - 1$, $y = 7 - 2x$, $y = \frac{1}{2}x + 100$ are all linear. Notice that on the right hand side there are only x terms and number terms. There are **no** terms involving x^2, x^3, $1/x$, etc. Linear graphs can be used to represent a wide variety of practical situations like those in the first exercise below. We begin by drawing linear graphs **accurately**.

Exercise 10

Draw the following graphs, using a scale of 2 cm to 1 unit on the x-axis and 1 cm to 1 unit on the y-axis.

1. $y = 2x + 1$ for $^-3 \leqslant x \leqslant 3$
2. $y = 3x - 4$ for $^-3 \leqslant x \leqslant 3$
3. $y = 8 - x$ for $^-2 \leqslant x \leqslant 4$
4. $y = 10 - 2x$ for $^-2 \leqslant x \leqslant 4$
5. $y = \dfrac{x + 5}{2}$ for $^-3 \leqslant x \leqslant 3$
6. $y = 3(x - 2)$ for $^-3 \leqslant x \leqslant 3$
7. $y = \frac{1}{2}x + 4$ for $^-3 \leqslant x \leqslant 3$
8. $v = 2t - 3$ for $^-2 \leqslant t \leqslant 4$
9. $z = 12 - 3t$ for $^-2 \leqslant t \leqslant 4$

10. Kendal Motors hires out vans.

Copy and complete the table where x is the number of miles travelled and C is the total cost in pounds.

x	0	50	100	150	200	250	300
C	35			65			95

Draw a graph of C against x, using scales of 2 cm for 50 miles on the x-axis and 1 cm for £10 on the C-axis.
   (a) Use the graph to find the number of miles travelled when the total cost was £71.
   (b) What is the formula connecting C and x?

11. Jeff sets up his own business as a plumber.

Copy and complete the table where C stands for his total charge and h stands for the number of hours he works. Draw a graph with h across the page and C up the page. Use scales of 2 cm to 1 hour for h and 2 cm to £10 for C.

h	0	1	2	3
C		33		

   (a) Use your graph to find how long he worked if his charge was £55.50.
   (b) What is the equation connecting C and h?

12. The equation connecting the annual mileage, M miles, of a certain car and the annual running cost, £C is $C = \dfrac{M}{20} + 200$.

Draw the graph for $0 \leqslant M \leqslant 10\,000$ using scales of 1 cm for 1000 miles for M and 2 cm for £100 for C.
   (a) From the graph find:
      (i) the cost when the annual mileage is 7200 miles
      (ii) the annual mileage corresponding to a cost of £320.
   (b) There is a '200' in the formula for C. Explain where this figure might come from.

13. Some drivers try to estimate their annual cost of repairs £c in relation to their average speed of driving s km/h using the equation $c = 6s + 50$.

Draw the graph for $0 \leqslant s \leqslant 160$.
From the graph find:
   (a) the estimated repair bill for a man who drives at an average speed of 23 km/h
   (b) the average speed at which a motorist drives if his annual repair bill is £1000.

Line segment

The graph shows a line segment from A(1, 2) to B(7, 4).

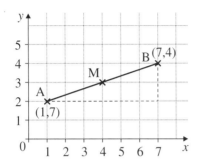

The coordinates of the mid-point M are the mean of the coordinates A and B i.e. M has coordinates

$$\left(\frac{1+7}{2}, \frac{2+4}{2}\right) = (4, 3).$$

The length of the line segment is found using Pythagoras.

$$AB = \sqrt{(7-1)^2 + (4-2)^2}$$
$$= \sqrt{40} \text{ units.}$$

Exercise 1P

1. Find the coordinates of the mid-point of the line joining the following pairs of points.
   (a) $A(3, 7), B(5, 13)$
   (b) $C(0, 10), D(12, 4)$
   (c) $P(^-3, 6), Q(5, 0)$
   (d) $X(4, 3), Y(^-2, 3)$

2. Find the length of the line segment joining the following pairs of points.
   (a) $E(7, 4), F(10, 8)$
   (b) $G(0, ^-2), H(5, 10)$
   (c) $S(3, ^-1), T(10, 4)$

3. The line $2x + y = 12$ cuts the axes at the points A and B.
   Find the coordinates of A and B and hence find:
   (a) the coordinates of the mid-point of AB
   (b) the length of the line AB.

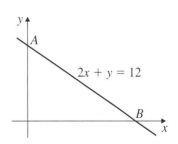

4. Find the coordinates of the mid-point of the line joining $P(a, 5a)$ and $Q(7a, ^-3a)$.

The general straight line, $y = mx + c$

Gradient

The gradient of a straight line is a measure of how steep it is.

Example

Gradient of line $AB = \dfrac{3-1}{6-1} = \dfrac{2}{5}$.

Gradient of line $AC = \dfrac{6-1}{3-1} = \dfrac{5}{2}$.

Gradient of line $BC = \dfrac{6-3}{3-6} = -1$.

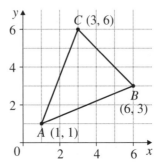

A line which slopes upwards to the right has a **positive** gradient.
A line which slopes upwards to the left has a **negative** gradient.

- Gradient $= \dfrac{\text{difference in } y\text{-coordinates}}{\text{difference in } x\text{-coordinates}}$

Exercise 1Q

1. Find the gradient of AB, BC, AC.

2. Find the gradient of PQ, PR, QR.

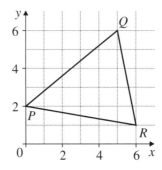

3. Find the gradient of the lines joining the following pairs of points:
   (a) $(5, 2) \to (7, 8)$
   (b) $(^-1, 3) \to (1, 6)$
   (c) $(\tfrac{1}{2}, 1) \to (\tfrac{3}{4}, 2)$
   (d) $(3{\cdot}1, 2) \to (3{\cdot}2, 2{\cdot}5)$

4. Find the value of a if the line joining the points $(3a, 4)$ and $(a, ^-3)$ has a gradient of 1.

5. (a) Write down the gradient of the line joining the points $(2m, n)$ and $(3, {}^-4)$.
(b) Find the value of n if the line is parallel to the x-axis
(c) Find the value of m if the line is parallel to the y-axis.

6. On the grid 1 square = 1 unit both across and up the page. Three pairs of intersecting lines are drawn. In each case AB is perpendicular to CD.

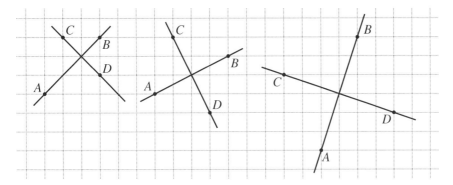

(a) In each case find the gradient of AB and the gradient of CD.
(b) In each case work out the product,

$$\text{(gradient of } AB) \times \text{(gradient of } CD)$$

(c) Copy and complete the following sentence:
'The product of the gradients of lines which are perpendicular is always ☐.'
(d) A more straightforward result is true for **parallel** lines. Copy and complete this sentence:
'Parallel lines have _____ gradient.'

7. Copy and complete the table.

Gradient of line 1	Gradient of line 2	Information
5	☐	Lines are parallel
2	☐	Lines are perpendicular
☐	$-\frac{2}{3}$	Lines are parallel
$^-5$	☐	Lines are perpendicular
☐	$\frac{3}{4}$	Lines are perpendicular

8. PQR is a right-angled triangle with vertices at $P(2, 1)$, $Q(4, 3)$, and $R(8, t)$. Given that angle PQR is $90°$, find t.

Intercept constant c

Here are two straight lines.

For $y = 2x - 3$, the gradient is 2 and the y-intercept is $^-3$.

For $y = -\frac{1}{2}x + 5$ the gradient is $-\frac{1}{2}$ and the y-intercept is 5.

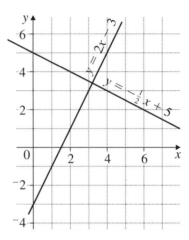

These two lines illustrate a general rule. When the equation of a straight line is written in the form

$$y = mx + c$$

the gradient of the line is m and the intercept on the y-axis is c.

Example 1

Draw the line $y = 2x + 3$ on a **sketch** graph. The word 'sketch' implies that we do not plot a series of points but simply show the position and slope of the line.

The line $y = 2x + 3$ has a gradient of 2 and cuts the y-axis at (0, 3).

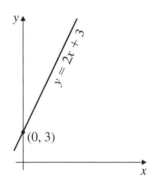

Example 2

Draw the line $x + 2y - 6 = 0$ on a sketch graph.
(a) Rearrange the equation to make y the subject.

$$x + 2y - 6 = 0$$
$$2y = {}^-x + 6$$
$$y = -\frac{1}{2}x + 3.$$

(b) The line has a gradient of $-\frac{1}{2}$ and cuts the y-axis at (0, 3).

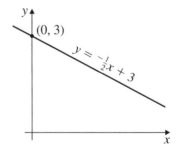

Exercise 1R

In Questions **1** to **20**, find the gradient of the line and the intercept on the y-axis. Hence draw a small sketch graph of each line.

1. $y = x + 3$
2. $y = x - 2$
3. $y = 2x + 1$
4. $y = 2x - 5$
5. $y = 3x + 4$
6. $y = \frac{1}{2}x + 6$
7. $y = 3x - 2$
8. $y = 2x$
9. $y = \frac{1}{4}x - 4$
10. $y = -x + 3$
11. $y = 6 - 2x$
12. $y = 2 - x$
13. $y + 2x = 3$
14. $3x + y + 4 = 0$
15. $2y - x = 6$
16. $3y + x - 9 = 0$
17. $4x - y = 5$
18. $3x - 2y = 8$
19. $10x - y = 0$
20. $y - 4 = 0$

21. Find the equations of the lines A and B.

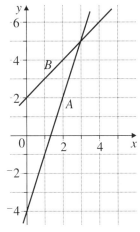

22. Find the equations of the lines C and D.

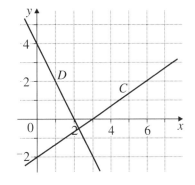

23. Here are the equations of several straight lines

A $y = -x + 7$
B $2y = x - 7$
C $4x + 12y = 5$
D $x + 3y = 10$
E $y = x - 3$
F $y = 5 - 5x$
G $5x + y = 12$
H $y = 3 - 2x$
I $y = 2x + 4$

(a) Find **two** pairs of lines which are parallel.
(b) Find **two** pairs of lines which are perpendicular.
(c) Find **one** line which is neither parallel nor perpendicular to any other line.

24. The sketch represents a section of the curve $y = x^2 - 2x - 8$. Calculate:
   (a) the coordinates of A and of B,
   (b) the gradient of the line AB,
   (c) the equation of the straight line AB.

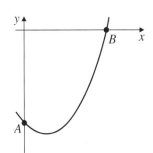

Finding the equation of a line

Example

Find the equation of the straight line which passes through (1, 3) and (3, 7).

(a) Let the equation of the line take the form $y = mx + c$.

$$\text{The gradient, } m = \frac{7-3}{3-1} = 2$$

so we may write the equation as

$$y = 2x + c \ldots [1]$$

(b) Since the line passes through (1, 3), substitute 3 for y and 1 for x in [1].

$$\therefore \quad 3 = 2 \times 1 + c$$
$$1 = c$$

The equation of the line is $y = 2x + 1$.

Exercise 1S

In Questions **1** to **10** find the equation of the line which passes through:

1. (0, 7) at a gradient of 3

2. (0, ¯9) at a gradient of 2

3. (0, 5) at a gradient of ¯1

4. (2, 3) at a gradient of 2

5. (2, 11) at a gradient of 3

6. (4, 3) at a gradient of ¯1

7. (6, 0) at a gradient of $\frac{1}{2}$

8. (2, 1) and (4, 5)

9. (5, 4) and (6, 7)

10. (0, 5) and (3, 2).

Finding a law

A ball is released from the top of a building. Its position is recorded every second using an electronic photoflash camera. Here are the results, giving distance fallen d and time taken t.

t	0	1	2	3
d	0	4·9	19·6	44·1

By plotting d against t, a curve is obtained.

It is not easy to find the equation (the law) connecting d and t from a curve.

Try plotting d against t^2.

t	0	1	2	3
t^2	0	1	4	9
d	0	4·9	19·6	44·1

This is a straight line with gradient 4·9 and intercept 0 on the d-axis.

∴ the law connecting d and t is
$$d = 4{\cdot}9t^2 + 0$$
or $d = 4{\cdot}9t^2$

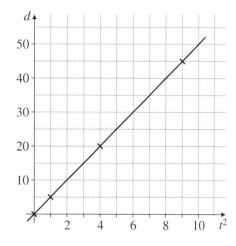

Note: We can only find the law from a straight line graph. It is not easy to discover an equation from a curve.

Exercise 1T

1. Two variables Z and X are thought to be connected by an equation of the form $Z = aX + c$. Here are some values of Z and X.

X	1	2	3·6	4·2	6·4
Z	4	6·5	10·5	12	17·5

   Draw a graph with X on the horizontal axis using a scale of 2 cm to 1 unit and Z on the vertical axis with a scale of 1 cm to 1 unit.
   Use your graphs to find a and c and hence write down the equation relating Z and X.

2. In an experiment, the following measurements of the variables q and t were taken.

q	0·5	1·0	1·5	2·0	2·5	3·0
t	3·85	5·0	6·1	7·0	7·75	9·1

A scientist suspects that q and t are related by an equation of the form $t = mq + c$, (m and c constants). Plot the values obtained from the experiment and draw the line of best fit through the the points. Plot q on the horizontal axis with a scale of 4 cm to 1 unit, and t on the vertical axis with a scale of 2 cm to 1 unit. Find the gradient and intercept on the t-axis and hence estimate the values of m and c.

3. In an experiment, the following measurements of p and z were taken:

z	1·2	2·0	2·4	3·2	3·8	4·6
p	11·5	10·2	8·8	7·0	6·0	3·5

Plot the points on a graph with z on the horizontal axis and draw the line of best fit through the points. Hence estimate the values of n and k if the equation relating p and z is of the form $p = nz + k$.

4. In an experiment the following measurements of t and z were taken:

t	1·41	2·12	2·55	3·0	3·39	3·74
z	3·4	3·85	4·35	4·8	5·3	5·75

Draw a graph, plotting t^2 on the horizontal axis and z on the vertical axis, and hence confirm that the equation connecting t and z is of the form $z = mt^2 + c$. Find approximate values for m and c.

5. Two variables T and V are thought to be related by an equation which is either.
$V = a\sqrt{T} + b$ or $V = \dfrac{a}{T} + b$.
Here are some values of T and V.

T	1	2	4	10
V	15·5	9·5	6·5	4·7

Draw a suitable graph (or graphs) to determine which equation is the correct one.
What are the values of a and b?

6. Two engineers measured the power P (kilowatts) required for a car to travel at various speeds V (km/h). Here are their results.

V	40	60	80	100	120	140
P	13·5	30·5	54	84·5	122	166

John thinks the formula connecting P and V is of the form $P = mV + c$.
Steve thinks the formula is $\sqrt{P} = kV$.
Steve says that John must be wrong, just by looking at the results. Explain why John's formula cannot be correct.
Draw a graph with V on the horizontal axis (1 cm = 20 units) and $\sqrt{P}$ on the vertical axis (1 cm = 1 unit).
Draw a line of best fit to confirm that Steve's formula is correct and estimate the value of the constant k.

1.6 Real-life graphs

Travel graphs and sketch graphs

Exercise 1U

Travel graphs

1. The graph shows the journeys made by a van and a car starting at York, travelling to Durham and returning to York.
   (a) For how long was the van stationary during the journey?
   (b) At what time did the car first overtake the van?
   (c) At what speed was the van travelling between 09:30 and 10:00?
   (d) What was the greatest speed attained by the car during the entire journey?
   (e) What was the average speed of the car over its entire journey?

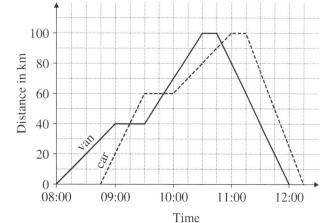

2. The graph shows the journeys of a bus and a car along the same road. The bus goes from Leeds to Darlington and back to Leeds. The car goes from Darlington to Leeds and back to Darlington.

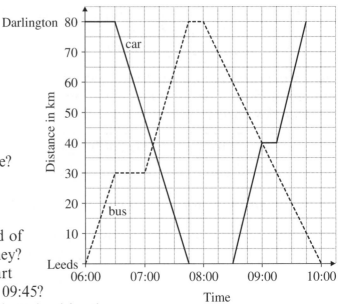

   (a) When did the bus and the car meet for the second time?
   (b) At what speed did the car travel from Darlington to Leeds?
   (c) What was the average speed of the bus over its entire journey?
   (d) Approximately how far apart were the bus and the car at 09:45?
   (e) What was the greatest speed attained by the car during its entire journey?

In Questions 3, 4, 5, draw a travel graph to illustrate the journey described. Draw axes with the same scales as in Question 2.

3. Mrs Chuong leaves home at 08:00 and drives at a speed of 50 km/h. After $\frac{1}{2}$ hour she reduces her speed to 40 km/h and continues at this speed until 09:30. She stops from 09:30 until 10:00 and then returns home at a speed of 60 km/h.
   Use a graph to find the approximate time at which she arrives home.

4. Mr Coe leaves home at 09:00 and drives at a speed of 20 km/h. After $\frac{3}{4}$ hour he increases his speed to 45 km/h and continues at this speed until 10:45. He stops from 10:45 until 11:30 and then returns home at a speed of 50 km/h.
   Use the graph to find the approximate time at which he arrives home.

5. At 10:00 Akram leaves home and cycles to his grandparents' house which is 70 km away. He cycles at a speed of 20 km/h until 11:15, at which time he stops for $\frac{1}{2}$ hour. He then completes the journey at a speed of 30 km/h.
   At 11:45 Akram's sister, Hameeda, leaves home and drives her car at 60 km/h. Hameeda also goes to her grandparents' house and uses the same road as Akram.
   At approximately what time does Hameeda overtake Akram?

6. A boat can travel at a speed of 20 km/h in still water. The current in a river flows at 5 km/h so that downstream the boat travels at 25 km/h and upstream it travels at only 15 km/h.

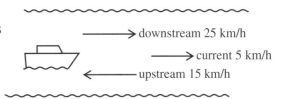

The boat has only enough fuel to last 3 hours. The boat leaves its base and travels downstream. Draw a distance–time graph and draw lines to indicate the outward and return journeys. After what time must the boat turn round so that it can get back to base without running out of fuel?

Sketch graphs

7. Which of the graphs A to D below best fits each of the following statements?
   (a) The birthrate was falling but is now steady.
   (b) Unemployment, which rose slowly until 1990, is now rising rapidly.
   (c) Inflation, which has been rising steadily, is now beginning to fall.
   (d) The price of gold has fallen steadily over the last year.

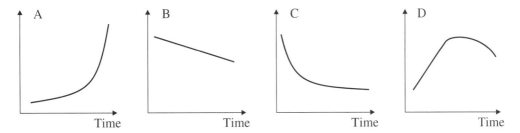

8. Which of the graphs A to D best fits the following statement: 'The price of oil was rising more rapidly in 1999 than at any time in the previous ten years.'?

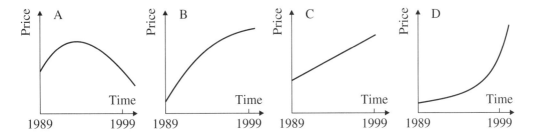

9. A car hire firm makes the charges shown. Draw a sketch graph showing 'miles travelled' across the page and 'total cost' up the page. Assume the car is hired for just one day.

10. Here are two car racing circuits.

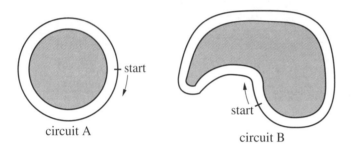

Sketch a speed–time graph to show the speed of a racing car as it goes around one lap of each circuit. [Not the first lap. Why?]

11. Water is poured at a constant rate into each of the containers A, B and C.

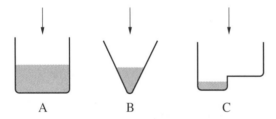

The graphs X, Y and Z show how the water level rises. Decide which graph fits each container.

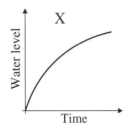
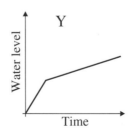
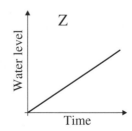

12. A car travels along a motorway and the amount of petrol in its tank is monitored as shown on the graph.
    (a) How much petrol was bought at the first stop?
    (b) What was the petrol consumption in miles per gallon:
        (i) before the first stop,
        (ii) between the two stops?
    (c) What was the average petrol consumption over the 200 miles?

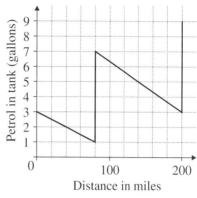

After it leaves the second service station the car encounters road works and slow traffic for the next 20 miles. Its petrol consumption is reduced to 20 m.p.g. After that, the road clears and the car travels a further 75 miles during which time the consumption is 30 m.p.g. Draw the graph above and extend it to show the next 95 miles. How much petrol is in the tank at the end of the journey?

1.7 Simultaneous linear equations

Graphical solution

Louise and Philip are two children; Louise is 5 years older than Philip. The sum of their ages is 12 years. How old is each child?

Let Louise be x years old and Philip be y years old.
The sum of their ages is 12,

so $x + y = 12$.

The difference of their ages is 5,

so $x - y = 5$.

Because both equations relate to the same information, both can be plotted on the same axes.

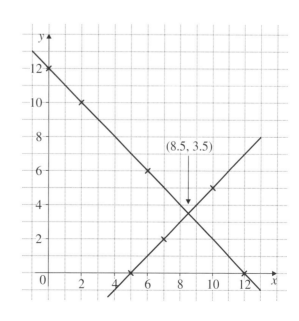

$x + y = 12$ goes through (0, 12), (2, 10), (6, 6), (12, 0).
$x - y = 5$ goes through (5, 0), (7, 2), (10, 5).

The values of x and y are found from the point where the two lines intersect. The point (8·5, 3·5) lies on both lines.

The solution is $x = 8·5$, $y = 3·5$

So Louise is $8\frac{1}{2}$ years old and Philip is $3\frac{1}{2}$ years old.

Note that only three points are required to define a straight line.

Exercise 1V

1. Use the graph to solve the simultaneous equations.

   (a) $x + y = 10$
   $y - 2x = 1$
   (b) $2x + 5y = 17$
   $y - 2x = 1$
   (c) $x + y = 10$
   $2x + 5y = 17$

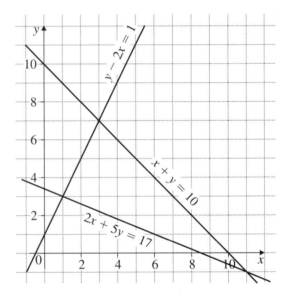

In Questions **2** to **6**, solve the simultaneous equations by first drawing graphs.

2. $x + y = 6$
   $2x + y = 8$
   Draw axes with x and y from 0 to 8.

3. $x + 2y = 8$
   $3x + y = 9$
   Draw axes with x and y from 0 to 9.

4. $x + 3y = 6$
   $x - y = 2$
   Draw axes with x from 0 to 8 and y from ⁻2 to 4.

5. $5x + y = 10$
   $x - y = {}^-4$
   Draw axes with x from ${}^-4$ to 4 and y from 0 to 10.

6. $a + 2b = 11$
   $2a + b = 13$
   Here, the unknowns are a and b. Draw the a-axis across the page from 0 to 13 and the b-axis up the page also from 0 to 13.

7. There are four lines drawn here. Write down the solutions to the following:

   (a) $x - 2y = 4$
       $x + 4y = 4$
   (b) $x + y = 7$
       $y - 3x = 3$
   (c) $y - 3x = 3$
       $x - 2y = 4$
   (d) $x + 4y = 4$
       $x + y = 7$
   (e) $x + 4y = 4$
       $y - 3x = 3$

   [Here, x and y should be given correct to 1 d.p.]

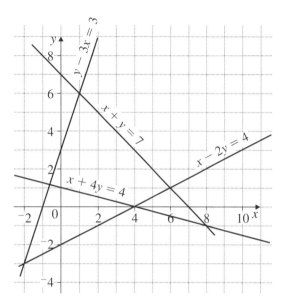

Algebraic solution

Simultaneous equations can be solved without drawing graphs. There are two methods: substitution and elimination. Choose the one which seems suitable.

Substitution method

This method is used when one equation contains a single x or y as in equation [2] of the example below.

Example
$$3x - 2y = 0 \quad \ldots [1]$$
$$2x + y = 7 \quad \ldots [2]$$

(a) Label the equations so that the working is made clear.
(b) In **this** case, write y in terms of x from equation [2].
(c) Substitute this expression for y in equation [1] and solve to find x.

(d) Find y from equation [2] using this value of x.
$$2x + y = 7$$
$$y = 7 - 2x$$

Substituting in [1]
$$3x - 2(7 - 2x) = 0$$
$$3x - 14 + 4x = 0$$
$$7x = 14$$
$$x = 2$$

Substituting in [2]
$$2 \times 2 + y = 7$$
$$y = 3$$

The solution is $x = 2$, $y = 3$.

Exercise 1W

Use the substitution method to solve the following:

1. $2x + y = 5$
   $x + 3y = 5$
2. $x + 2y = 8$
   $2x + 3y = 14$
3. $3x + y = 10$
   $x - y = 2$
4. $2x + y = {}^-3$
   $x - y = 2$
5. $4x + y = 14$
   $x + 5y = 13$
6. $x + 2y = 1$
   $2x + 3y = 4$
7. $2x + y = 5$
   $3x - 2y = 4$
8. $2x + y = 13$
   $5x - 4y = 13$
9. $7x + 2y = 19$
   $x - y = 4$
10. $3m = 2n - 6\frac{1}{2}$
    $4m + n = 6$
11. $2w + 3x - 13 = 0$
    $x + 5w - 13 = 0$
12. $x + 2(y - 6) = 0$
    $3x + 4y = 30$
13. $2x = 4 + z$
    $6x - 5z = 18$
14. $3m - n = 5$
    $2m + 5n = 7$
15. $5c - d - 11 = 0$
    $4d + 3c = {}^-5$

Elimination method

Use this method when the first method is unsuitable (some prefer to use it for every question).

Example

$$2x + 3y = 5 \quad \ldots [1]$$
$$5x - 2y = {}^-16 \quad \ldots [2]$$

$[1] \times 5$ $\quad 10x + 15y = 25 \quad \ldots [3]$
$[2] \times 2$ $\quad 10x - 4y = {}^-32 \quad \ldots [4]$
$[3] - [4]$ $\quad 15y - ({}^-4y) = 25 - ({}^-32)$
$\quad\quad\quad\quad\quad 19y = 57$
$\quad\quad\quad\quad\quad y = 3$

Substitute in [1] $\quad 2x + 3 \times 3 = 5$
$\quad\quad\quad\quad\quad\quad 2x = 5 - 9 = {}^-4$
$\quad\quad\quad\quad\quad\quad x = {}^-2$

The solution is $x = {}^-2$, $y = 3$.

Exercise 1X

Use the elimination method to solve the following:

1. $2x + 5y = 24$
   $4x + 3y = 20$
2. $5x + 2y = 13$
   $2x + 6y = 26$
3. $3x + y = 11$
   $9x + 2y = 28$
4. $x + 2y = 17$
   $8x + 3y = 45$
5. $3x + 2y = 19$
   $x + 8y = 21$
6. $2a + 3b = 9$
   $4a + b = 13$
7. $2x + 7y = 17$
   $5x + 3y = {}^-1$
8. $5x + 3y = 23$
   $2x + 4y = 12$
9. $3x + 2y = 11$
   $2x - y = {}^-3$
10. $3x + 2y = 7$
    $2x - 3y = {}^-4$
11. $x - 2y = {}^-4$
    $3x + y = 9$
12. $5x - 7y = 27$
    $3x - 4y = 16$
13. $x + 3y - 7 = 0$
    $2y - x - 3 = 0$
14. $3a - b = 9$
    $2a + 2b = 14$
15. $2x - y = 5$
    $\dfrac{x}{4} + \dfrac{y}{3} = 2$
16. $3x - y = 17$
    $\dfrac{x}{5} + \dfrac{y}{2} = 0$
17. $4x - 0{\cdot}5y = 12{\cdot}5$
    $3x + 0{\cdot}8y = 8{\cdot}2$
18. $0{\cdot}4x + 3y = 2{\cdot}6$
    $x - 2y = 4{\cdot}6$

Solving problems using simultaneous equations

As with linear equations, solving problems involves four steps.

(a) Let the two unknown quantities be x and y.
(b) Write the problem in the form of two equations.
(c) Solve the equations and give the answers in words.
(d) Check your solution using the problem and not your equations.

Example

A motorist buys 24 litres of petrol and 5 litres of oil for £26·75, while another motorist buys 18 litres of petrol and 10 litres of oil for £31.
Find the cost of 1 litre of petrol and 1 litre of oil at this garage.

(a) Let the cost of 1 litre of petrol be x pence and the cost of 1 litre of oil be y pence.

(b) $\quad 24x + 5y = 2675 \ldots$ [1]
   $\quad 18x + 10y = 3100 \ldots$ [2]

(c) Solve the equations: $x = 75$, $y = 175$

   1 litre of petrol costs 75 pence.
   1 litre of oil costs 175 pence.

(d) Check: $24 \times 75 + 5 \times 175 = 2675\text{p} = £26{\cdot}75$
   $18 \times 75 + 10 \times 175 = 3100\text{p} = £31$ ✓

Exercise 1Y

Solve each problem by forming a pair of simultaneous equations.

1. Find two numbers with a sum of 15 and a difference of 4.

2. Twice one number added to three times another gives 21. Find the numbers, if the difference between them is 3.

3. The average of two numbers is 7, and three times the difference between them is 18. Find the numbers.

4. Here is a puzzle from a newspaper. The ? and * stand for numbers which are to be found. The totals for the rows and columns are given. Write down two equations involving ? and * and solve them to find the values of ? and *.

?	*	?	*	36
?	*	*	?	36
*	?	*	*	33
?	*	?	*	36
39	33	36	33	

5. The line, with equation $y + ax = c$, passes through the points (1, 5) and (3, 1). Find a and c.

Hint:
For the point (1, 5) put $x = 1$ and $y = 5$ into $y + ax = c$, etc.

6. The line $y = mx + c$ passes through (2, 5) and (4, 13). Find m and c.

7. A stone is thrown into the air and its height, h metres above the ground, is given by the equation
$h = at - bt^2$.
From an experiment we know that $h = 40$ when $t = 2$ and that $h = 45$ when $t = 3$.
Show that, $a - 2b = 20$
and $a - 3b = 15$.
Solve these equations to find a and b.

8. A television addict can buy either two televisions and three video-recorders for £1750 or four televisions and one video-recorder for £1250. Find the cost of one of each.

9. A tortoise makes a journey in two parts; it can either walk at 2 m/min or crawl at 1 m/min.

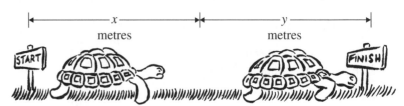

walks at 2 m/min   crawls at 1 m/min

If the tortoise walks the first part and crawls the second, the journey takes 110 minutes.
If it crawls the first part and walks the second, the journey takes 100 minutes.
Let x metres be the length of the first part and y metres be the length of the second part.
Write down two simultaneous equations and solve them to find the lengths of the two parts of the journey.

For questions **9** and **10** use the formula,
Time = Distance ÷ Speed

10. A cyclist completes a journey of 500 m in 22 seconds, part of the way at 10 m/s and the remainder at 50 m/s. How far does she travel at each speed?

11. A bag contains forty coins, all of them either 2p or 5p coins. If the value of the money in the bag is £1·55, find the number of each kind.

12. Thirty tickets were sold for a concert, some at 60p and the rest at £1. If the total raised was £22, how many had the cheaper tickets?

13. If the numerator and denominator of a fraction are both decreased by one the fraction becomes $\frac{2}{3}$. If the numerator and denominator are both increased by one the fraction becomes $\frac{3}{4}$. Find the original fraction.

14. In three years' time a pet mouse will be as old as his owner was four years ago. Their present ages total 13 years. Find the age of each now.

15. The curve $y = ax^2 + bx + c$ passes through the points (1, 8), (0, 5) and (3, 20). Find the values of a, b and c and hence the equation of the curve.

Linear/quadratic simultaneous equations

A pair of simultaneous equations can contain equations which are linear, quadratic, cubic and so on.
Consider the pair of simultaneous equations
$$y = 2x + 2$$
$$\text{and } y = x^2 - 1$$

The diagram shows that the line and the curve intersect at (¯1, 0) and (3, 8).

These solutions can be obtained algebraically
using the method of substitution as follows:

$$y = 2x + 2 \quad \ldots [A]$$
$$y = x^2 - 1 \quad \ldots [B]$$

Substitute $y = 2x + 2$ into equation [B]
$$2x + 2 = x^2 - 1$$
$$\Rightarrow x^2 - 2x - 3 = 0$$
$$(x - 3)(x + 1) = 0$$
$$x = 3 \text{ or } x = {}^-1$$

when $x = 3$, $y = 8$
when $x = {}^-1$, $y = 0$ } These are the solutions.

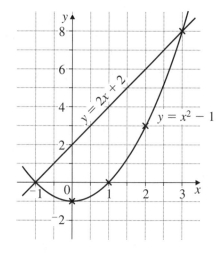

Exercise 1Z

This work requires a knowledge of the solution of quadratic equations.

1. Solve the simultaneous equations:
   (a) $y = x^2 - 2x$
   $y = x + 4$

   (b) $y = 7x - 8$
   $y = x^2 - x + 7$

   > Quadratic equations are explained on page 421.

2. Solve the simultaneous equations:
   (a) $y = x^2 - 3x + 7$
   $5x - y = 8$

   (b) $y = 9x - 4$
   $y = 2x^2$

3. A circle with centre $(0, 0)$ and radius r has equation $x^2 + y^2 = r^2$.
   Find the coordinates of the points of intersection of the line $y = x + 1$ and the circle $x^2 + y^2 = 13$.

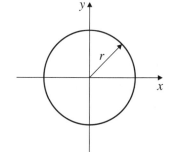

4. Solve the simultaneous equations:
   $x^2 + y^2 = 20$, $y = x - 2$

5. Find the coordinates of the two points where the line $y = 4x - 8$ intersects the curve $y^2 = 16x$.

6. Explain why there are no solutions to the simultaneous equations $x^2 + y^2 = 1$ and $y = x + 10$.

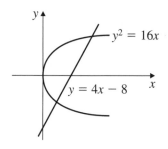

7. Find the coordinates of the point of intersection of the curve $y = x^3$ and the line $x + y = 10$.

Summary

1. You can expand brackets.

2. You can factorise expressions.

3. You can solve linear equations.

4. You can use trial and improvement to find appropriate solutions.

5. You know how to find the nth term of a sequence.

6. You can draw a straight line graph.

7. You can find the gradient and intercept of a line graph.

Check out AS1

1. (a) Expand the brackets: $2x(3x^2 - 7)$
   (b) Expand the brackets and simplify:
   $5(2x + 3) - 5(7 - 4x)$

2. Factorise: (a) $x^2 - 8x$
   (b) $16x^2 - 25y^2$

3. Solve: (a) $3x + 7 = x - 11$
   (b) $\dfrac{x+5}{4} = \dfrac{2x-7}{5}$

4. If $x^2 + x = 7$, use trial and improvement to find x to 1 decimal place.

5. Find the nth term of the sequence
   4, 11, 18, 25, 32 …

6. Draw the line $y = 2x - 5$.

7. Write down the gradient and intercept of the line shown below.

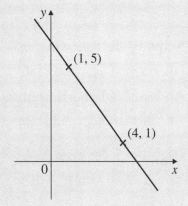

8. You can draw and interpret a distance-time graph.	8. Kim leaves home at 9 am and cycles at 10 mph for one hour. She then has a 30 minute rest before cycling another 6 miles at 8 mph. She stops again for 30 minutes before cycling home at a constant 7 mph. Draw a distance-time graph to show her journey and use your graph to find when Kim arrived home.
9. You can solve simultaneous equations using a variety of methods.	9. Solve the simultaneous equations $x + 2y = 8$ and $y = 3x - 3$ using: (a) a graphical method (b) a substitution method (c) an elimination method.

Revision exercise AS1

1. (a) Solve the equation
$$3x - 11 = 2(x - 5).$$
   (b) Solve the simultaneous equations
$$2x - 3y = 17$$
$$3x + y = 9.$$
   [AQA]

2. (a) Simplify the following:
   (i) $e^4 \times e$ (ii) $f^7 \div f^3$ (iii) $(3hk^3)^2$
   (b) Multiply out and simplify.
   (i) $3p(p + 2)$ (ii) $(x - 7y)(x - y)$ [SEG]

3. (a) Simplify these expressions.
   (i) $\dfrac{a^2}{a^2}$ (ii) $2x^4 \times 3x^3$ (iii) $(2pq^2)^5$
   (b) Multiply out and simplify $(y - 7)^2$.
   (c) Calculate the values of p and q in the identity
$$x^2 - 6x + 14 = (x - p)^2 + q.$$
   [AQA]

4. A solution of the equation $x^3 - x = 77$ lies between 4 and 5. Use a trial and improvement method to find this solution. Give your answer to one decimal place. You **must** show all your trials. [SEG]

5. (a) Multiply out $3p(6 - 2p)$.
   (b) Multiply out and simplify $(p - 1)(2p + 3)$.
   (c) Factorise the expression $6a^2b - 3a^2$. [SEG]

6. A solution of the equation $x^3 - x^2 = 7$ lies between $x = 2$ and $x = 3$. Use trial and improvement to find this solution. Give your answer correct to one decimal place. You **must** show all your working. [SEG]

7. The graph shows the line $4x + 5y = 20$. By copying the graph and drawing another line, use the graph to solve the simultaneous equations:

   $4x + 5y = 20$
   and $3x - 2y = 6$ [AQA]

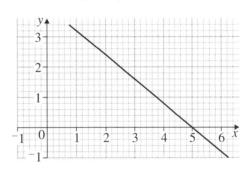

8. (a) Simplify
   (i) $p^6 \div p^2$ (ii) $(q^5)^2$
   (b) (i) Make c the subject of the formula
      $P = 2a + 2b + 2c$.
   (ii) Find the value of c when $P = 46$, $a = 7.7$ and $b = 10.8$. [SEG]

9. Points A, B and C are shown on the grid.
   (a) Write down the equation of the line AB.
   (b) Use the grid to work out the gradient of the line CB.
   (c) Write down the equation of the line CB. [SEG]

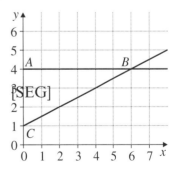

10. Part of a sequence of numbers is shown below.

    2   5   10   17   26   ...

    (a) Find an expression for the nth term of this sequence.
    (b) Will the number 403 be in this sequence? Explain your answer. [SEG]

AS2 Shape, Space and Measures 1

This unit will show you how to:

- Calculate with angles
- Know the properties of 2-D and 3-D solids
- Draw geometrical constructions and loci
- Understand congruence and similarity
- Use Pythagoras' theorem
- Calculate areas, volumes and surface areas
- Calculate areas and perimeters of circles, arc lengths, areas of sectors and segments
- Find areas and volumes of similar shapes

Before you start:

You should know how to...	Check in AS2
1. Use BODMAS.	1. Find the value of $\frac{1}{2} \times 25(17 + 8)$
2. Divide numbers mentally. For example, $360 \div 8 = 45$	2. Without using a calculator, find: (a) $\frac{360}{6}$ (b) $\frac{360}{9}$
3. Use ratios.	3. Divide 36 in the ratio of $4:5$.
4. Substitute in formulae. For example, $V = lwb$ Find V where $l = 6$, $w = 8$ and $b = 4$. $V = 6 \times 8 \times 4 = 192$	4. (a) Using $A = 4\pi r^2$, find A when $r = 5$. (b) $A = \frac{1}{2}h(a+b)$ Find A when $h = 6$, $a = 8$, $b = 7$.
5. Rearrange formulae. For example, $V = \pi r^2 h$. Find r. $r = \sqrt{\dfrac{V}{\pi h}}$	5. (a) Find r when $A = \pi r^2$. (b) Find h when $A = 2\pi r^2 + 2\pi rh$.
6. Use proportionality. For example, $y \propto x$. When $y = 5$, $x = 7$. Find y when $x = 11$. $y \propto x \Rightarrow y = kx$ When $y = 5$, $x = 7 \Rightarrow k = \frac{5}{7}$ $\therefore y = \frac{5}{7}x \therefore y = \frac{5}{7} \times 11 = 7 \cdot 86$	6. (a) $h \propto l$, $h = 7$ when $l = 4$. Find h when $l = 10$. (b) $y \propto x^2$, when $y = 3$, $x = 5$. Find y when $x = 9$.

2.1 Angle facts

- The angles at a point add up to 360°.
  The angles on a straight line add up to 180°.

Example 1

(a)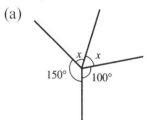

$$x + x + 150 + 100 = 360$$
$$\therefore \ 2x = 360 - 250$$
$$x = 55°$$

(b)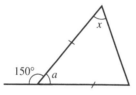

$$3a + 90 = 180$$
$$3a = 90$$
$$a = 30°$$

- The angles in a triangle add up to 180°.

Example 2

$a = 180° - 150° = 30°$
The triangle is isosceles

$$\therefore \ 2x + 30 = 180$$
$$2x = 150$$
$$x = 75°$$

- When a line cuts a pair of parallel lines all the acute angles are equal and all the obtuse angles are equal.

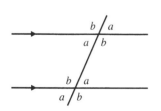

Corresponding angles

Alternate angles

Some people remember 'F angles'

and 'Z angles'

Exercise 2A

Find the angles marked with letters. The lines *AB* and *CD* are straight.

1.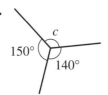
2.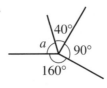
3.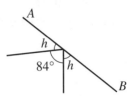
4.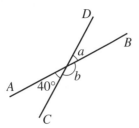
5.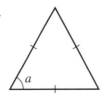
6.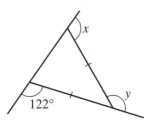
7.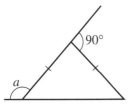
8.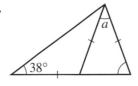
9.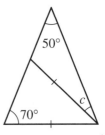
10.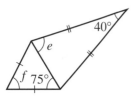
11.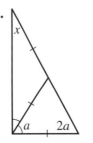
12.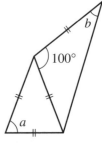
13.
14.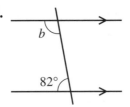
15.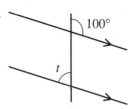
16.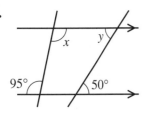
17.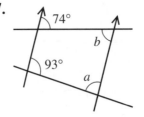
18.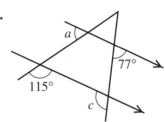

19. 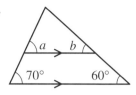   **20.** 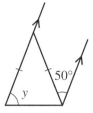   **21.**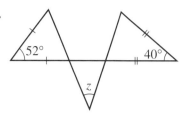

22. Here is a triangle.   A line is drawn parallel to one side.

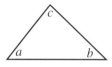   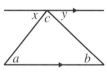

(a) What is angle x? What is angle y?
(b) Prove that the sum of the angles in a triangle is $180°$.

23. (a) Write down the sum of the angles p, q and r.
(b) Write down the sum of the angles q and x.
(c) Hence show that the exterior angle of a triangle is equal to the sum of the interior angles at the other two vertices.

24. The diagram shows two equal squares joined to a triangle.
Find the angle x.

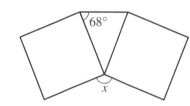

25. Find the angle a between the diagonals of the parallelogram.

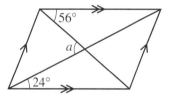

26. The diagram shows a series of isosceles triangles drawn between two lines. Find the value of x.

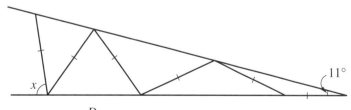

27. In the diagram $AB = BD = DC$ and $A\widehat{D}C - D\widehat{A}C = 70°$

Find $A\widehat{D}B$.

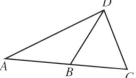

Quadrilaterals

Square: Four equal sides;
All angles 90°;
Four lines of symmetry;
Rotational symmetry of order 4.

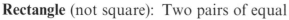

Rectangle (not square): Two pairs of equal and parallel sides;
All angles 90°;
Two lines of symmetry;
Rotational symmetry of order 2.

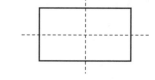

Rhombus: Four equal sides;
Opposite sides parallel;
Diagonals bisect at right angles;
Diagonals bisect angles of rhombus;
Two lines of symmetry;
Rotational symmetry of order 4.

Parallelogram: Two pairs of equal and parallel sides;
Opposite angles equal;
No lines of symmetry (in general);
Rotational symmetry of order 2.

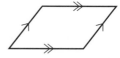

Trapezium: One pair of parallel sides;
Rotational symmetry of order 1.

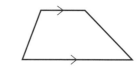

Kite: $AB = AD$, $CB = CD$;
Diagonals meet at 90°;
One line of symmetry;
Rotational symmetry of order 1.

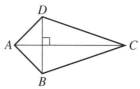

- For all quadrilaterals the sum of the interior angles is 360°.

Exercise 2B

1. *ABCD* is a rhombus whose diagonals intersect at *M*.
   Find the coordinates of *C* and *D*.

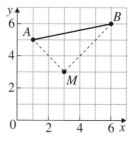

2. Quadrilateral *PQRS* is split into two triangles.
   (a) Write down the sum of the angles *a*, *b* and *c*.
   (b) Write down the sum of the angles *d*, *e* and *f*.
   (c) Show that the sum of the angles in the quadrilateral is 360°.

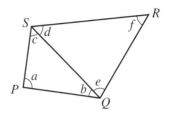

In Questions **3** to **11** begin by drawing a diagram and remember to put the letters around the shape in alphabetical order.

3. In a parallelogram *WXYZ*, $W\hat{X}Y = 72°$, $Z\hat{W}Y = 80°$. Calculate:
   (a) $W\hat{Z}Y$ (b) $X\hat{W}Z$ (c) $W\hat{Y}X$

4. In a kite *ABCD*, $AB = AD$, $BC = CD$, $C\hat{A}D = 40°$ and $C\hat{B}D = 60°$. Calculate:
   (a) $B\hat{A}C$ (b) $B\hat{C}A$ (c) $A\hat{D}C$

5. In a rhombus *ABCD*, $A\hat{B}C = 64°$. Calculate:
   (a) $B\hat{C}D$ (b) $A\hat{D}B$ (c) $B\hat{A}C$

6. In the parallelogram *PQRS* line *QA* bisects (cuts in half) angle *PQR*. Calculate the size of angle *RAQ*.

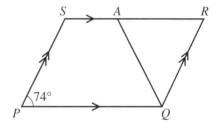

7. In a trapezium *ABCD*, *AB* is parallel to *DC*, $AB = AD$, $BD = DC$ and $B\hat{A}D = 128°$. Find:
   (a) $A\hat{B}D$ (b) $B\hat{D}C$ (c) $B\hat{C}D$

8. *ABE* is an equilateral triangle drawn inside the square *ABCD*. Calculate the size of angle *DEC*.

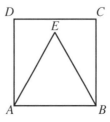

9. In a kite *PQRS* with $PQ = PS$ and $RQ = RS$, $Q\hat{R}S = 40°$ and $Q\hat{P}S = 100°$. Find $P\hat{Q}R$.

10. In a rhombus $PQRS$, $R\hat{P}Q = 54°$. Find:
    (a) $P\hat{R}Q$ (b) $P\hat{S}R$ (c) $R\hat{Q}S$

11. In a kite $PQRS$, $R\hat{P}S = 2 P\hat{R}S$, $PQ = QS = PS$ and $QR = RS$. Find:
    (a) $Q\hat{P}S$ (b) $P\hat{R}S$ (c) $Q\hat{S}R$

Angles in polygons

Exterior angles of a polygon

The exterior angle of a polygon is the angle between a produced side and the adjacent side of the polygon. The word 'produced' in this context means 'extended'.

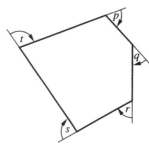

If we put all the exterior angles together we can see that the sum of the angles is 360°. This is true for any polygon.

- The sum of the exterior angles of a polygon $= 360°$

Note:
(a) In a **regular** polygon all exterior angles are equal.
(b) For a regular polygon with n sides, each exterior angle $= \dfrac{360}{n}$.

Example

The diagram shows a regular octagon (8 sides).

(a) Calculate the size of each exterior angle (marked e).
(b) Calculate the size of each interior angle (marked i).

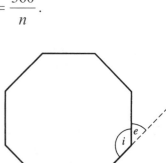

(a) There are 8 exterior angles and the sum of this angle is 360°

$\therefore$ angle $e = \dfrac{360}{8} = 45°$

(b) $e + i = 180°$ (angles on a straight line)

$\therefore$ $i = 135°$

Exercise 2C

1. Look at the polygon shown.
   (a) Calculate each exterior angle
   (b) Check that the total of the exterior angles is 360°.

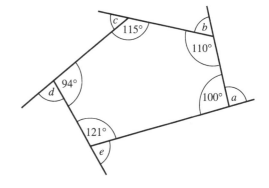

2. The diagram shows a regular decagon.
   (a) Calculate the angle *a*.
   (b) Calculate the interior angle of a regular decagon.

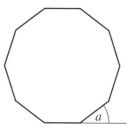

3. Find: (a) the exterior angle
         (b) the interior angle of a regular polygon with
   (i) 9 sides   (ii) 18 sides   (iii) 45 sides   (iv) 60 sides.

4. Find the angles marked with letters.

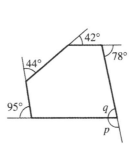
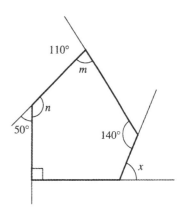

5. Each exterior angle of a regular polygon is 15°. How many sides has the polygon?

6. Each interior angle of a regular polygon is 140°. How many sides has the polygon?

7. Each exterior angle of a regular polygon is 18°. How many sides has the polygon?

Sum of interior angles

pentagon, 5 sides

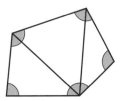

There are 3 triangles.
Sum of interior angles = 3 × 180°
= (5 − 2) × 180°

hexagon, 6 sides

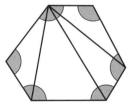

There are 4 triangles.
Sum of interior angles = 4 × 180°
= (6 − 2) × 180°

- In general: The sum of the interior angles of a polygon with n sides is $(n − 2) \times 180°$.

Exercise 2D

1. Use the formula above to find the sum of the interior angles in:
   (a) an octagon (8 sides)
   (b) a decagon (10 sides).

2. (a) Work out the sum of the interior angles in a polygon with 20 sides.
   (b) What is the size of each interior angle in a **regular** polygon with 20 sides?

3. Find the angles marked with letters.

   (a)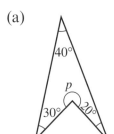

   (b)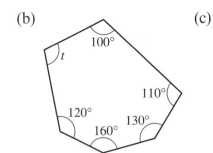

   (c)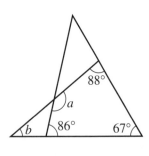

4. A regular dodecagon has 12 sides.
   (a) Calculate the size of each interior angle, i.
   (b) Use your answer to find the size of each exterior angle, e.

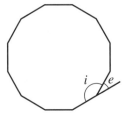

5. The sum of the interior angles of a polygon with n sides is 3600°. Find the value of n.

6. The sides of a regular polygon subtend angles of 18° at the centre of the polygon. How many sides has the polygon?

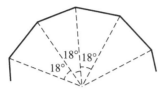

2.2 Congruent triangles

Two plane figures are congruent if one fits exactly on the other. The four types of congruence for triangles are as follows:

(a) Two sides and the included angle (SAS)

(b) Two angles and a corresponding side (AAS)

(c) Three sides (SSS)

(d) Right angle, hypotenuse and one other side (RHS)

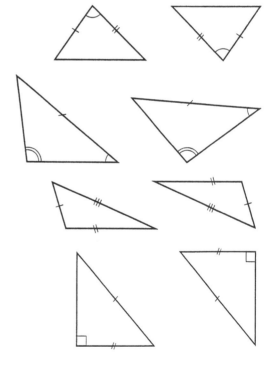

Example

$PQRS$ is a square. M is the mid-point of PQ, N is the mid-point of RQ.
Prove that $SN = RM$.

$S\hat{R}Q = P\hat{Q}R$ (Both right angles)
$SR = RQ$ (Sides of a square)
$RN = MQ$ (Given)

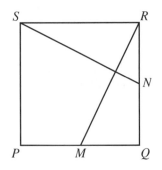

∴ Triangle SRN is congruent to triangle RQM. (SAS)
We can write $\triangle SRN \equiv \triangle RQM$
The sign $\equiv$ means 'is identically equal to'.
[Note that the order of letters shows corresponding vertices in the triangles.]

Since the triangles are congruent, it follows that $SN = RM$.

Exercise 2E

For Questions **1** to **6**, decide whether the pair of triangles are congruent. If they are congruent, state which conditions for congruency are satisfied.

1.   **2.**   **3.**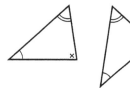

4.   **5.**   **6.**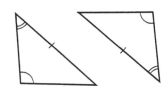

7. $ABCDE$ is a regular pentagon.
List all the triangles which are congruent to triangle BCE.

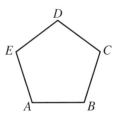

8. By construction show that it is possible to draw two **different** triangles with the angle and sides shown.

This shows that 'ASS' is not a condition for triangles to be congruent.

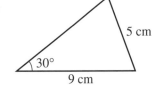

9. Draw triangle ABC with $AB = BC$ and point D at the mid-point of AC.
Prove that triangles ABD and CBD are congruent and state which case of congruency applies.
Hence prove that angles A and C are equal.

10. Tangents TA and TB are drawn to touch the circle with centre O. Use congruent triangles to prove that the two tangents are the same length.

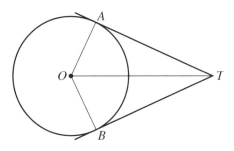

11. Triangle LMN is isosceles with $LM = LN$; X and Y are points on LM, LN respectively such that $LX = LY$. Prove that triangles LMY and LNX are congruent.

12. $ABCD$ is a quadrilateral and a line through A parallel to BC meets DC at X. If $\hat{D} = \hat{C}$, prove that $\triangle ADX$ is isosceles.

13. XYZ is a triangle with $XY = XZ$. The bisectors of angles Y and Z meet the opposite sides in M and N respectively. Prove that $YM = ZN$.

14. In the diagram, $DX = XC$, $DV = ZC$ and the lines AB and DC are parallel. Prove that:
    (a) $AX = BX$
    (b) $AC = BD$
    (c) triangles DBZ and CAV are congruent.

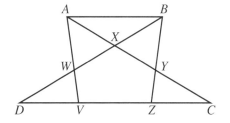

15. Draw a parallelogram $ABCD$ and prove that its diagonals bisect each other.

16. PQRS is a square and equilateral triangles PQA and RQB are drawn with A inside and B outside the square.

    By considering triangles PQR and AQB, prove that AB is equal to PR.

2.3 Locus

In mathematics, the word **locus** describes the position of points which obey a certain rule. The locus can be the path traced out by a moving point.

Three important loci

Circle

The locus of points which are equidistant from a fixed point O is shown. It is a **circle** with centre O.

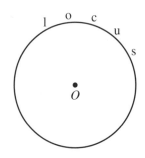

Perpendicular bisector

The locus of points which are equidistant from two fixed points A and B is shown.

The locus is the **perpendicular bisector** of the line AB. Use compasses to draw arcs, as shown, or use a ruler and a protractor.

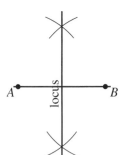

Angle bisector

The locus of points which are equidistant from two fixed lines AB and AC is shown.

The locus is the line which bisects the angle BAC. Use compasses to draw arcs or use a protractor to construct the locus.

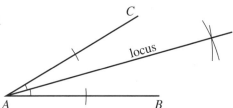

Exercise 2F

1. Draw the locus of a point P which moves so that it is always 3 cm from a fixed point X.

2. Mark two points P and Q which are 10 cm apart. Draw the locus of points which are equidistant from P and Q.

3. Draw two lines AB and AC of length 8 cm, where $B\hat{A}C = 40°$. Draw the locus of points which are equidistant from AB and AC.

4. A sphere rolls along a surface from *A* to *B*. Sketch the locus of the centre of the sphere in each case.

(a) (b) (c)

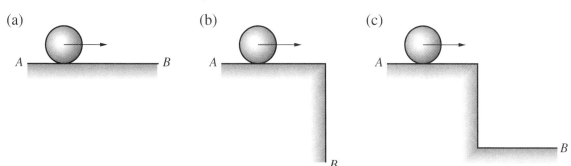

5. A rectangular slab is rotated around corner *B* from position 1 to position 2. Draw a diagram, on squared paper, to show:
   (a) the locus of corner *A*
   (b) the locus of corner *C*.

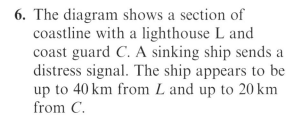

6. The diagram shows a section of coastline with a lighthouse L and coast guard C. A sinking ship sends a distress signal. The ship appears to be up to 40 km from *L* and up to 20 km from *C*.

   Copy the diagram and show the region in which the sinking ship could be.

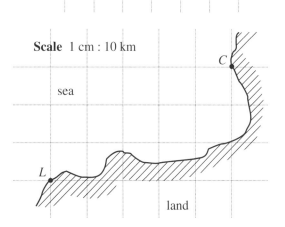

7. (a) Draw the triangle *LMN* full size.
   (b) Draw the locus of the points which are:
       (i) equidistant from *L* and *N*
       (ii) equidistant from *LN* and *LM*
       (iii) 4 cm from *M*.
   [Draw the three loci in different colours.]

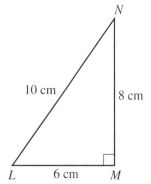

8. Draw a line AB of length 6 cm. Draw the locus of a point P so that angle $ABP = 90°$.

9. The diagram shows a garden with a fence on two sides and trees at two corners.
   A sand pit is to be placed so that it is:
   (a) equidistant from the 2 fences
   (b) equidistant from the 2 trees.

   Make a scale drawing (1 cm = 1 m) and mark where the sand pit goes.

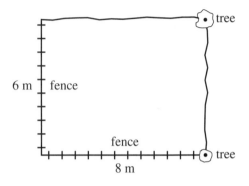

10. Channel 9 in Australia is planning the position of a new TV mast to send pictures all over the country. The new mast is to be placed:
    (a) an equal distance from Darwin and Adelaide
    (b) not more than 2000 km from Perth
    (c) not more than 1800 km from Brisbane.
    Make a copy of the map on squared paper using the grid lines as reference.
    Show clearly where the mast could be placed so that it satisfies the conditions (a), (b), (c) above.

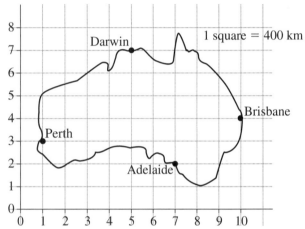

Exercise 2G

1. The perpendicular from a point to a line is obtained as follows.

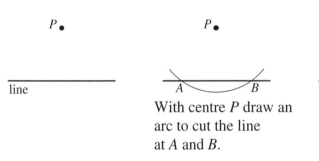

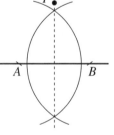

Construct the perpendicular of bisector of AB.

Draw a line and a point P about 5 cm away.
Construct the line which passes through P and which is perpendicular to the line.

2. Draw a straight line and mark a point A on the line. Construct the perpendicular at A.

3. A circle, centre O, radius 5 cm, is inscribed inside a square $ABCD$. Point P moves so that $OP \leqslant 5$ cm and $BP \leqslant 5$ cm. Shade the set of points indicating where P can be.

4. Construct a triangle ABC where $AB = 9$ cm, $BC = 7$ cm and $AC = 5$ cm.
   (a) Sketch and describe the locus of points within the triangle which are equidistant from AB and AC.
   (b) Shade the set of points within the triangle which are less than 5 cm from B and are also nearer to AC than to AB.

5. A goat is tied to one corner on the outside of a barn. The diagram shows a plan view.
   Sketch two plan views of the barn and show the locus of points where the goat can graze if:
   (a) the rope is 4 m long
   (b) the rope is 7 m long

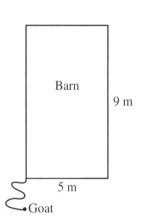

6. Rod AB rotates about the fixed point A. B is joined to C which slides along a fixed rod. $AB = 15$ cm and $BC = 25$ cm.

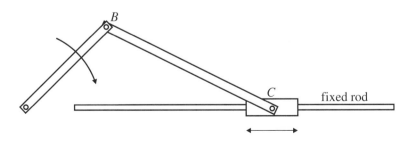

Describe the locus of C as AB rotates clockwise about A.
Find the smallest and the largest distance between C and A.

7. Draw two points M and N 16 cm apart. Draw the locus of a point P which moves so that the area of triangle MNP is 80 cm².

8. Describe the locus of a point which moves in three dimensional space and is equidistant from two fixed points.

9. Draw two points A and B 10 cm apart.

   Place the corner of a piece of paper (or a set square) so that the edges of the paper pass through A and B. Mark the position of corner C.
   Slide the paper around so the edge still passes through A and B and mark the new position of C. Repeat several times and describe the locus of the point C which moves so that angle ACB is always 90°.

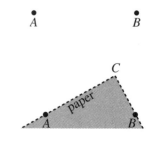

10. Draw any triangle ABC and construct the bisectors of angles B and C to meet at point Y. With centre at Y draw a circle which just touches the sides of the triangle.
    This is the **inscribed** circle of the triangle.

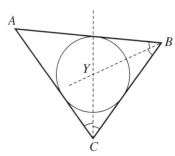

11. Copy the diagram shown. $OC = 4$ cm.
    Sketch the locus of P which moves so that PC is equal to the perpendicular distance from P to the line AB.
    This locus is called a **parabola.**

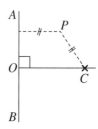

12. A rod OA of length 60 cm rotates about O at a constant rate of 1 revolution per minute. An ant, with good balance, walks along the rod at a speed of 1 cm per second. Sketch the locus of the ant for 1 minute after it leaves O.

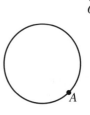

13. Here is a circle with point A on its circumference. The centre of the circle is unknown.
    Using ruler and compasses, construct a diameter to the circle through the point A.

2.4 Pythagoras' theorem

In a right-angled triangle the square on the hypotenuse is equal to the sum of the squares on the other two sides.

$a^2 + b^2 = c^2$

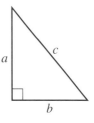

Exercise 2H

In Questions **1** to **8**, find x. All the lengths are in cm.

1. 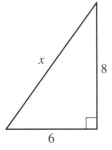   **2.** 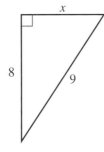   **3.** 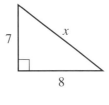   **4.**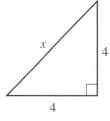

5. 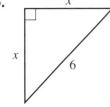   **6.** 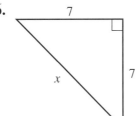   **7.** 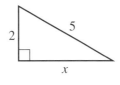   **8.**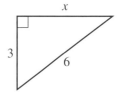

9. Find the length of a diagonal of a rectangle of length 9 cm and width 4 cm.

10. An isosceles triangle has sides 10 cm, 10 cm and 4 cm. Find the height of the triangle.

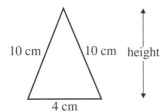

11. A 4 m ladder rests against a vertical wall with its foot 2 m from the wall. How far up the wall does the ladder reach?

12. A ship sails 20 km due North and then 35 km due East. How far is it from its starting point?

13. Find the length of a diagonal of a square of side 9 cm.

14. The square and the rectangle have the same length diagonal. Find x.

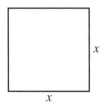 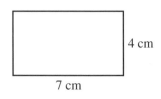

15. (a) Find the height of the triangle, h.
    (b) Find the area of the triangle ABC.

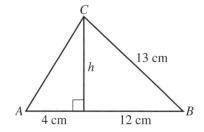

16. A thin wire of length 18 cm is bent in the shape shown.

    Calculate the length from A to B.

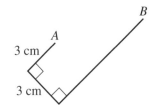

17. A paint tin is a cylinder of radius 12 cm and height 22 cm. Leonardo, the painter, drops his stirring stick into the tin and it disappears.
    Work out the maximum length of the stick.

18. A square of side 10 cm is drawn inside a circle.
    (a) Find the diameter of the circle.
    (b) Find the shaded area.

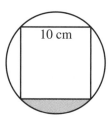

Exercise 21

1. In the diagram A is $(1, 2)$ and B is $(6, 4)$. Work out the length AB. (First find the length of AN and BN).

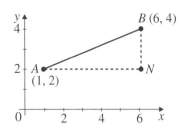

2. On squared paper plot $P(1, 3)$, $Q(6, 0)$, $R(6, 6)$. Find the lengths of the sides of triangle PQR. Is the triangle isosceles?

In Questions **3** to **8** find x.

3.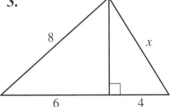

4.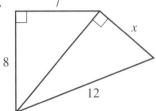

5.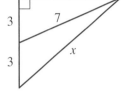

6.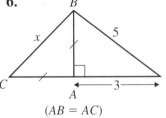

7.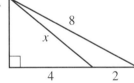

8.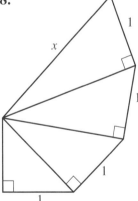

9. The diagram shows a rectangular block. Calculate: (a) AC (b) AY.

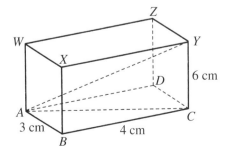

10. Find the length of a diagonal of a rectangular room of length 5 m, width 3 m and height 2·5 m.

11. TC is a vertical pole whose base lies at a corner of the horizontal rectangle $ABCD$.
    The top of the pole T is connected by straight wires to points A, B and D.

    Calculate: (a) TC
    (b) TD
    (c) (harder) TA.

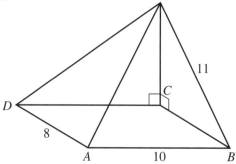

12. The diagonal of a rectangle exceeds the length by 2 cm. If the width of the rectangle is 10 cm, find the length.

13. A ladder reaches H when held vertically against a wall. When the base is 6 feet from the wall, the top of the ladder is 2 feet lower than H.
    How long is the ladder?

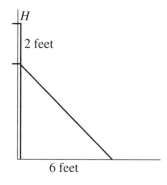

14. (a) Sketch axes in three dimensions as shown.
    (b) Find the length of the line from $O(0, 0, 0)$ to $A(3, 4, 7)$.
    (c) Find the length of the line from $O(0, 0, 0)$ to $B(1, 10, 13)$.

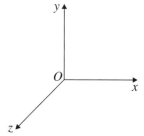

15. The diagram represents the starting position (AB) and the finishing position (CD) of a ladder as it slips. The ladder is leaning against a vertical wall.

    Given: $AC = x$, $OC = 4AC$, $BD = 2AC$
    and $OB = 5$ m.
    Form an equation in x, find x and hence find the length of the ladder.

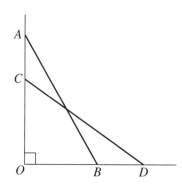

16. The most well known right-angled triangle is the 3, 4, 5 triangle $[3^2 + 4^2 = 5^2]$.
    It is interesting to look at other right-angled triangles where all the sides are whole numbers.

    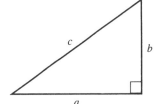

    (a) (i) Find c if $a = 5$, $b = 12$.
        (ii) Find c if $a = 7$, $b = 24$.
        (iii) Find a if $c = 41$, $b = 40$.

    (b) Write the results in a table.

a	b	c
3	4	5
5	12	?
7	24	?
?	40	41

    (c) Look at the sequences in the 'a' column and in the 'b' column. Also write down the connection between b and c for each triangle.

    (d) Predict the next three sets of values of a, b, c. Check to see if they really do form right-angled triangles.

17. The net of a square-based pyramid is shown. The base is 8 cm × 8 cm and the vertical height is 10 cm. Find x.

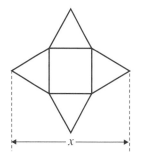

2.5 Area

For a **triangle,** area = $\frac{1}{2}$ × base × height **or** area = $\frac{1}{2} ab \sin C$.

Use the second formula when you know two sides and the included angle.

For a **rectangle,** area = base × height.

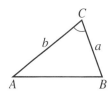

Trapezium

A trapezium has two parallel sides.
area $= \frac{1}{2}(a+b) \times h$

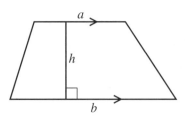

Parallelogram

area $= b \times h$

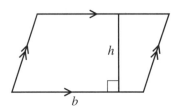

Exercise 2J

For Questions **1** to **6,** find the area of each shape.
All lengths are in cm.

1. 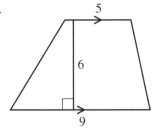 **2.** 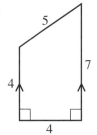 **3.**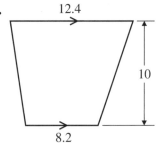

4. 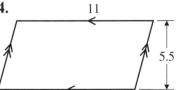 **5.** 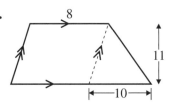 **6.**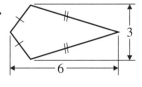

7. A rectangle has an area of $117 \, m^2$ and a width of 9 m. Find its length.

8. A trapezium of area $105 \, cm^2$ has parallel sides of length 5 cm and 9 cm. How far apart are the parallel sides?

9. A floor 5 m by 20 m is covered by square tiles of side 20 cm. How many tiles are needed?

10. Find the area of each triangle. All lengths are in cm.

(a)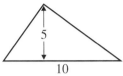
(b)
(c)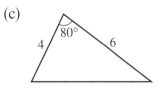

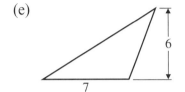

(d)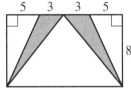
(e)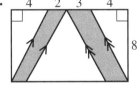
(f)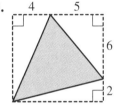

In Questions **11** to **14**, find the area shaded.

11.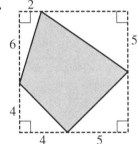
12.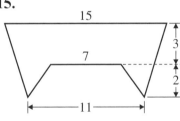
13.

14.
15.
16.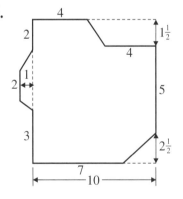

Exercise 2K

1. A rectangular pond measuring 10 m by 6 m is surrounded by a path which is 2 m wide. Find the area of the path.

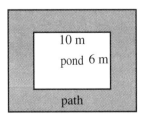

2. Find the length x. The area is shown inside the shape.
   (a) 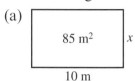 (b) 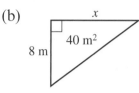 (c)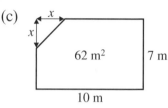

3. Work out the area of this shape. All lengths are in cm.

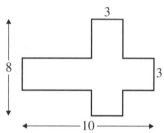

4. The arrowhead has an area of $3{\cdot}6\,\text{cm}^2$. Find the length x.

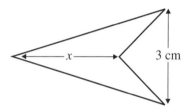

5. The diagram shows a square $ABCD$ in which $DX = XY = YC = AW$. The area of the square is $45\,\text{cm}^2$.
   (a) What is the fraction $\dfrac{DX}{DC}$?
   (b) What fraction of the square is shaded?
   (c) Find the area of the unshaded part.

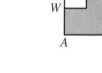

6. On squared paper draw a 7×7 square. Design a pattern which divides it up into nine smaller squares.

7. A rectangular field, 400 m long, has an area of 6 hectares. Calculate the perimeter of the field. [1 hectare $= 10\,000\,\text{m}^2$]

8. On squared paper draw the triangle with vertices at (1, 1), (5, 3), (3, 5). Find the area of the triangle.

9. Draw the quadrilateral with vertices at (1, 1), (6, 2), (5, 5), (3, 6). Find the area of the quadrilateral.

10. The side of the small square is half the length of the side of the large square. The L-shape has an area of $75\,\text{cm}^2$. Find the side length of the large square.

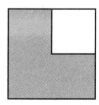

11. A regular hexagon is circumscribed by a circle of radius 3 cm with centre O.
    (a) What is angle EOD?
    (b) Find the area of triangle EOD and hence find the area of the hexagon $ABCDEF$.

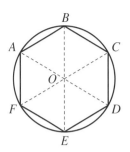

12. Find the area of a parallelogram $ABCD$ with $AB = 7$ m, $AD = 20$ m and $B\hat{A}D = 62°$.

13. In the diagram if $AE = \frac{1}{3}AB$, find the area shaded.

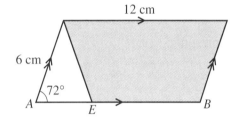

14. The area of an equilateral triangle ABC is $50\,\text{cm}^2$. Find AB.

15. The area of a triangle XYZ is $11\,\text{m}^2$. Given $YZ = 7$ m and $X\hat{Y}Z = 130°$, find XY.

16. A rhombus has an area of $40\,\text{cm}^2$ and adjacent angles of $50°$ and $130°$. Find the length of a side of the rhombus.

17. The diagram shows a part of the perimeter of a regular polygon with n sides. The centre of the polygon is at O and $OA = OB = 1$ unit.
    (a) What is the angle AOB in terms of n?
    (b) Work out an expression in terms of n for:
        (i) the area of triangle OAB
        (ii) the area of the whole polygon.
    (c) Find the area of the polygons where $n = 100$ and $n = 1000$.
        What do you notice?

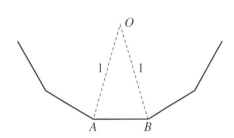

18. The area of a regular pentagon is $600\,\text{cm}^2$. Calculate the length of one side of the pentagon.

2.6 Circles, arcs and sectors

For any circle, the ratio $\left(\dfrac{\text{circumference}}{\text{diameter}}\right)$ is equal to π.

The value of π is usually taken to be 3·14, but this is not an exact value. Through the centuries, mathematicians have been trying to obtain a better value for π. Electronic computers are now able to calculate the value of π to many millions of figures, but its value is still not exact. It was shown in 1761 that π is an **irrational number** which, like $\sqrt{2}$ or $\sqrt{3}$ cannot be expressed exactly as a fraction.

The following formulae should be memorised

- circumference $= \pi d$
  $\qquad\qquad\quad\; = 2\pi r$
- area $= \pi r^2$

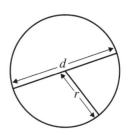

Example 1
Find the circumference and area of a circle of diameter 8 cm. Take π from a calculator.

$$\begin{aligned}
\text{Circumference} &= \pi d \\
&= \pi \times 8 \\
&= 25\cdot1 \text{ cm (3 s.f.)}
\end{aligned}$$

$$\begin{aligned}
\text{Area} &= \pi r^2 \\
&= \pi \times 4^2 \\
&= 50\cdot3 \text{ cm}^2 \text{ (3 s.f.)}
\end{aligned}$$

Example 2
Find the perimeter and area of this shape.

$$\begin{aligned}
\text{Perimeter} &= \text{arc } AB + \text{arc } BC + CD + DA \\
&= \left(\dfrac{\pi \times 12}{2}\right) + \left(\dfrac{\pi \times 7}{2}\right) + 12 + 7 \\
&= 9\tfrac{1}{2}\pi + 19 \\
&= 48\cdot8 \text{ m (3 s.f.)}
\end{aligned}$$

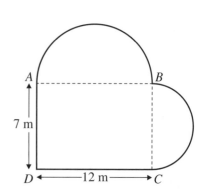

Area = large semicircle + small semicircle + rectangle
$$= \left(\frac{\pi \times 6^2}{2}\right) + \left(\frac{\pi \times 3\cdot 5^2}{2}\right) + (12 \times 7)$$
$$= 160 \, \text{m}^2 \, (3 \text{ s.f.})$$

Note:
Do not approximate **early** in questions of this type. If your final answer is given to 3 s.f., work to 5 s.f. in the working. Better still, leave the working in the calculator memory.

Exercise 2L

For each shape find (a) the perimeter, (b) the area. All lengths are in cm unless otherwise stated. All the arcs are either semicircles or quarter circles.

1.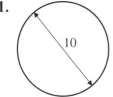

2.

3.

4.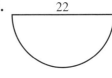

5.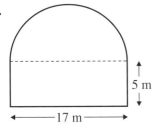

6.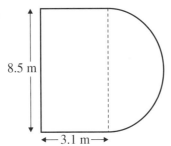

7.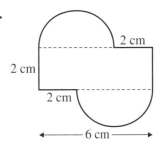

8.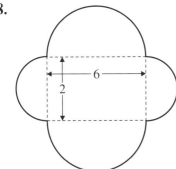

9.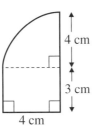

10.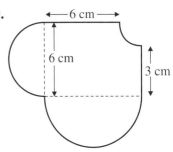

11.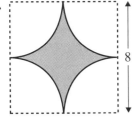

12.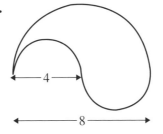

13.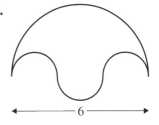

Example

(a) A circle has a circumference of 20 m. Find the radius of the circle.
Let the radius of the circle be r m.

$$\text{Circumference} = 2\pi r$$
$$\therefore \quad 2\pi r = 20$$
$$\therefore \quad r = \frac{20}{2\pi}$$
$$r = 3.18$$

The radius of the circle is 3·18 m (3 s.f.).

(b) A circle has an area of 45 cm². Find the radius of the circle.
Let the radius of the circle be r cm.

$$\pi r^2 = 45$$
$$r^2 = \frac{45}{\pi}$$
$$r = \sqrt{\left(\frac{45}{\pi}\right)} = 3.78$$

The radius of the circle is 3·78 cm (3 s.f.).

Exercise 2M

1. A circle has an area of 15 cm². Find its radius.

2. An odometer is a wheel used by surveyors to measure distances along roads. The circumference of the wheel is one metre. Find the diameter of the wheel.

3. Find the radius of a circle of area 22 km².

4. Find the radius of a circle of circumference 58·6 cm.

5. The handle of a paint tin is a semicircle of wire which is 28 cm long. Calculate the diameter of the tin.

6. A circle has an area of 16 mm². Find its circumference.

7. A circle has a circumference of 2500 km. Find its area.

8. A circle of radius 5 cm is inscribed inside a square as shown.
Find the area shaded.

9. Discs of radius 4 cm are cut from a rectangular plastic sheet of length 84 cm and width 24 cm. How many complete discs can be cut out?
Find:
   (a) the total area of the discs cut
   (b) the area of the sheet wasted.

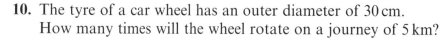

10. The tyre of a car wheel has an outer diameter of 30 cm.
How many times will the wheel rotate on a journey of 5 km?

11. A golf ball of diameter 1·68 inches rolls a distance of 4 m in a straight line. How many times does the ball rotate completely? (1 inch ≈ 2·54 cm)

12. A circular pond of radius 6 m is surrounded by a path of width 1 m.
    (a) Find the area of the path.
    (b) The path is resurfaced with astroturf which is bought in packs each containing enough to cover an area of 7 m^2. How many packs are required?

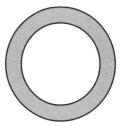

13. Calculate the radius of a circle whose area is equal to the sum of the areas of three circles of radii 2 cm, 3 cm and 4 cm respectively.

14. The diagram below shows a lawn (unshaded) surrounded by a path of uniform width (shaded). The curved end of the lawn is a semicircle of diameter 10 m.

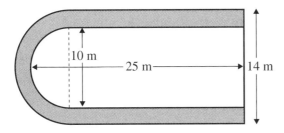

Calculate the total area of the path.

15. The diameter of a circle is given as 10 cm, correct to the nearest cm. Calculate:
    (a) the maximum possible circumference
    (b) the minimum possible area of the circle consistent with this data.

16. The governor of a prison has 100 m of wire fencing. What area can he enclose if he makes a circular compound?

17. In 'equable shapes' the numerical value of the area is equal to the numerical value of the perimeter.
    Find the dimensions of the following equable shapes:
    (a) square   (b) circle   (c) equilateral triangle.

18. The semicircle and the isosceles triangle have the same base AB and the same area.
    Find the angle x.

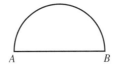
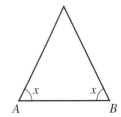

19. Mr Gibson decided to measure the circumference of the earth using a very long tape measure. For reasons best known to himself he held the tape measure 1 m from the surface of the (perfectly spherical) earth all the way round. When he had finished Mrs Gibson told him that his measurement gave too large an answer. She suggested taking off 6 m. Was she correct? [Take the radius of the earth to be 6400 km.]

Exercise 2N Mixed questions

1. The sloping line divides the area of the square in the ratio 1 : 5. What is the ratio $a : b$?

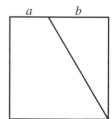

2. *ABCD* is a square of side 14 cm. *P*, *Q*, *R*, *S* are the mid-points of the sides. Semicircles are drawn with centres *P*, *Q*, *R*, *S* as shown on the diagram.

   Find the shaded area, using $\pi = \dfrac{22}{7}$.

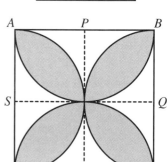

3. What fraction of the area of a circle of radius 4 cm is more than 3 cm from the centre?

4. Two equal circles are cut from a square piece of paper of side 2 m. Calculate the radius of the largest possible circles. Give your answer correct to 3 significant figures.

5. These circles have radii 2, 3, 5 cm. Express the shaded area as a percentage of the area of the largest circle.

6. A square of side x cm is drawn in the middle of a square of side 10 cm. The corners of the smaller square are 2 cm from the corners of the larger square and are on the diagonals of the larger square. Find x correct to 4 s.f.

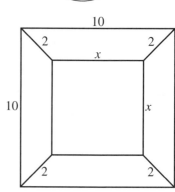

7. $ABCDEFGH$ is a regular octagon. Calculate angle ACD.

8. The diagram shows two diagonals drawn on the faces of a cube. Calculate the angle between the diagonals.

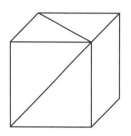

9. The radius of the circle inscribed in a regular hexagon has length 4 cm. Find the area of the hexagon, correct to 3 s.f.

10. In $\triangle ABC$, $AC = BC = 12$ cm.
    CD is perpendicular to AB.
    $MD = 1$ cm.
    M is equidistant from A, B and C.
    Calculate the length CM.

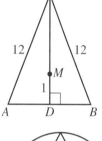

11. The circumcircle and the inscribed circle of an equilateral triangle are shown. Calculate the ratio of the area of the circumcircle to the area of the inscribed circle.

12. $ABCDEFGH$ is a regular octagon of side 1 unit.
    (a) Calculate the length AF
    (b) Calculate the length AC
    (c) (Much harder) You probably used a calculator to work out AF. Can you calculate the **exact** value of AF?
        Leave your answer in a form involving a square root.

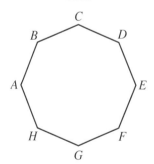

Arcs and sectors

- Arc length, $l = \dfrac{\theta}{360} \times 2\pi r$

    We take a fraction of the whole circumference depending on the angle at the centre of the circle.

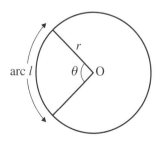

- Sector area, $A = \dfrac{\theta}{360} \times \pi r^2$

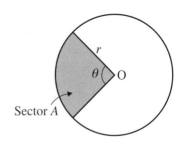

Sector A

We take a fraction of the whole area depending on the angle at the centre of the circle.

Example

(a) Find the length of an arc which subtends an angle of 140° at the centre of a circle of radius 12 cm. (Take $\pi = \tfrac{22}{7}$.)

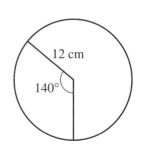

$$\text{Arc length} = \dfrac{140}{360} \times 2 \times \dfrac{22}{7} \times 12$$

$$= \dfrac{88}{3}$$

$$= 29\tfrac{1}{3} \text{ cm}$$

(b) A sector of a circle of radius 10 cm has an area of 25 cm². Find the angle at the centre of the circle.

Let the angle at the centre of the circle be θ.

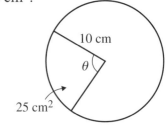

$$\dfrac{\theta}{360} \times \pi \times 10^2 = 25$$

$$\therefore \quad \theta = \dfrac{25 \times 360}{\pi \times 100}$$

$$\theta = 28\cdot 6° \text{ (1 d.p.)}$$

The angle at the centre of the circle is 28·6°.

Exercise 20

1. Find the length of the arc AB.

(a)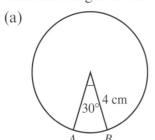

(b)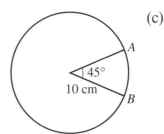

(c)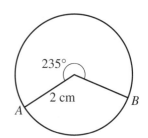

2. A pendulum of length 55 cm swings through an angle of 16°. Through what distance does the tip of the pendulum swing?

3. Find the area of the sector shaded.

(a) 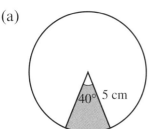 (b) 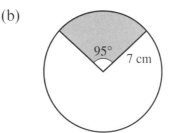 (c)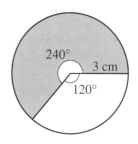

4. Find the shaded area.

(a) 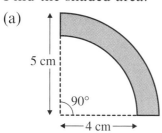 (b)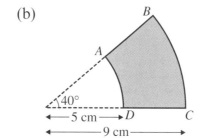

5. Find the length of the perimeter of the shaded area in Question **4** part (b).

6. A sector is cut from a round cheese as shown. Calculate the volume of the piece of cheese.

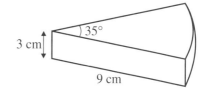

7. Work out the shaded area.

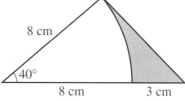

8. The lengths of the minor and major arcs of a circle are 5·2 cm and 19·8 cm respectively. Find:
   (a) the radius of the circle,
   (b) the angle subtended at the centre by the minor arc.

9. The length of the minor arc AB of a circle, centre O, is 2π cm and the length of the major arc is 22π cm. Find:
   (a) the radius of the circle,
   (b) the acute angle AOB.

10. A wheel of radius 10 cm is turning at a rate of 5 revolutions per minute. Calculate:
    (a) the angle through which the wheel turns in 1 second,
    (b) the distance moved by a point on the rim in 2 seconds.

11. In the diagram the arc length is l and the sector area is A.
    (a) Find θ, when $r = 5$ cm and $l = 7 \cdot 5$ cm.
    (b) Find θ, when $r = 2$ m and $A = 2$ m^2.
    (c) Find r, when $\theta = 55°$ and $l = 6$ cm.

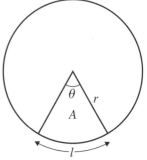

12. Two parallel lines are drawn 2 cm from the centre of a circle of radius 4 cm. Calculate the area shaded.

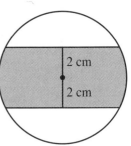

13. Find the angle θ so that the arc length is equal to the radius. The angle you have found is called a radian. Your calculator may have a RAD mode, where RAD stands for radian.

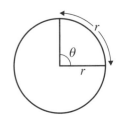

14. (a) *ABCDE is a regular pentagon.* Calculate the size of angle x.

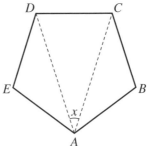

   (b) In this diagram, arcs are drawn with centres at the opposite corners of the pentagon, e.g. arc CD has centre at A. The radius of each arc is 5 cm. Show that the perimeter of the shape is 5π cm.

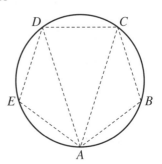

15. In the diagram, the arc length is l and the sector area is A.
    (a) Find l, when $\theta = 72°$ and $A = 15\,\text{cm}^2$.
    (b) Find l, when $\theta = 135°$ and $A = 162\,\text{m}^2$.
    (c) Find A, when $l = 11\,\text{cm}$ and $r = 5.2\,\text{cm}$.

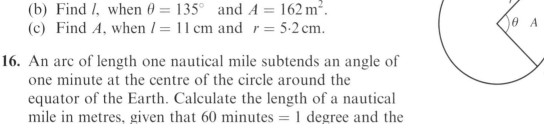

16. An arc of length one nautical mile subtends an angle of one minute at the centre of the circle around the equator of the Earth. Calculate the length of a nautical mile in metres, given that 60 minutes = 1 degree and the radius of the Earth is about 6370 km.

17. The length of an arc of a circle is 12 cm. The corresponding sector area is $108\,\text{cm}^2$. Find:
    (a) the radius of the circle
    (b) the angle subtended at the centre of the circle by the arc.

18. In the diagram, AB is a tangent to the circle at A. Straight lines BCD and ACE pass through the centre of the circle C. Angle $ACB = x°$ and the radius of the circle is 1 unit.

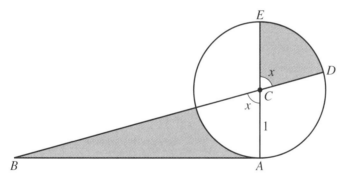

The two shaded areas are equal.
(a) Show that x satisfies the equation $x = \left(\dfrac{90}{\pi}\right) \tan x$.
(b) Use trial and improvement to find a solution for x correct to 1 decimal place.

19. (Hard, unless you can find a neat solution!)
    In the diagram:
    BC is a diameter of the semicircle,
    $A\widehat{B}C = 90°$,
    shaded area ① = shaded area ② and angle $ACB = x$.
    Show that $\tan x = \dfrac{\pi}{4}$.
    Hence find angle x.

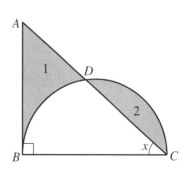

Segments

The line AB is a chord. The area of a circle cut off by a chord is called a **segment.** In the diagram the **minor** segment is shaded and the **major** segment is unshaded.

(a) The line from the centre of a circle to the mid-point M of a chord **bisects** the chord at **right angles.**

(b) The line from the centre of a circle to the mid-point of a chord bisects the angle subtended by the chord at the centre of the circle.

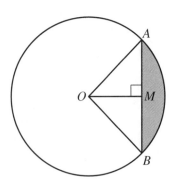

Example

XY is a chord of length 12 cm of a circle of radius 10 cm, centre O. Calculate

(a) the angle XOY

(b) the area of the minor segment cut off by the chord XY.

Let the mid-point of XY be M

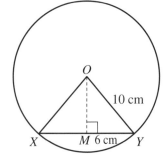

$\therefore \quad MY = 6 \text{ cm}$

$\sin M\hat{O}Y = \dfrac{6}{10}$

$\therefore \quad M\hat{O}Y = 36\cdot 87°$

$\therefore \quad X\hat{O}Y = 2 \times 36\cdot 87$

$\qquad \qquad = 73\cdot 74°$

area of minor segment = area of sector XOY − area of $\triangle XOY$

$\qquad$ area of sector $XOY = \dfrac{73\cdot 74}{360} \times \pi \times 10^2$

$\qquad \qquad \qquad \qquad \quad = 64\cdot 32 \text{ cm}^2$

$\qquad$ area of $\triangle XOY = \tfrac{1}{2} \times 10 \times 10 \times \sin 73\cdot 74°$

$\qquad \qquad \qquad \quad = 48\cdot 00 \text{ cm}^2$

$\therefore$ area of minor segment $= 64\cdot 32 - 48\cdot 00$

$\qquad \qquad \qquad \qquad \quad = 16\cdot 3 \text{ cm}^2$ (3 s.f.)

Circles, arcs and sectors 289

Exercise 2P
Use the $\boxed{\pi}$ button on a calculator.

1. The chord AB subtends an angle of $130°$ at the centre O. The radius of the circle is 8 cm. Find:
   (a) the length of AB,
   (b) the area of sector OAB,
   (c) the area of triangle OAB,
   (d) the area of the minor segment (shown shaded).

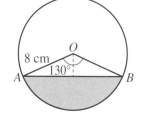

2. Find the shaded area when:
   (a) $r = 6$ cm, $\theta = 70°$
   (b) $r = 14$ cm, $\theta = 104°$
   (c) $r = 5$ cm, $\theta = 80°$

3. Find θ and hence the shaded area when:
   (a) $AB = 10$ cm, $r = 10$ cm
   (b) $AB = 8$ cm, $r = 5$ cm

4. How far is a chord of length 8 cm from the centre of a circle of radius (a) 5 cm (b) 6 cm?

5. The diagram shows the cross section of a cylindrical pipe with water lying in the bottom.
   (a) If the maximum depth of the water is 2 cm and the radius of the pipe is 7 cm, find the area shaded.
   (b) What is the **volume** of water in a length of 30 cm?

6. An equilateral triangle is inscribed in a circle of radius 10 cm. Find:
   (a) the area of the triangle,
   (b) the area shaded.

7. A regular hexagon is circumscribed by a circle of radius 6 cm. Find the area shaded.

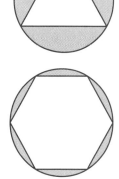

8. A regular octagon is circumscribed by a circle of radius r cm. Find the area enclosed between the circle and the octagon. (Give the answer in terms of r.)

9. Find the radius of the circle:
   (a) when $\theta = 90°$, $A = 20\,\text{cm}^2$
   (b) when $\theta = 30°$, $A = 35\,\text{cm}^2$
   (c) when $\theta = 150°$, $A = 114\,\text{cm}^2$

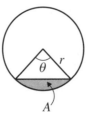

10. The diagram shows a regular pentagon of side 10 cm with a star inside. Calculate the area of the star.

2.7 Volume and surface area

A **prism** is an object with the same cross section throughout its length.

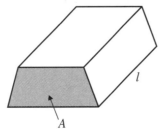

- Volume of prism = (area of cross section) × length
  $= A \times l$

A **cuboid** is a prism whose six faces are all rectangles. A cube is a special case of a cuboid in which all six faces are squares.

A **cylinder** is a prism whose cross section is a circle.

radius = r
height = h

- Volume of cylinder = area of cross section × length
  $= \pi r^2 h$

Example

Calculate the height of a cylinder of volume 500 cm³ and base radius 8 cm.

Let the height of the cylinder be h cm.

$$\pi r^2 h = 500$$

$$3 \cdot 14 \times 8^2 \times h = 500$$

$$h = \frac{500}{3 \cdot 14 \times 64}$$

$$h = 2 \cdot 49 \text{ (3 s.f.)}$$

The height of the cylinder is 2·49 cm.

Exercise 2Q

1. Calculate the volume of the prisms. All lengths are in cm.

   (a)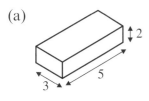

   (b)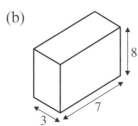

   (c)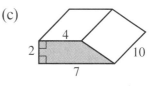

   (d)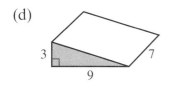

   (e)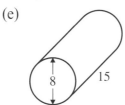

   (f)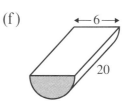

2. Calculate the volume of the following cylinders:
   (a) $r = 4$ cm, $h = 10$ cm
   (b) $r = 11$ m, $h = 2$ m

3. A gas cylinder has diameter 18 cm and length 40 cm. Calculate the capacity of the cylinder, correct to the nearest litre.

4. A solid cylinder of radius 5 cm and length 15 cm is made from material of density 6 g/cm³. Calculate the mass of the cylinder.

5. The two solid cylinders shown have the same mass. Calculate the density, x g/cm³, of cylinder B.

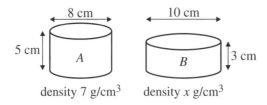

6. The diagram shows a view of the water in a swimming pool.

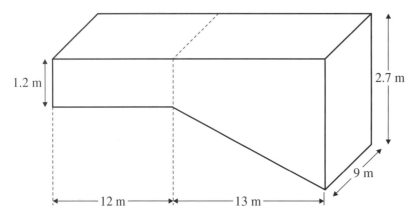

Calculate the volume of water in the pool.

7. Find the height of a cylinder of volume 200 cm³ and radius 4 cm.

8. Find the length of a cylinder of volume 2 litres and radius 10 cm.

9. Find the radius of a cylinder of volume 45 cm³ and length 4 cm.

10. When 3 litres of oil is removed from an upright cylindrical can, the level falls by 10 cm. Find the radius r of the can.

11. A solid cylinder of radius 4 cm and length 8 cm is melted down and recast into a solid cube. Find the side of the cube.

12. Water flows through a circular pipe of internal diameter 3 cm at a speed of 10 cm/s. If the pipe is full, how much water issues from the pipe in one minute? (answer in litres)

13. Water issues from a hose-pipe of internal diameter 1 cm at a rate of 5 litres per minute. At what speed is the water flowing through the pipe?

14. A cylindrical metal pipe has external diameter of 6 cm and internal diameter of 4 cm. Calculate the volume of metal in a pipe of length 1 m. If the density of the metal is 8 g/cm^3, find the weight of the pipe.

15. A cylindrical can of internal radius 20 cm stands upright on a flat surface. It contains water to a depth of 20 cm. Calculate the rise h in the level of the water when a brick of volume 1500 cm^3 is immersed in the water.

16. Rain which falls onto a flat rectangular surface of length 6 m and width 4 m is collected in a cylinder of internal radius 20 cm. What is the depth of water in the cylinder after a storm in which 1 cm of rain fell?

Pyramid, sphere, cone

Pyramid
Volume = $\frac{1}{3}$(base area) × height

Sphere
Volume = $\frac{4}{3}\pi r^3$

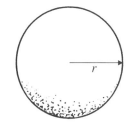

Cone
Volume = $\frac{1}{3}\pi r^2 h$

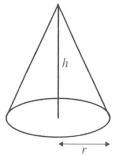

The formula for the volume of a pyramid is demonstrated below:

Figure 1 shows a cube of side $2a$ broken down into six pyramids of height a as shown in figure 2.

If the volume of each pyramid is V,

then $\quad 6V = 2a \times 2a \times 2a$

$\quad\quad V = \frac{1}{6} \times (2a)^2 \times 2a$

so $\quad\quad V = \frac{1}{3} \times (2a)^2 \times a$

$\quad\quad V = \frac{1}{3}$(base area) × height.

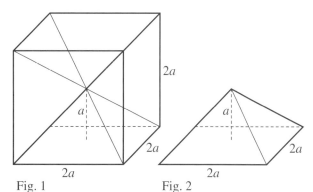

Fig. 1   Fig. 2

Example

(a) Calculate the volume of the cone.

$$\text{Volume} = \tfrac{1}{3} \times \pi \times 2^2 \times 5$$

$$= \frac{\pi \times 20}{3}$$

$$= 20 \cdot 9 \text{ cm}^3 \quad (3 \text{ s.f.})$$

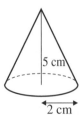

(b) A solid sphere of radius 4 cm is made of metal of density 8 g/cm³. Calculate the mass of the sphere.

$$\text{Volume} = \tfrac{4}{3} \times \pi \times 4^3 \text{ cm}^3$$

$$\text{Mass} = \tfrac{4}{3} \times \pi \times 4^3 \times 8 = 2140 \text{ g} \quad (3 \text{ s.f.})$$

Exercise 2R

In Questions **1** to **6** find the volume of each object.
All lengths are in cm.

1.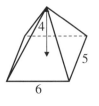

2.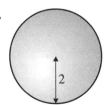

3.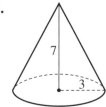

4.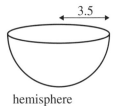
   hemisphere

5.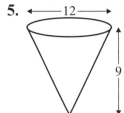

6.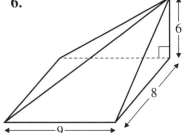

7. A solid sphere is made of metal of density 9 g/cm³. Calculate the mass of the sphere if the radius is 5 cm.

8. Find the volume of a hemisphere of radius 5 cm.

9. Calculate the volume of a cone of height and radius 7 m.

10. A cone is attached to a hemisphere of radius 4 cm. If the total length of the object is 10 cm, find its volume.

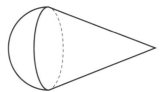

11. Find the height of a pyramid of volume 20 m³ and base area 12 m².

12. A single drop of oil is a sphere of radius 3 mm. The drop of oil falls on water to produce a thin circular film of radius 100 mm. Calculate the thickness of this film in mm.

13. Gold is sold in solid spherical balls.
Which is worth more: 10 balls of radius 2 cm or 1 ball of radius 4 cm?

$r = 2$ cm $r = 4$ cm

14. A toy consists of a cylinder of diameter 6 cm 'sandwiched' between a hemisphere and a cone of the same diameter. If the cone is of height 8 cm and the cylinder is of height 10 cm, find the total volume of the toy.

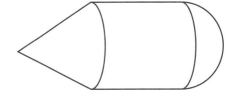

15. Water is flowing into an inverted cone, of diameter and height 30 cm, at a rate of 4 litres per minute. How long, in seconds, will it take to fill the cone?

16. Metal spheres of radius 2 cm are packed into a rectangular box of internal dimensions 16 cm × 8 cm × 8 cm. When 16 spheres are packed the box is filled with a preservative liquid. Find the volume of this liquid.

17. The Pyramids in Egypt are seriously large! One Pyramid has a square base of side 100 m and height 144 m. An average slave could just manage to carry a brick measuring 40 cm by 20 cm by 16 cm. Assuming no spaces were left for mummies or treasure, work out how many bricks would be needed to make this Pyramid.

Example

Calculate the radius of a sphere of volume 500 cm³.

Let the radius of the sphere be r cm

$$\tfrac{4}{3}\pi r^3 = 500$$

$$r^3 = \frac{3 \times 500}{4\pi}$$

$$r = \sqrt[3]{\left(\frac{3 \times 500}{4\pi}\right)} = 4\cdot 92 \text{ (3 s.f.)}$$

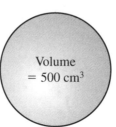

The radius of the sphere is 4·92 cm.

Exercise 2S

1. Calculate the radius of each sphere.
   (a) (b) (c)

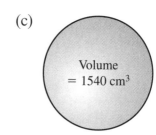

2. A solid metal sphere of radius 4 cm is recast into a solid cube. Find the length of a side of the cube.

3. Find the height of a cone of volume 2500 cm³ and radius 10 cm.

4. A solid metal sphere is recast into many smaller spheres. Calculate the number of the smaller spheres if the initial and final radii are as follows:
   (a) initial radius = 10 cm, final radius = 2 cm
   (b) initial radius = 1 m, final radius = $\tfrac{1}{3}$ cm.

5. A spherical ball is immersed in water contained in a vertical cylinder.
   Assuming the water covers the ball, calculate the rise in the water level if:
   (a) sphere radius = 3 cm, cylinder radius = 10 cm
   (b) sphere radius = 2 cm, cylinder radius = 5 cm.

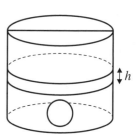

6. A spherical ball is immersed in water contained in a vertical cylinder. The rise in water level is measured in order to calculate the radius of the spherical ball. Calculate the radius of the ball in the following cases:
   (a) cylinder of radius 10 cm, water level rises 4 cm
   (b) cylinder of radius 100 cm, water level rises 8 cm.

7. The diagram shows the cross section of an inverted cone of height $MC = 12$ cm. If $AB = 6$ cm and $XY = 2$ cm, use similar triangles to find the length NC.
   Hence find the volume of the cone of height NC.

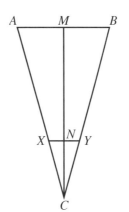
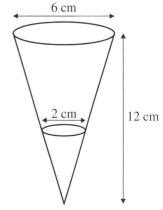

8. An inverted cone of height 10 cm and base radius 6·4 cm contains water to a depth of 5 cm, measured from the vertex. Calculate the volume of water in the cone.

9. An inverted cone of height 15 cm and base radius 4 cm contains water to a depth of 10 cm. Calculate the volume of water in the cone.

10. A frustum is a cone with 'the end chopped off'. A bucket in the shape of a frustum as shown has diameters of 10 cm and 4 cm at its ends and a depth of 3 cm. Calculate the volume of the bucket.

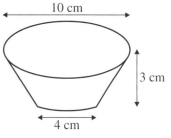

11. The diagram shows a sector of a circle of radius 10 cm.
    (a) Find, as a multiple of π, the arc length of the sector. The straight edges are brought together to make a cone. Calculate
    (b) the radius of the base of the cone.
    (c) the vertical height of the cone.

12. A sphere passes through the eight vertices of a cube of side 10 cm. Find the volume of the sphere.

Surface area

We are concerned here with the surface areas of the **curved** parts of cylinders, spheres and cones. The areas of the plane faces are easier to find.

Cylinder
Curved surface area = $2\pi rh$

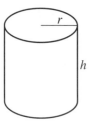

Sphere
Surface area = $4\pi r^2$

Cone
Curved surface area = πrl, where l is the slant height

Example
Calculate the **total** surface area of a solid cylinder of radius 3 cm and height 8 cm.

Curved surface area = $2\pi rh$
$= 2 \times \pi \times 3 \times 8$
$= 48\pi \text{ cm}^2$

Area of two ends = $2 \times \pi r^2$
$= 2 \times \pi \times 3^2$
$= 18\pi \text{ cm}^2$

Total surface area = $(48\pi + 18\pi) \text{ cm}^2$
$= 207 \text{ cm}^2$ (3 s.f.)

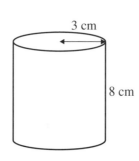

Exercise 2T

1. Work out the **curved** surface area of these objects. Leave π in your answers. All lengths are in cm.

(a)
$r = 3$

(b)
$r = 4$
$h = 5$

(c)
$l = 10$
$r = 6$

(d)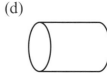
$r = 0.7$
$h = 1$

2. Work out the **total** surface area of these objects. Leave π in your answers. All lengths are in cm.

(a) $l = 6$, $r = 4$

(b) $r = 2$, $h = 5$

(c) hemisphere $r = 4$

(d) $r = 3$, vertical height $= 4$

3. A solid cylinder of height 10 cm and radius 4 cm is to be plated with material costing £11 per cm². Find the cost of the plating.

4. A tin of paint covers a surface area of 60 m² and costs £4·50. Find the cost of painting the outside surface of a cylindrical gas holder of height 30 m and radius 18 m. The top of the gas holder is a flat circle.

5. A solid wooden cylinder of height 8 cm and radius 3 cm is cut in two along a vertical axis of symmetry. Calculate the total surface area of the two pieces.

6. Calculate the total surface area of the combined cone/cylinder/hemisphere.

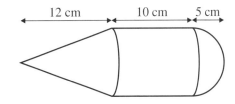

12 cm   10 cm   5 cm

7. Find the radius of a sphere of surface area 34 cm².

8. Find the slant height of a cone of curved surface area 20 cm² and radius 3 cm.

9. Find the height of a solid cylinder of radius 1 cm and **total** surface area 28 cm².

10. Find the volume of a sphere of surface area 100 cm².

11. Find the surface area of a sphere of volume 28 cm³.

12. An inverted cone of vertical height 12 cm and base radius 9 cm contains water to a depth of 4 cm. Find the area of the interior surface of the cone not in contact with the water.

13. A circular paper of radius 20 cm is cut in half and each half is made into a hollow cone by joining the straight edges. Find the slant height and base radius of each cone.

14. A solid metal cube of side 6 cm is recast into a solid sphere.
    (a) Find the radius of the sphere.
    (b) By how much is the surface area of the original cube greater than the surface area of the new sphere?

15. A solid cuboid, measuring 5 cm × 5 cm × 3 cm, has a hole of radius 2 cm drilled right through. Calculate the surface area of the object.

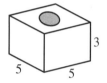

2.8 Similar shapes

If one shape is an enlargement of another, the two shapes are mathematically **similar**.

The two triangles A and B are similar if they have the same angles.

For other shapes to be similar, not only must corresponding angles be equal, but also corresponding edges must be in the same proportion.

The two quadrilaterals C and D are similar. All the edges of shape D are twice as long as the edges of shape C.

The two rectangles E and F are not similar even though they have the same angles.

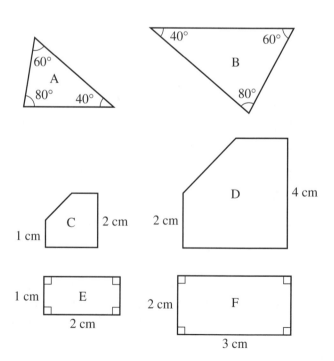

Example

(a) These triangles are similar. Find x.
(b) Triangle B is an enlargement of triangle A. Corresponding sides are in the same ratio.

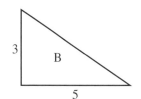

$$\frac{x}{5} = \frac{2}{3}$$

$x = \frac{2}{3} \times 5 \Rightarrow x = 3\frac{1}{3}$

Exercise 2U

1. Which of the shapes B, C, D is/are similar to shape A?

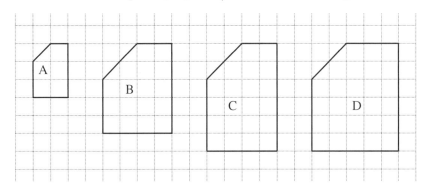

In Questions **2** to **7**, find the sides marked with letters; all lengths are given in cm. The pairs of shapes are similar.

2.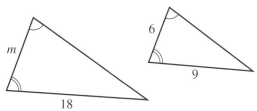

3.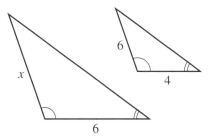

4.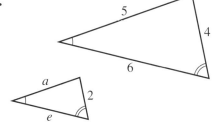

5.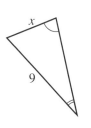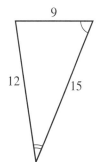

6.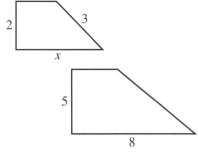

7.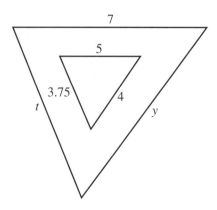

8. Picture B is an enlargement of picture A. Calculate the length x.

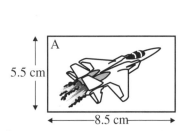
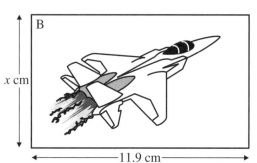

9. The drawing shows a rectangular picture 16 cm × 8 cm surrounded by a border of width 4 cm.
   Are the two rectangles similar?

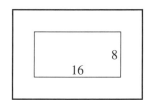

10. Which of the following **must** be similar to each other?
    (a) two equilateral triangles   (b) two rectangles
    (c) two isosceles triangles   (d) two squares
    (e) two regular pentagons   (f) two kites
    (g) two rhombuses   (h) two circles

11. (a) Explain why triangles ABC and EBD are similar.
    (b) Given that $EB = 7$ cm, calculate the length AB.
    (c) Write down the length AE.

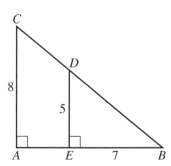

In Questions **12**, **13**, **14** use similar triangles to find the sides marked with letters. All lengths are in cm.

12.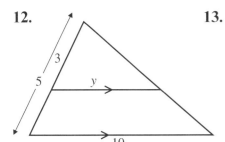

13.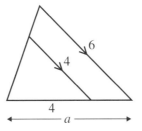

14.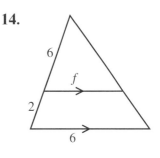

Exercise 2V

1. A tree of height 4 m casts a shadow of length 6·5 m. Find the height of a house casting a shadow 26 m long.

2. A small cone is cut from a larger cone. Find the radius of the smaller cone.

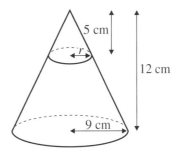

3. The diagram shows the side view of a swimming pool being filled with water. Calculate the length x.

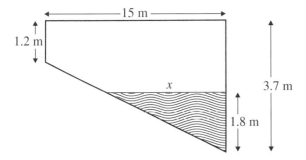

4. The diagonals of a trapezium $ABCD$ intersect at O. AB is parallel to DC, $AB = 3$ cm and $DC = 6$ cm. Show that triangles ABO and CDO are similar. If $CO = 4$ cm and $OB = 3$ cm, find AO and DO.

In Questions **5** and **6** find the sides marked with letters.

5. $B\widehat{A}C = D\widehat{B}C$
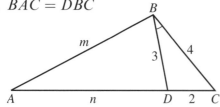

6.
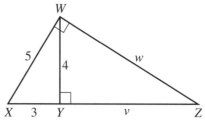

7. A rectangle 11 cm by 6 cm is similar to a rectangle 2 cm by x cm. Find the two possible values of x.

8. From the rectangle $ABCD$ a square is cut off to leave rectangle $BCEF$.
Rectangle $BCEF$ is similar to $ABCD$. Find x and hence state the ratio of the sides of rectangle $ABCD$. $ABCD$ is called the Golden Rectangle and is an important shape in architecture.

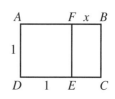

9. Triangles ABC and EBD are similar but DE is **not** parallel to AC. Work out the length x.

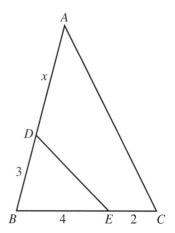

10. In the diagram $A\widehat{B}C = A\widehat{D}B = 90°$, $AD = p$ and $DC = q$.
   (a) Use similar triangles ABD and ABC to show that $x^2 = pz$.
   (b) Find a similar expression for y^2.
   (c) Add the expressions for x^2 and y^2 and hence prove Pythagoras' theorem.

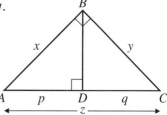

11. In a triangle ABC, a line is drawn parallel to BC to meet AB at D and AC at E. DC and BE meet at X. Prove that:
   (a) the triangles ADE and ABC are similar
   (b) the triangles DXE and BXC are similar
   (c) $\dfrac{AD}{AB} = \dfrac{EX}{XB}$

Areas of similar shapes

The two rectangles shown are similar.
The ratio of their corresponding sides is k.

area of $ABCD = ab$
area of $WXYZ = ka \times kb = k^2 ab$

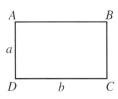

$$\therefore \quad \frac{\text{area of } WXYZ}{\text{area } ABCD} = \frac{k^2 ab}{ab} = k^2$$

This illustrates an important general rule for all similar shapes:

If two figures are similar and the ratio of corresponding sides is k, then the ratio of their areas is k^2.

Note: k may be called the **linear scale factor**.

This result also applies for the surface areas of similar three-dimensional objects.

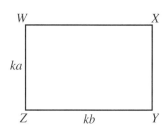

Example 1

In triangle ABC, XY is parallel to BC and $\dfrac{AB}{AX} = \dfrac{3}{2}$.

If the area of triangle AXY is $4\,\text{cm}^2$, find the area of triangle ABC.
The triangles ABC and AXY are similar.

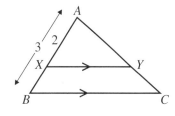

Ratio of corresponding sides $(k) = \dfrac{3}{2}$

$\therefore$ Ratio of areas $(k^2) = \dfrac{9}{4}$

$\therefore$ Area of $\triangle ABC \;= \dfrac{9}{4} \times (\text{area of } \triangle AXY)$

$\qquad\qquad\qquad\quad = \dfrac{9}{4} \times (4) = 9\,\text{cm}^2$

Example 2

Two similar triangles have areas of $18\,\text{cm}^2$ and $32\,\text{cm}^2$ respectively. If the base of the smaller triangle is $6\,\text{cm}$, find the base of the larger triangle.

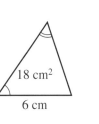 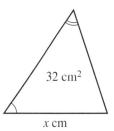

Ratio of areas $(k^2) = \dfrac{32}{18} = \dfrac{16}{9}$

$\therefore$ Ratio of corresponding sides $(k) = \sqrt{\left(\dfrac{16}{9}\right)} = \dfrac{4}{3}$

$\therefore$ Base of larger triangle $= 6 \times \dfrac{4}{3} = 8\,\text{cm}$

Exercise 2W

In this exercise, a number written inside a figure represents the area of the shape in cm². Numbers on the outside give linear dimensions in cm. In each case the shapes are similar.
In Questions **1** to **6**, find the unknown area A.

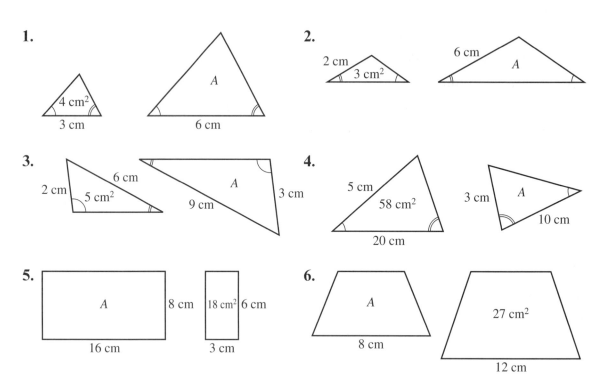

In Questions **7** to **10**, find the lengths marked for each pair of similar shapes.

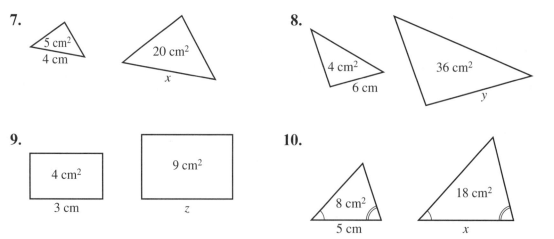

11. A triangle with sides 3, 4 and 5 cm has an area of 6 cm². A similar triangle has an area of 54 cm². Find the sides of the larger triangle.

12. A giant ball is made to promote the sales of a new make of golf ball. The surface area of an ordinary ball is 50 cm².
   The diameter of the giant ball is 100 times as great as a normal ball.
   Work out the surface area of the giant ball:
   (a) in cm²
   (b) in m².

13. It takes 30 minutes to cut the grass in a square field of side 20 m. How long will it take to cut the grass in a square field of side 60 m?

14. A floor is covered by 600 tiles which are 10 cm by 10 cm. How many 20 cm by 20 cm tiles are needed to cover the same floor?

15. A wall is covered by 160 tiles which are 15 cm by 15 cm. How many 10 cm by 10 cm tiles are needed to cover the same wall?

16. A ball of radius r cm has a surface area of 20 cm². Find, in terms of r, the radius of a ball with a surface area of 500 cm².

17. Given $AD = 3$ cm, $AB = 5$ cm and area of $\triangle ADE = 6$ cm². Find:
   (a) area of $\triangle ABC$
   (b) area of $DECB$.

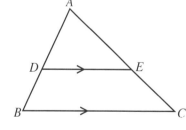

18. Given $XY = 5$ cm, $MY = 2$ cm and area of $\triangle MYN = 4$ cm². Find:
   (a) area of $\triangle XYZ$
   (b) area of $MNZX$.

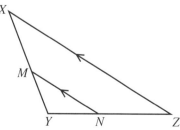

19. The triangles ABC and EBD are similar (AC and DE are **not** parallel). If $AB = 8$ cm, $BE = 4$ cm and the area of $\triangle DBE = 6$ cm^2, find the area of $\triangle ABC$.

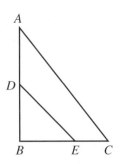

20. When potatoes are peeled do you lose more peel or less when big potatoes are used as opposed to small ones?

21. A supermarket offers cartons of cheese with 10% extra for the same price. The old carton has a radius of 6 cm and a thickness of 1·2 cm. The new carton has the same thickness. Calculate its radius.

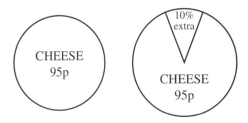

22. An A3 sheet of paper can be cut into two sheets of A4. The A3 and A4 sheets are mathematically similar.

    Find the scale factor: $\left(\dfrac{\text{long side of A3 sheet}}{\text{long side of A4 sheet}}\right)$ $\left[\text{i.e. } \dfrac{x}{y}\right]$

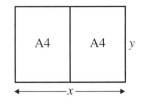

23. Rectangle $ABCD$ is inscribed in $\triangle PQR$. The length of AD is one third the length of the perpendicular from Q to PR.

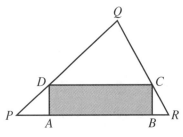

Calculate the ratio $\left(\dfrac{\text{area } ABCD}{\text{area } PQR}\right)$.

Volumes of similar objects

When solid objects are similar, one is an accurate enlargement of the other. If two objects are similar and the ratio of corresponding sides is k, then the ratio of their volumes is k^3. A line has one dimension, and the scale factor is used once. An area has two dimensions, and the scale factor is used twice. A volume has three dimensions, and the scale factor is used three times.

Example 1

Two similar cylinders have heights of 3 cm and 6 cm respectively. If the volume of the smaller cylinder is 30 cm³, find the volume of the larger cylinder.

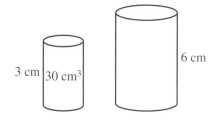

$$\text{ratio of heights } (k) = \tfrac{6}{3} \text{ (linear scale factor)}$$
$$= 2$$
$$\therefore \quad \text{ratio of volumes } (k^3) = 2^3$$
$$= 8$$

and volume of larger cylinder $= 8 \times 30$
$$= 240 \text{ cm}^3$$

Example 2

Two similar spheres made of the same material have weights of 32 kg and 108 kg respectively. If the radius of the larger sphere is 9 cm, find the radius of the smaller sphere.

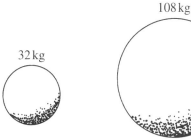

We may take the ratio of weights to be the same as the ratio of volumes.

$$\text{ratio of volumes } (k^3) = \frac{32}{108} = \frac{8}{27}$$

$$\text{ratio of corresponding lengths } (k) = \sqrt[3]{\left(\frac{8}{27}\right)} = \frac{2}{3}$$

$$\therefore \quad \text{radius of smaller sphere} = \frac{2}{3} \times 9 = 6 \text{ cm}$$

Exercise 2X

In this exercise, the objects are similar and a number written inside a figure represents the volume of the object in cm³. Numbers on the outside give linear dimensions in cm.
In Questions **1** to **8**, find the unknown volume *V*.

1.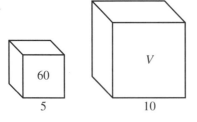

2.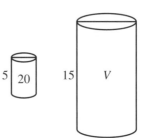

3.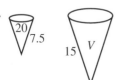

4.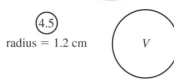

radius = 1.2 cm   radius = 12 cm

5.

6.

7.

8.

In Questions **9** to **12**, find the lengths marked by a letter.

9.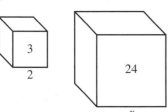

10.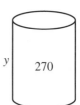

11.

12.

13. Two similar jugs have heights of 4 cm and 6 cm respectively. If the capacity of the smaller jug is 50 cm³, find the capacity of the larger jug.

14. Two similar cylindrical tins have base radii of 6 cm and 8 cm respectively. If the capacity of the larger tin is 252 cm³, find the capacity of the small tin.

15. Two solid metal spheres have masses of 5 kg and 135 kg respectively. If the radius of the smaller one is 4 cm, find the radius of the larger one.

16. Two similar cones have surface areas in the ratio 4 : 9. Find the ratio of:
    (a) their lengths,   (b) their volumes.

17. The area of the bases of two similar glasses are in the ratio 4 : 25. Find the ratio of their volumes.

18. A model of a treasure chest is 3 cm long and has a volume of 11 cm³. Work out the length of the real treasure chest if its volume is 88 000 cm³.

19. The ingredients of a standard Toblerone bar cost 20p. A giant-sized bar is similar in shape and four times as long. What should be the cost of the ingredients of the giant bar?

20. Sam has a model of a pirate ship which is made to a scale of 1 : 50.

    Copy and complete the table using the given units.

	On model	On actual ship
Length	42 cm	☐ m
Capacity of hold	500 cm³	☐ m³
Area of sails	☐ cm²	175 m²
Number of cannon	12	☐
Deck area	370 cm²	☐ m²

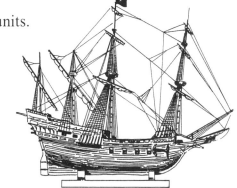

21. Two solid spheres have surface areas of 5 cm² and 45 cm² respectively and the mass of the smaller sphere is 2 kg. Find the mass of the larger sphere.

22. The masses of two similar objects are 24 kg and 81 kg respectively. If the surface area of the larger object is 540 cm², find the surface area of the smaller object.

Summary

1. You know the angle properties of parallel lines.

2. You know the properties of specific quadrilaterals.

3. You can find the exterior and the interior angles of a regular polygon.

4. You know the cases for congruency of two triangles and can prove that triangles are congruent.

5. You can make geometrical constructions.

6. You can use Pythagoras' theorem.

7. You can find the area of a polygon.

Check out AS2

1. What is meant by corresponding angles?

2. What are the properties of:
   (a) a square
   (b) a trapezium?

3. Calculate the interior angles of a hexagon.

4.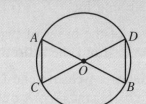

   O is the centre of the circle $ABCD$. Prove that triangles AOC and BOD are congruent.

5. (a) Draw a triangle with sides 4·8 cm, 7·2 cm and 3·9 cm.
   (b) Measure the smallest angle.

6.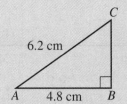

   Find the length of BC.

7.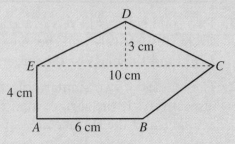

   Find the area $ABCDE$.

8. You can find the circumference and area of a circle.

8. A circle has a radius of 6·2 cm.
   Find: (a) the circumference
   (b) the area of the circle.

9. You know how to find arc lengths, areas of sectors and segments.

9.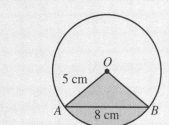

Calculate:

(a) the arc length AB

(b) the area of the shaded sector OAB

(c) the area of the minor segment cut off by the chord AB.

10. You can find the surface areas and volumes of cuboids, cylinders, pyramids, spheres and cones.

10. Find: (a) the curved surface area and
    (b) the volume of:

    (i) a sphere of radius 6 cm
    (ii) a cone of with base radius 7 cm and vertical height 3 cm
    (iii) a cylinder of radius 2 cm and height 15 cm.

11. Knowing the area or volume of a shape you can find the area or volume of a similar shape.

11.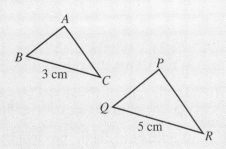

The two triangles are similar.
The area of triangle PQR is 13 cm^2.
Find the area of triangle ABC.

Revision exercise AS2

1. (a) These two triangles are similar. Calculate the value of *a*.

   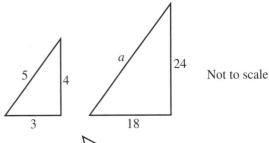

   (b) Are these two triangles similar? You **must** explain your answer.

   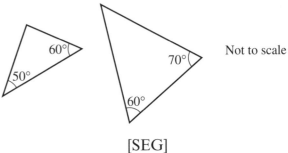

   [SEG]

2. (a) The two triangles shown are congruent.

   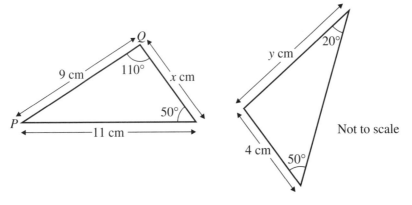

   Write down the values of *x* and *y*.

   (b) The length of the side *PQ* was given as 9 centimetres, measured to the nearest millimetre.

   What is the maximum length of *PQ* in millimetres? [SEG]

3. (a) Calculate the size of an interior angle of a regular octagon.

   (b) Explain why a regular polygon cannot have an interior angle equal to 110°. [SEG]

4. The positions of three villages, Oldacre (*O*), Adchester (*A*) and Byetoft (*B*), are shown on the diagram.

   Angle *OAB* is 90°.
   The distance from Oldacre to Adchester is 8 km.

   The distance from Oldacre to Byetoft is 10 km.

   Calculate the distance from Adchester to Byetoft. [SEG]

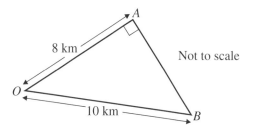

5. The diagram shows Kim's garden.

   The line *AB* represents the wall of Kim's house.

   PQRS is a rectangular shed and there is a tree at *T*.

   Kim wants to plant another tree in the garden.

   The new tree must be at least 2 m away from the wall, *AB*. It must also be at least 3 m away from the tree, *T*, and at least 2 m away from the shed, *PQRS*.

   By accurate drawing, shade in the area where the new tree can be planted. [SEG]

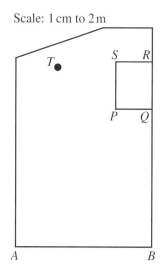

6. Pamela has a fish-tank which is 30 cm long and has a capacity of 4·8 litres.

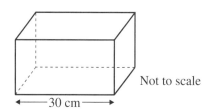

   Rani has a similar fish-tank which is 45 cm long.

   What is the capacity of Rani's fish-tank?

   Give your answer in litres.

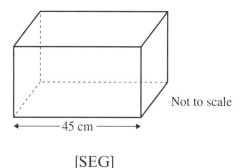

[SEG]

7. The descriptions of three different quadrilaterals are given. Choose the correct names for each one from this list:

   kite   rectangle   rhombus   square   trapezium

   (You may find it helpful to sketch the quadrilaterals.)

   (a) A quadrilateral with one pair of sides parallel but not equal in length.

   (b) A quadrilateral with diagonals of different lengths that cross at 90°.
   Each diagonal is a line of symmetry of the quadrilateral.

   (c) A quadrilateral with diagonals equal in length that bisect each other, but do not cross at right-angles.   [SEG]

8. The Gnome factory makes garden gnomes in two sizes.

   The gnomes are similar in shape.

   The larger gnome is 50 cm high and the smaller one is 30 cm high.

   It takes 45 400 cm³ of plaster to make a large gnome.

   How much plaster is needed to make a small gnome?   [SEG]

9. A sector, OPQ, is cut from a circle of radius 8 cm.

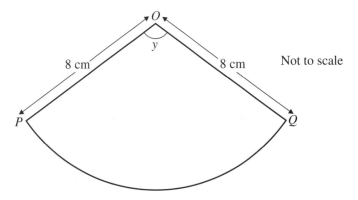

   The perimeter of sector OPQ is 30 cm.

   Calculate the angle y.   [SEG]

AS3 Algebra 2

This unit will show you how to:

- Rearrange equations
- Solve inequalities
- Know the general shape of curves such as x^3, $y = \dfrac{1}{x}$, $y = 2^x$
- Use the rules of indices to solve equations

Before you start:

You should know how to...	Check in AS3
1. Expand brackets. For example, $3x(7x - 2) = 21x^2 - 6x$	1. Expand: (a) $2x(3x + 5)$ (b) $4x^2(5 - 2x)$
2. Square terms. For example, $(x - 3y)^2 = x^2 - 6xy + 9y^2$	2. Evaluate: (a) $(3x + y)^2$ (b) $(7x - 2)^2$
3. Draw the graph of a straight line. For example, $y = 2x + 1$	3. Draw the lines: $y = x - 4$ $2y = 5x + 1$
4. Find the gradient and intercept of a graph. For example, find the gradient and intercept of $3y + 4x = 4$.	4. Write down the gradient and intercept of $y = 2x - 5$.
5. Use the rules of indices: $a^m \times a^n = a^{m+n}$ $a^m \div a^n = a^{m-n}$ $(a^m)^n = a^{mn}$ $a^0 = 1 \ (a \neq 0)$ $a^{-n} = \dfrac{1}{a^n}$ $a^{\frac{1}{n}} = \sqrt[n]{a}$	5. Write down the values of: (a) $x^7 \times x^3$ (b) $x^5 \div x^9$ (c) $(x^4)^3$ (d) 7^0 (e) $8^{-\frac{1}{3}}$ (f) 2^{-3}

3.1 Changing the subject of a formula

The operations involved in solving ordinary linear equations are exactly the same as the operations required in changing the subject of a formula.

Example
(a) Solve the equation $3x + 1 = 12$.
(b) Make x the subject of the formula $Mx + B = A$.
(c) Make y the subject of the formula $x(y - a) = e$.

(a) $3x + 1 = 12$
$3x = 12 - 1$
$x = \dfrac{12 - 1}{3} = \dfrac{11}{3}$

(b) $Mx + B = A$
$Mx = A - B$
$x = \dfrac{A - B}{M}$

(c) $x(y - a) = e$
$xy - xa = e$
$y = \dfrac{e + xa}{x}$

Exercise 3A
Make x the subject of the following:

1. $Ax = B$
2. $Nx = T$
3. $Mx = K$
4. $xy = 4$
5. $9x = T + N$
6. $Ax = B - R$
7. $Cx = R + T$
8. $Lx = N - R^2$
9. $R - S^2 = Nx$
10. $x + 5 = 7$
11. $x + B = S$
12. $N = x + D$
13. $N^2 + x = T$
14. $L + x = N + M$
15. $x - R = A$
16. $x - A = E$
17. $F = x - B$
18. $F^2 = x - B^2$

Make y the subject of the following:

19. $L = y - B$
20. $N = y - T$
21. $3y + 1 = 7$
22. $2y - 4 = 5$
23. $Ay + C = N$
24. $By + D = L$
25. $Dy + E = F$
26. $Ny - F = H$
27. $Vy + m = Q$
28. $ty - m = n + a$
29. $V^2y + b = c$
30. $r = ny - 6$
31. $3(y - 1) = 5$
32. $A(y + B) = C$
33. $D(y + E) = F$
34. $h(y + n) = a$
35. $b(y - d) = q$
36. $n = r(y + t)$

Example
(a) Solve the equation $\dfrac{3a + 1}{2} = 4$.
(b) Make a the subject of the formula $\dfrac{na + b}{m} = n$.
(c) Make a the subject of the formula $x - na = y$.

(a) $\dfrac{3a + 1}{2} = 4$
$3a + 1 = 8$
$3a = 7$
$a = \tfrac{7}{3}$

(b) $\dfrac{na + b}{m} = n$
$na + b = mn$
$na = mn - b$
$a = \dfrac{mn - b}{n}$

(c) $x - na = y$
Make the 'a' term positive
$x - y = na$
$\dfrac{x - y}{n} = a$

Exercise 3B

Make a the subject.

1. $\dfrac{a}{4} = 3$
2. $\dfrac{a}{5} = 2$
3. $\dfrac{a}{D} = B$
4. $\dfrac{a}{B} = T$
5. $\dfrac{a}{N} = R$
6. $b = \dfrac{a}{m}$
7. $\dfrac{a-2}{4} = 6$
8. $\dfrac{a-A}{B} = T$
9. $\dfrac{a-D}{N} = A$
10. $\dfrac{a+Q}{N} = B^2$
11. $g = \dfrac{a-r}{e}$
12. $\dfrac{2a+1}{5} = 2$
13. $\dfrac{Aa+B}{C} = D$
14. $\dfrac{na+m}{p} = q$
15. $\dfrac{ra-t}{S} = v$
16. $\dfrac{za-m}{q} = t$
17. $\dfrac{m+Aa}{b} = c$
18. $A = \dfrac{Ba+D}{E}$
19. $n = \dfrac{ea-f}{h}$
20. $q = \dfrac{ga+b}{r}$
21. $6 - a = 2$
22. $7 - a = 9$
23. $5 = 7 - a$
24. $A - a = B$
25. $C - a = E$
26. $D - a = H$
27. $n - a = m$
28. $t = q - a$
29. $b = s - a$
30. $v = r - a$
31. $t = m - a$
32. $5 - 2a = 1$
33. $T - Xa = B$
34. $M - Na = Q$
35. $V - Ma = T$
36. $L = N - Ra$
37. $r = v^2 - ra$
38. $t^2 = w - na$
39. $n - qa = 2$
40. $\dfrac{3-4a}{2} = 1$
41. $\dfrac{5-7a}{3} = 2$
42. $\dfrac{B-Aa}{D} = E$
43. $\dfrac{D-Ea}{N} = B$
44. $\dfrac{h-fa}{b} = x$
45. $\dfrac{v^2-ha}{C} = d$
46. $\dfrac{M(a+B)}{N} = T$
47. $\dfrac{f(Na-e)}{m} = B$
48. $\dfrac{T(M-a)}{E} = F$
49. $\dfrac{y(x-a)}{z} = t$
50. $\dfrac{k^2(m-a)}{x} = x$

Formulae with fractions

Example

(a) Solve the equation $\dfrac{4}{z} = 7$.

(b) Make z the subject of the formula $\dfrac{n}{z} = k$.

(c) Make t the subject of the formula $\dfrac{x}{t} + m = a$.

(a) $\dfrac{4}{z} = 7$

$4 = 7z$

$\dfrac{4}{7} = z$

(b) $\dfrac{n}{z} = k$

$n = kz$

$\dfrac{n}{k} = z$

(c) $\dfrac{x}{t} + m = a$

$\dfrac{x}{t} = a - m$

$x = (a - m)t$

$\dfrac{x}{(a - m)} = t$

Exercise 3C

Make *a* the subject.

1. $\dfrac{7}{a} = 14$
2. $\dfrac{5}{a} = 3$
3. $\dfrac{B}{a} = C$
4. $\dfrac{T}{a} = X$
5. $t = \dfrac{v}{a}$
6. $\dfrac{n}{a} = \sin 20°$
7. $\dfrac{7}{a} = \cos 30°$
8. $\dfrac{B}{a} = x$
9. $\dfrac{v}{a} = \dfrac{m}{s}$
10. $\dfrac{t}{b} = \dfrac{m}{a}$
11. $\dfrac{B}{a + D} = C$
12. $\dfrac{Q}{a - C} = T$
13. $\dfrac{V}{a - T} = D$
14. $\dfrac{L}{Ma} = B$
15. $\dfrac{N}{Ba} = C$
16. $\dfrac{m}{ca} = d$
17. $t = \dfrac{b}{c - a}$
18. $x = \dfrac{z}{y - a}$

Make *x* the subject.

19. $\dfrac{2}{x} + 1 = 3$
20. $\dfrac{5}{x} - 2 = 4$
21. $\dfrac{A}{x} + B = C$
22. $\dfrac{V}{x} + G = H$
23. $\dfrac{r}{x} - t = n$
24. $q = \dfrac{b}{x} + d$
25. $t = \dfrac{m}{x} - n$
26. $h = d - \dfrac{b}{x}$
27. $C - \dfrac{d}{x} = e$
28. $r - \dfrac{m}{x} = e^2$
29. $t^2 = b - \dfrac{n}{x}$
30. $\dfrac{d}{x} + b = mn$
31. $3M = M + \dfrac{N}{P + x}$
32. $A = \dfrac{B}{c + x} - 5A$
33. $\dfrac{m^2}{x} - n = {}^-p$
34. $t = w - \dfrac{q}{x}$

Squares and square roots

Example

Make x the subject of the formulas.

(a) $\sqrt{(x^2 + A)} = B$
$\quad x^2 + A = B^2$ (square both sides)
$\quad x^2 = B^2 - A$
$\quad x = \pm\sqrt{(B^2 - A)}$

(b) $(Ax - B)^2 = M$
$\quad Ax - B = \pm\sqrt{M}$ (square root both sides)
$\quad Ax = B \pm \sqrt{M}$
$\quad x = \dfrac{B \pm \sqrt{M}}{A}$

Exercise 3D

Make x the subject.

1. $\sqrt{x} = 2$
2. $\sqrt{(x-2)} = 3$
3. $\sqrt{(x+C)} = D$
4. $\sqrt{(ax+b)} = c$
5. $b = \sqrt{(gx-t)}$
6. $\sqrt{(d-x)} = t$
7. $c = \sqrt{(n-x)}$
8. $g = \sqrt{(c-x)}$
9. $\sqrt{(Ax+B)} = \sqrt{D}$
10. $x^2 = g$
11. $x^2 = B$
12. $x^2 - A = M$

Make k the subject.

13. $C - k^2 = m$
14. $mk^2 = n$
15. $\dfrac{kz}{a} = t$
16. $n = a - k^2$
17. $\sqrt{(k^2 - A)} = B$
18. $t = \sqrt{(m + k^2)}$
19. $A\sqrt{(k+B)} = M$
20. $\sqrt{\left(\dfrac{N}{k}\right)} = B$
21. $\sqrt{(a^2 - k^2)} = t$
22. $2\pi\sqrt{(k+t)} = 4$
23. $\sqrt{(ak^2 - b)} = C$
24. $k^2 + b = x^2$

Collecting from both sides

Example

Make x the subject of each formula.

(a) $Ax - B = Cx + D$
$\quad Ax - Cx = D + B$ (x terms on one side)
$\quad x(A - C) = D + B$ (factorise)
$\quad x = \dfrac{D + B}{A - C}$

(b) $\quad x+a=\dfrac{x+b}{c}$

$c(x+a) = x+b$

$cx + ca = x + b$

$cx - x = b - ca$ (x terms on one side)

$x(c-1) = b - ca$ (factorise)

$$x = \dfrac{b-ca}{c-1}$$

Exercise 3E

Make y the subject.

1. $5(y-1) = 2(y+3)$
2. $7(y-3) = 4(3-y)$
3. $Ny + B = D - Ny$
4. $My - D = E - 2My$
5. $ay + b = 3b + by$
6. $my - c = e - ny$
7. $xy + 4 = 7 - ky$
8. $Ry + D = Ty + C$
9. $ay - x = z + by$
10. $m(y+a) = n(y+b)$
11. $x(y-b) = y + d$
12. $\dfrac{a-y}{a+y} = b$
13. $\dfrac{1-y}{1+y} = \dfrac{c}{d}$
14. $\dfrac{M-y}{M+y} = \dfrac{a}{b}$
15. $m(y+n) = n(n-y)$
16. $y + m = \dfrac{2y-5}{m}$
17. $y - n = \dfrac{y+2}{n}$
18. $y + b = \dfrac{ay+e}{b}$
19. $\dfrac{ay+x}{x} = 4 - y$
20. $c - dy = e - ay$
21. $y(a-c) = by + d$
22. $y(m+n) = a(y+b)$
23. $t - ay = s - by$
24. $\dfrac{y+x}{y-x} = 3$
25. $\dfrac{v-y}{v+y} = \dfrac{1}{2}$
26. $y(b-a) = a(y+b+c)$
27. $\sqrt{\left(\dfrac{y+x}{y-x}\right)} = 2$
28. $\sqrt{\left(\dfrac{z+y}{z-y}\right)} = \dfrac{1}{3}$
29. $\sqrt{\left[\dfrac{m(y+n)}{y}\right]} = p$
30. $n - y = \dfrac{4y-n}{m}$

Exercise 3F Mixed questions

Make the letter in square brackets the subject.

1. $ax + by + c = 0$ [x]
2. $\sqrt{\{a(y^2 - b)\}} = e$ [y]
3. $\dfrac{\sqrt{(k-m)}}{n} = \dfrac{1}{m}$ [k]
4. $a - bz = z + b$ [z]
5. $\dfrac{x+y}{x-y} = 2$ [x]
6. $\sqrt{\left(\dfrac{a}{z} - c\right)} = e$ [z]
7. $lm + mn + a = 0$ [n]
8. $t = 2\pi\sqrt{\left(\dfrac{d}{g}\right)}$ [d]

9. $t = 2\pi\sqrt{\left(\dfrac{d}{g}\right)}$ [g]

10. $\sqrt{(x^2 + a)} = 2x$ [x]

11. $\sqrt{\left[\dfrac{b(m^2 + a)}{e}\right]} = t$ [m]

12. $\sqrt{\left(\dfrac{x+1}{x}\right)} = a$ [x]

13. $a + b - mx = 0$ [m]

14. $\sqrt{(a^2 + b^2)} = x^2$ [a]

15. $\dfrac{a}{k} + b = \dfrac{c}{k}$ [k]

16. $a - y = \dfrac{b + y}{a}$ [y]

17. $G = 4\pi\sqrt{(x^2 + T^2)}$ [x]

18. $M(ax + by + c) = 0$ [y]

19. $x = \sqrt{\left(\dfrac{y - 1}{y + 1}\right)}$ [y]

20. $a\sqrt{\left(\dfrac{x^2 - n}{m}\right)} = \dfrac{a^2}{b}$ [x]

21. $\dfrac{M}{N} + E = \dfrac{P}{N}$ [N]

22. $\dfrac{Q}{P - x} = R$ [x]

23. $\sqrt{(z - ax)} = t$ [a]

24. $e + \sqrt{(x + f)} = g$ [x]

3.2 Inequalities and regions

Symbols

Here are some examples of inequalities.

$x < 4$ means 'x is *less than* 4'
$y > 7$ means 'y is *greater than* 7'
$z \leqslant 10$ means 'z is *less than or equal to* 10'
$t \geqslant {}^-3$ means 't is *greater than or equal to* ${}^-3$'

With two symbols in one statement look at each part separately.
For example, if n is an **integer** and $3 < n \leqslant 7$, n has to be greater than 3 but at the same time it has to be less than or equal to 7.
So n could be 4, 5, 6 or 7 only.

Example

Illustrate on a number line the range of values of x stated.

(a) $x > 1$ The circle at the left hand end of the range is open. This means that 1 is not included.

(b) $x \leqslant {}^-2$

The circle at ${}^-2$ is filled in to indicate that ${}^-2$ is included.

(c) $1 \leqslant x < 4$

Exercise 3G

1. Write down each statement with either > or < in the box.
   (a) 3 ☐ 7
   (b) 0 ☐ ${}^-2$
   (c) 3·1 ☐ 3·01
   (d) ${}^-3$ ☐ ${}^-5$
   (e) 100 mm ☐ 1 m
   (f) 1 kg ☐ 1 lb

2. Write down the inequality displayed. Use x for the variable.

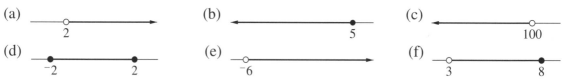

3. Draw a number line to display these inequalities:
   (a) $x \geqslant 7$
   (b) $x < 2\cdot5$
   (c) $1 < x < 7$
   (d) $0 \leqslant x \leqslant 4$
   (e) ${}^-1 < x \leqslant 5$

4. Write an inequality for each statement.
   (a) You must be at least 16 to get married. [Use A for age.]
   (b) Vitamin J1 is not recommended for people over 70 or for children 3 years or under.
   (c) To braise badger the oven temperature should be between 150°C and 175°C. [Use T for temperature.]
   (d) Applicants for training as paratroopers must be at least 1·75 m tall. [Use h for height.]

5. Answer 'true' or 'false':
   (a) n is an integer and $1 < n \leqslant 4$, so n can be 2, 3 or 4.
   (b) x is an integer and $2 \leqslant x < 5$, so x can be 2, 3 or 4.
   (c) p is an integer and $p \geqslant 10$, so p can be 10, 11, 12, 13 …

Inequalities and regions

6. Write two different inequalities involving x which are satisfied by the integers 3, 4, 5, 6, 7.

7. Write down one inequality to show the values of x which satisfy all three of the following inequalities.

   $\boxed{x < 5}$ $\boxed{0 < x < 6}$ $\boxed{3 \leqslant x < 10}$

Solving inequalities

We follow the same procedure used for solving equations except that when we multiply or divide by a **negative** number the inequality is **reversed**.

e.g. $4 > {}^-2$ but multiplying by ${}^-2$, ${}^-8 < 4$

It is best to avoid dividing by a negative number as in the following example.

Example

(a) $2x - 1 > 5$
$2x > 5 + 1$ [add 1]
$x > \dfrac{6}{2}$ [divide by 2]
$x > 3$

(b) $x + 1 < 2x < 3x + 2$
Solve the two inequalities separately:
$x + 1 < 2x \qquad 2x < 3x + 2$
$1 < x \qquad\qquad x < 2$
$\therefore$ The solution is $1 < x < 2$.

Exercise 3H

Solve the following inequalities:

1. $x - 3 > 10$
2. $x + 1 < 0$
3. $5 > x - 7$
4. $2x + 1 \leqslant 6$
5. $3x - 4 > 5$
6. $10 \leqslant 2x - 6$
7. $5x < x + 1$
8. $2x \geqslant x - 3$
9. $4 + x < {}^-4$
10. $3x + 1 < 2x + 5$
11. $2(x + 1) > x - 7$
12. $7 < 15 - x$
13. $9 > 12 - x$
14. $4 - 2x \leqslant 2$
15. $3(x - 1) < 2(1 - x)$
16. $7 - 3x < 0$
17. $\dfrac{x}{3} < {}^-1$
18. $\dfrac{2x}{5} > 3$

19. The height of the picture has to be greater than the width.

Find the range of possible values of x.

height
$2(x + 1)$

width
$(x + 7)$

20. $10 \leqslant 2x \leqslant x + 9$

21. $x < 3x + 2 < 2x + 6$

22. $10 \leqslant 2x - 1 \leqslant x + 5$

23. $3 < 3x - 1 < 2x + 7$

24. $x - 10 < 2(x - 1) < x$

25. $4x + 1 < 8x < 3(x + 2)$

Hint:
In Questions **20** to **25**, solve the two inequalities separately.

26. Sumitra said 'I think of an integer.
        I subtract 14.
        I multiply the result by 5.
        I divide by 2.
        The answer is greater than the number I thought of.'

Write an inequality and solve it to find the smallest number Sumitra thought of.

Squares and square roots in inequalities need care.

The equation $x^2 = 4$ becomes $x = \pm 2$, which is correct.

For the inequality $x^2 < 4$, we might wrongly write $x < \pm 2$.
Consider $x = {}^-3$, say.
${}^-3$ is less than ${}^-2$ and is also less than ${}^+2$.
But $({}^-3)^2$ is not less than 4 and so
$x = {}^-3$ does not satisfy the inequality $x^2 < 4$.
The correct solution for $x^2 < 4$ is ${}^-2 < x < 2$.

Example
Solve the inequality $2x^2 - 1 > 17$.

$2x^2 - 1 > 17$
$\quad 2x^2 > 18$
$\quad\quad x^2 > 9$
$\quad\quad\quad x > 3 \text{ or } x < -3$

[Avoid the temptation to write $x > \pm 3$!]

Exercise 3I

1. The area of the rectangle must be greater than the area of the triangle.
   Find the range of possible values of x.

   For Questions **2** to **8**, list the solutions which satisfy the given condition.

2. $3a + 1 < 20$; a is a positive integer.

3. $b - 1 \geqslant 6$; b is a prime number less than 20.

4. $1 < z < 50$; z is a square number.

5. $2x > {}^-10$; x is a negative integer.

6. $x + 1 < 2x < x + 13$; x is an integer.

7. $0 \leqslant 2z - 3 \leqslant z + 8$; z is a prime number.

8. $\dfrac{a}{2} + 10 > a$; a is a positive even number.

9. Given that $4x > 1$ and $\dfrac{x}{3} \leqslant 1\tfrac{1}{3}$, list the possible integer values of x.

10. State the smallest integer n for which $4n > 19$.

11. Given that ${}^-4 \leqslant a \leqslant 3$ and ${}^-5 \leqslant b \leqslant 4$, find:
    (a) the largest possible value of a^2
    (b) the smallest possible value of ab
    (c) the largest possible value of ab
    (d) the value of b if $b^2 = 25$.

12. For any shape of triangle ABC, complete the statement AB + BC ☐ AC, by writing $<, >$ or $=$ inside the box.

13. Find a simple fraction r such that $\tfrac{1}{3} < r < \tfrac{2}{3}$.

14. Find the largest prime number p such that $p^2 < 400$.

15. Find the integer n such that $n < \sqrt{300} < n + 1$.

16. If $f(x) = 2x - 1$ and $g(x) = 10 - x$ for what values of x is $f(x) > g(x)$?

17. (a) The solution of $x^2 < 9$ is $^-3 < x < 3$.
    [The square roots of 9 are $^-3$ and 3.]
    (b) Copy and complete:
       (i) If $x^2 < 100$, then $\square < x < \square$
       (ii) If $x^2 < 81$, then $\square < x < \square$
       (iii) If $x^2 > 36$, then $x > \square$ or $x < \square$

Solve the inequalities.

18. $x^2 < 25$ 19. $x^2 \leqslant 16$ 20. $x^2 > 1$

21. $2x^2 \geqslant 72$ 22. $3x^2 + 5 > 5$ 23. $5x^2 - 2 < 18$

24. Given $2 \leqslant p \leqslant 10$ and $1 \leqslant q \leqslant 4$, find the range of values of:
    (a) pq (b) $\dfrac{p}{q}$ (c) $p - q$ (d) $p + q$

25. If $2^r > 100$, what is the smallest integer value of r?

26. Given $\left(\dfrac{1}{3}\right)^x < \dfrac{1}{200}$, what is the smallest integer value of x?

27. Find the smallest integer value of x which satisfies $x^x > 10\,000$.

28. What integer values of x satisfy $100 < 5^x < 10\,000$?

29. If x is an acute angle and $\sin x > \frac{1}{2}$, write down the range of values that x can take.

30. If x is an acute angle and $\cos x > \frac{1}{4}$, write down the range of values that x can take.

Shading regions

It is useful to represent inequalities on a graph, particularly where two variables (x and y) are involved.

Example

Draw a sketch graph and shade the area which represents the set of points that satisfy each of these inequalities.

(a) $x > 2$ (b) $1 \leqslant y \leqslant 5$ (c) $x + y \leqslant 8$

 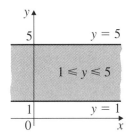 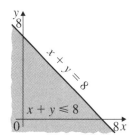

In each graph, the required region is shaded.

In (a), the line $x = 2$ is shown as a broken line to indicate that the points on the line are not included.

In (b) and (c) points on the line **are** included 'in the region' and the lines are drawn unbroken.

To decide which side to shade when the line is sloping, we take a **trial point**. This can be any point which is not actually on the line.

In (c) above, the trial point could be (1, 1).

Is (1, 1) in the region $x + y \leqslant 8$?
It satisfies $x + y < 8$ because $1 + 1 = 2$, which is less than 8.
So below the line is $x + y < 8$. We have shaded the required region.

Exercise 3J

In Questions **1** to **6**, describe the region shaded.

1.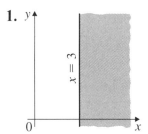

2.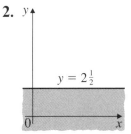

3.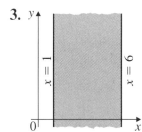

4.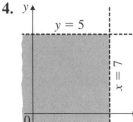
5.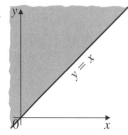
6.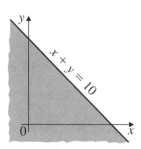

7. The point (1, 1), marked *, lies in the shaded region. Use this as a trial point to describe the unshaded region.

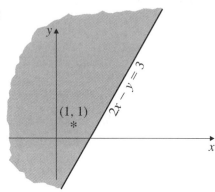

8. The point (3, 1), marked *, lies in the shaded triangle. Use this as a trial point to write down the three inequalities which describe the unshaded region.

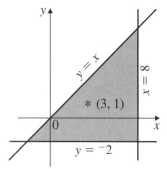

For Questions 9 to 26, draw a sketch graph similar to those above and indicate the set of points which satisfy the inequalities by shading the required region.

9. $2 < x < 7$
10. $0 < y < 3\frac{1}{2}$
11. $^-2 < x < 2$
12. $x < 6$ and $y < 4$
13. $0 < x < 5$ and $y < 3$
14. $1 < x < 6$ and $2 < y < 8$
15. $^-3 < x < 0$ and $^-4 < y < 2$
16. $y < x$
17. $x + y < 5$
18. $y > x + 2$ and $y < 7$
19. $x > 0$ and $y > 0$ and $x + y < 7$
20. $x > 0$ and $x + y < 10$ and $y > x$
21. $8 > y > 0$ and $x + y > 3$
22. $x + 2y < 10$ and $x > 0$ and $y > 0$
23. $3x + 2y < 18$ and $x > 0$ and $y > 0$
24. $x > 0$, $y > x - 2$, $x + y < 10$
25. $3x + 5y < 30$ and $y > \dfrac{x}{2}$
26. $y > \dfrac{x}{2}$, $y < 2x$ and $x + y < 8$

27. The two lines $y = x + 1$ and $x + y = 5$ divide the graph into four regions A, B, C, D. Write down the two inequalities which describe each of the regions A, B, C, D.

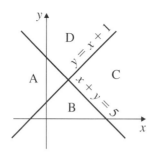

3.3 Curved graphs

Common curves

It is helpful to know the general shape of some of the more common curves.

Quadratic curves have an x^2 term as the highest power of x.

e.g. $y = 2x^2 - 3x + 7$ and $y = 5 + 2x - x^2$

(a) When the x^2 term is positive, the curve is ∪-shaped.

(b) When the x^2 term is negative the curve is an inverted ∩.

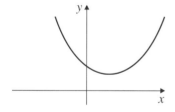
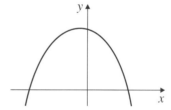

Cubic curves have an x^3 term as the highest power of x.

e.g. $y = x^3 + 7x - 4$ and $y = 8x - 4x^3$

(a) When the x^3 term is positive, the curve can be like one of the two shown below. Notice that as x gets larger, so does y.

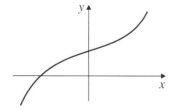
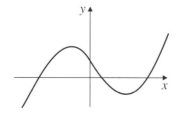

(b) When the x^3 term is negative, the curve can be like one of the two shown below. Notice that as x gets large, y is large but negative.

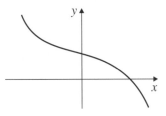 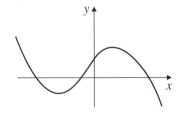

Reciprocal curves have a $\frac{1}{x}$ term.

e.g. $y = \dfrac{12}{x}$ and

$y = \dfrac{6}{x} + 5$

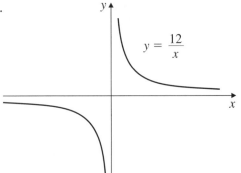

The curve has a break at $x = 0$. The x-axis and the y-axis are called **asymptotes** to the curve. The curve gets very near but never actually touches the asymptotes.

Exponential curves have a term involving a^x, where a is a constant.

e.g. $y = 3^x$

$y = \left(\frac{1}{2}\right)^x$

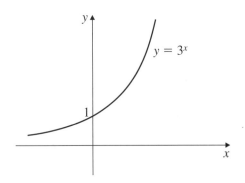

The x-axis is an asymptote to the curve.

Exercise 3K

1. What sort of curves are these? State as much information as you can.

   (a)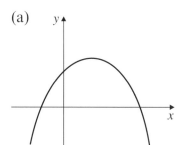
   (b)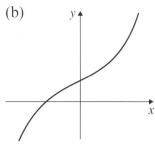
   (c)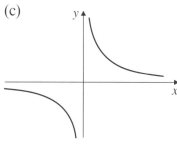
   (d)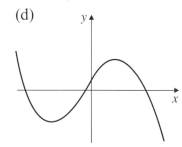
   (e)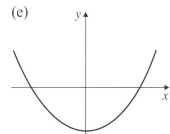
   (f)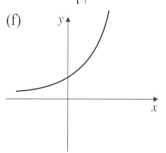

2. Draw the general shape of the following curves.

   (a) $y = 3x^2 - 7x + 11$
   (b) $y = 2^x$
   (c) $y = \dfrac{100}{x}$
   (d) $y = 8x - x^2$
   (e) $y = 10x^3 + 7x - 2$
   (f) $y = \dfrac{1}{x^2}$

3. Here are the equations of the six curves in Question **1**, but not in the correct order.

   (i) $y = \dfrac{8}{x}$
   (ii) $y = 2x^3 + x + 2$
   (iii) $y = 5 + 3x - x^2$
   (iv) $y = x^2 - 6$
   (v) $y = 5^x$
   (vi) $y = 12 + 11x - 2x^2 - x^3$

   Decide which equations fit the curves (a) to (f).

4. Sketch the two curves given and state the number of times the curves intersect.

   (a) $y = x^3$
   $y = 10 - x$
   (b) $y = x^2$
   $y = 10 - x^2$
   (c) $y = 3^x$
   $y = x^3$
   (d) $y = x^3$
   $y = x$

Plotting curved graphs

Example

Draw the graph of the function

$y = 2x^2 + x - 6$, for $^-3 \leqslant x \leqslant 3$.

(a)

x	$^-3$	$^-2$	$^-1$	0	1	2	3
$2x^2$	18	8	2	0	2	8	18
x	$^-3$	$^-2$	$^-1$	0	1	2	3
$^-6$	$^-6$	$^-6$	$^-6$	$^-6$	$^-6$	$^-6$	$^-6$
y	9	0	$^-5$	$^-6$	$^-3$	4	15

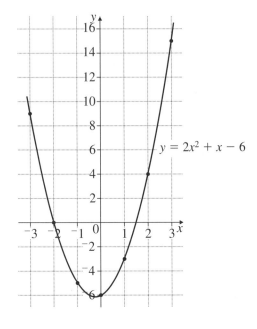

(b) Draw and label axes using suitable scales.
(c) Plot the points and draw a smooth curve through them with a pencil.
(d) Check any points which interrupt the smoothness of the curve.
(e) Label the curve with its equation.

Common errors with graphs

Avoid the following.

A series of 'mini curves'

Flat bottom

Wrong point

Exercise 3L

Draw the graphs of these functions using a scale of 2 cm for 1 unit on the x-axis and 1 cm for 1 unit on the y-axis.

1. $y = x^2 + 2x$, for $^-3 \leqslant x \leqslant 3$
2. $y = x^2 + 4x$, for $^-3 \leqslant x \leqslant 3$
3. $y = x^2 - 3x$, for $^-3 \leqslant x \leqslant 3$
4. $y = x^2 + 2$, for $^-3 \leqslant x \leqslant 3$

5. $y = x^2 - 7$, for $^-3 \leqslant x \leqslant 3$

6. $y = x^2 + x - 2$, for $^-3 \leqslant x \leqslant 3$

7. $y = x^2 + 3x - 9$, for $^-4 \leqslant x \leqslant 3$

8. $y = x^2 - 3x - 4$, for $^-2 \leqslant x \leqslant 4$

9. $y = x^2 - 5x + 7$, for $0 \leqslant x \leqslant 6$

10. $y = 2x^2 - 6x$, for $^-1 \leqslant x \leqslant 5$

11. $y = 2x^2 + 3x - 6$, for $^-4 \leqslant x \leqslant 2$

12. $y = 3x^2 - 6x + 5$, for $^-1 \leqslant x \leqslant 3$

Example

Draw the graph of $y = \dfrac{12}{x} + x - 6$, for $1 \leqslant x \leqslant 8$.

Use the graph to find approximate values for:

(a) the minimum value of $\dfrac{12}{x} + x - 6$

(b) the value of $\dfrac{12}{x} + x - 6$, when $x = 2 \cdot 25$.

Here is the table of values.

x	1	2	3	4	5	6	8
$\dfrac{12}{x}$	12	6	4	3	2·4	2	1·5
x	1	2	3	4	5	6	8
$^-6$	$^-6$	$^-6$	$^-6$	$^-6$	$^-6$	$^-6$	$^-6$
y	7	2	1	1	1·4	2	3·5

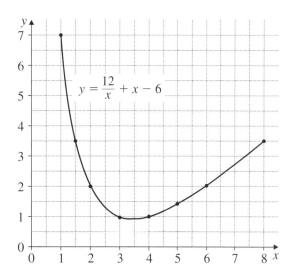

(a) From the graph, the minimum value of $\dfrac{12}{x} + x - 6$ (i.e. y) is approximately 0·9.

(b) At $x = 2 \cdot 25$, y is approximately 1·6.

Exercise 3M

Draw the following curves. The scales given are for one unit of x and y.

1. $y = \dfrac{12}{x}$, for $1 \leqslant x \leqslant 10$ (Scales: 1 cm for x and y)

2. $y = \dfrac{9}{x}$, for $1 \leqslant x \leqslant 10$ (Scales: 1 cm for x and y)

3. $y = \dfrac{12}{x+1}$, for $0 \leqslant x \leqslant 8$ (Scales: 2 cm for x, 1 cm for y)

4. $y = \dfrac{8}{x-4}$, for $^-4 \leqslant x \leqslant 3{\cdot}5$ (Scales: 2 cm for x, 1 cm for y)

5. $y = \dfrac{x}{x+4}$, for $^-3{\cdot}5 \leqslant x \leqslant 4$ (Scales: 2 cm for x and y)

6. $y = 3^x$, for $^-3 \leqslant x \leqslant 3$ (Scales: 2 cm for x, $\tfrac{1}{2}$ cm for y)

7. $y = \left(\tfrac{1}{2}\right)^x$, for $^-4 \leqslant x \leqslant 4$ (Scales: 2 cm for x, 1 cm for y)

8. $y = 5 + 3x - x^2$, for $^-2 \leqslant x \leqslant 5$ (Scales: 2 cm for x, 1 cm for y)
   Find:
   (a) the maximum value of the function $5 + 3x - x^2$,
   (b) the two values of x for which $y = 2$.

9. $y = \dfrac{15}{x} + x - 7$, for $1 \leqslant x \leqslant 7$. (Scales: 2 cm for x and y)
   From your graph find
   (a) the minimum value of y,
   (b) the y-value when $x = 5{\cdot}5$.

10. $y = x^3 - 2x^2$, for $0 \leqslant x \leqslant 4$. (Scales: 2 cm for x, $\tfrac{1}{2}$ cm for y)
    From your graph find:
    (a) the y-value at $x = 2{\cdot}5$,
    (b) the x-value at $y = 15$.

Exercise 3N Mixed questions

1. A farmer has 60 m of wire fencing which he uses to make a rectangular pen for his sheep. He uses a stone wall as one side of the pen so the wire is used for only 3 sides of the pen.

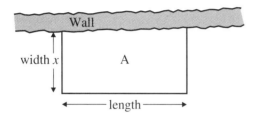

   If the width of the pen is x m, what is the length (in terms of x)?
   What is the area A of the pen?
   Draw a graph with area A on the vertical axis and the width x on the horizontal axis. Take values of x from 0 to 30.
   What dimensions should the pen have if the farmer wants to enclose the largest possible area?

2. A ball is thrown in the air so that t seconds after it is thrown, its height h metres above its starting point is given by the function $h = 25t - 5t^2$. Draw the graph of the function of $0 \leqslant t \leqslant 6$, plotting t on the horizontal axis with a scale of 2 cm to 1 second, and h on the vertical axis with a scale of 2 cm for 10 metres.
   Use the graph to find:
   (a) the time when the ball is at its greatest height,
   (b) the greatest height reached by the ball,
   (c) the interval of time during which the ball is at a height of more than 30 m.

3. Draw the graph of $y = x + \dfrac{1}{x}$ for
   $x = {}^-4, {}^-3, {}^-2, {}^-1, {}^-0{\cdot}5, {}^-0{\cdot}25, 0{\cdot}25, 0{\cdot}5, 1, 2, 3, 4$
   (Scales: 2 cm for x and y)

4. Draw the graph of $y = \dfrac{2^x}{x}$, for ${}^-4 \leqslant x \leqslant 7$, including $x = {}^-0{\cdot}5,\ x = 0{\cdot}5$.
   (Scales: 1 cm to 1 unit for x and y)

5. Consider the equation $y = \dfrac{1}{x}$.

   When $x = \frac{1}{2}$, $y = \frac{1}{\frac{1}{2}} = 2$.

   When $x = \frac{1}{100}$, $y = \frac{1}{\frac{1}{100}} = 100$.

   Draw the graph of $y = \dfrac{1}{x}$ for
   $x = {}^-4, {}^-3, {}^-2, {}^-1, {}^-0.5, {}^-0.25, 0.25, 0.5, 1, 2, 3, 4$
   (Scales: 2 cm for x and y)

   > As the denominator of $\frac{1}{x}$ gets smaller, the answer gets larger. An 'infinitely small' denominator gives an 'infinitely large' answer.
   > We write $\frac{1}{0} \to \infty$ '$\frac{1}{0}$ tends to an infinitely large number.'

6. A firm makes a profit of P thousand pounds from producing x thousand tiles.
   Corresponding values of P and x are given below

x	0	0·5	1·0	1·5	2·0	2·5	3·0
P	⁻1·0	0·75	2·0	2·75	3·0	2·75	2·0

   Using a scale of 4 cm to one unit on each axis, draw the graph of P against x. [Plot x on the horizontal axis.]
   Use your graph to find:
   (a) the number of tiles the firm should produce in order to make the maximum profit,
   (b) the minimum number of tiles that should be produced to cover the cost of production,
   (c) the range of values of x for which the profit is more than £2850.

7. At time $t = 0$, one bacteria is placed in a culture in a laboratory. The number of bacteria doubles every 10 minutes.
   (a) Draw a graph to show the growth of the bacteria from $t = 0$ to $t = 120$ mins.
      Use a scale of 1 cm to 10 minutes across the page and 5 cm to 1000 units up the page.
   (b) Use your graph to estimate the time taken to reach 800 bacteria.

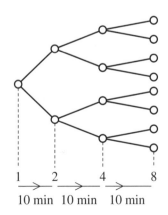

8. An economist estimates that the population of country A will be multiplied by 1·2 every 10 years and that the population of country B will be multiplied by 1·05 every 10 years. In 1980 the populations of A and B were 36 million and 100 million respectively.
   (a) Draw a graph to show the projected populations of the two countries from 1980 to 2060.
   Use a scale of 2 cm to 10 years across the page and 2 cm to 20 million up the page.
   (b) Estimate when the population of A will exceed the population of B for the first time.

9. Draw the graph of $\dfrac{x^4}{4^x}$, for $x = {}^-1, -\tfrac{3}{4}, -\tfrac{1}{2}, -\tfrac{1}{4}, 0, \tfrac{1}{4}, \tfrac{1}{2}, \tfrac{3}{4},$ 1, 1·5, 2, 2·5, 3, 4, 5, 6, 7.
   (Scales: 2 cm to 1 unit for x, 5 cm to 1 unit for y)
   (a) For what values of x is the gradient of the function zero?
   (b) For what values of x is $y = 0.5$?

3.4 Algebraic Indices

Example
Solve the equations (a) $2^x = 16$ (b) $5^y = \tfrac{1}{25}$
(c) $6^z = (\sqrt{6})^3$ (d) $3^{x-1} = 81$

We use the principle that if
$$a^x = a^y \text{ then } x = y.$$

(a) $2^x = 16 = 2^4$ (b) $5^y = \tfrac{1}{25} = 5^{-2}$
   $\therefore\ x = 4$ $\therefore\ y = {}^-2$
(c) $6^z = (\sqrt{6})^3 = 6^{\tfrac{3}{2}}$ (d) $3^{x-1} = 81 = 3^4$
   $\therefore\ z = \tfrac{3}{2}$ $\therefore\ x = 5$

Exercise 30
Solve these equations for x.

1. $2^x = 8$
2. $3^x = 81$
3. $5^x = \tfrac{1}{5}$
4. $10^x = \tfrac{1}{100}$
5. $3^{-x} = \tfrac{1}{27}$
6. $4^x = 64$
7. $6^{-x} = \tfrac{1}{6}$
8. $100\,000^x = 10$
9. $12^x = 1$
10. $10^x = 0.0001$
11. $2^x + 3^x = 13$
12. $(\tfrac{1}{2})^x = 32$
13. $5^{2x} = 25$
14. $1\,000\,000^{3x} = 10$

15. Here is a sketch of $y = 3^x$.

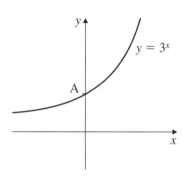

(a) Write down the coordinates of point A.
(b) Write down the solution of the equation $3^x = 5^x$.

16. Find a, b and c.
(a) $\frac{1}{8} = 2^a$ (b) $b^7 = 1$ (c) $2^{c-1} = 16$

17. Find p, q and r.
(a) $x^{\frac{1}{2}} \times x^2 = x^p$ (b) 10% of $100^2 = q$
(c) $x^{-\frac{1}{4}} \times x^2 = x^r$

18. Solve the equations.
(a) $10n^3 = 640$ (b) $10^n = 0\cdot 1$ (c) $2n^3 = 0$

19. Find two solutions of the equation $x^2 = 2^x$.

20. Use a calculator to find solutions correct to 3 significant figures.
(a) $x^x = 100$ (b) $x^x = 10\,000$

21. The last digit of 7^3 is 3 [$7^3 = 343$].
(a) Copy and complete the table below, which gives the last digit of 7^n.

n	1	2	3	4	5	6	7	8	9	10
Last digit of 7^n	7	9	3		7			1		

(b) Write down the last digit of:
(i) 7^{48} (ii) 7^{101} (iii) 49^{35}

22. It is given that $10^x = 3$ and $10^y = 7$. What is the value of 10^{x+y}?

23. In a laboratory we start with 2 cells in a dish. The number of cells in the dish doubles every 30 minutes. This is an example of **exponential growth**.
   (a) How many cells are in the dish after four hours?
   (b) After what time are there 2^{13} cells in the dish?
   (c) After $10\frac{1}{2}$ hours there are 2^{22} cells in the dish and an experimental fluid is added which eliminates half of the cells. How many cells are left?

24. Steve's bike is ill. Its computer-controlled ignition system has a virus. The doctor has advised Steve to keep the bike warm, in which case the number of germs in the bike will decay exponentially and will be $1\,000\,000 \times 2^{-n}$ after n hours.
   (a) How many germs will there be after 10 hours?
   (b) The bike will be cured when it contains less than one germ. After how many hours will it be cured?

25. A bank pays compound interest on money invested in an account. After n years a sum of £2000 will rise to £2000 × $1 \cdot 08^n$. This represents exponential growth.
   (a) How much money is in the account after three years?
   (b) After how many years will the original £2000 have nearly doubled?

Summary

1. You can change the subject of a formula.

2. You can solve inequalities.

Check out AS3

1. (a) Find s if $v^2 = u^2 + 2as$.
   (b) Find v if $\dfrac{1}{u} - \dfrac{1}{v} = \dfrac{1}{f}$.

2. Solve: (a) $10 \leqslant 2x - 5$ (b) $x^2 \geqslant 25$

3. You can recognise standard graphs.

3. (i) 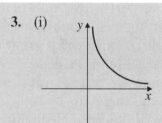 (ii)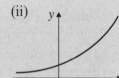

(iii)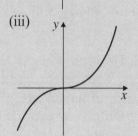

Which of the sketches shown is:
(a) $y = x^3$ (b) $y = \dfrac{1}{x}$ (c) $y = 2^x$?

4. You can draw graphs of curves.

4. Draw the graph of
$y = 2x^2 - 7x + 3$
for values of x from $^-1$ to $^+5$.

5. You can use indices to solve equations.

5. Solve: (a) $2^x = 16$
(b) $3^x = \dfrac{1}{81}$

Revision exercise AS3

1. (a) List all the solutions of the inequality
   $^-6 < 3n \leqslant 8$ where n is an integer.
   (b) Solve the inequality $y^2 > 4$. [SEG]

2. (a) A formula is given as $t = 7p - 50$.
   Rearrange the formula to make p the subject.
   (b) Make x the subject of $y = kx + 3x$. [SEG]

3. The graph of $y = x^3 - x^2 - 6x$ has been drawn on the grid below.

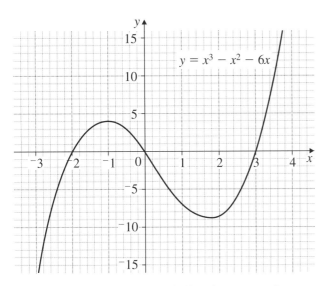

Use the graph to solve the following equations.

(a) $x^3 - x^2 - 6x = 2$
(b) $x^3 - x^2 - 7x - 10 = 0$ [SEG]

4. Rearrange this formula to make x the subject.
$$y = \frac{x}{5+x}$$ [AQA]

5. (a) Simplify $\dfrac{(x^{-\frac{1}{2}})^3}{x^{-2}}$ leaving your answer as a power of x.

   (b) Find the value of n such that $25^n = 5^{\frac{1}{4}}$. [NEAB]

6. Solve the inequality
$$(x+3)^2 < x^2 + 2x + 7.$$

Do **not** use a trial and improvement method. [NEAB]

AS4 Shape, Space and Measures 2

This unit will show you how to:

- Use isometric paper and draw nets
- Understand symmetry
- Use the trigonometric ratios
- Classify formulae by their dimensions
- Describe reflections, rotations, translations, enlargements
- Combine transformations
- Use vectors
- Use the sine and cosine rules
- Use the circle theorems and learn appropriate proofs

Before you start:

You should know how to...	Check in AS4
1. Transform equations. For example, find x when $28 = \dfrac{7}{x}$. $28 = \dfrac{7}{x}$ $28x = 7$ $x = \dfrac{7}{28} = \dfrac{1}{4}$	**1.** Find p when $37 = \dfrac{4}{p}$.
2. Substitute in formulae. For example, find V when $V = \dfrac{4}{3}\pi r^3$ and $r = 6\,\text{cm}$. $V = \dfrac{4}{3}\pi 6^3 \,\text{cm}^3 = \dfrac{4}{3}\pi \times 216\,\text{cm}^3$ $= 288\pi\,\text{cm}^3 = 905\,\text{cm}^3$	**2.** Find V when $V = 2\pi r^2 h$ and $r = 7\,\text{cm}$, $h = 8\,\text{cm}$.

4.1 Drawing 3-D shapes, symmetry

Isometric drawing

When we draw a solid on paper we are making a 2-D representation of a 3-D object. Here are two pictures of the same cuboid, measuring 4 × 3 × 2 units.

(a) On ordinary squared paper

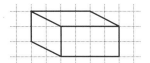

(b) On isometric paper
[a grid of equilateral triangles]

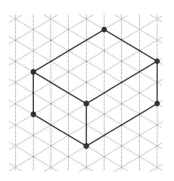

The dimensions of the cuboid cannot be taken from the first picture but they can be taken from the picture drawn on isometric paper. Instead of isometric paper you can also use 'triangular dotty' paper like this:
Be careful to use it the right way round (as shown here).

Exercise 4A

In Questions **1** to **3** the objects consist of 1 cm cubes joined together. Draw each object on isometric paper (or 'triangular dotty' paper).
Questions **1** and **2** are already drawn on isometric paper.

1. 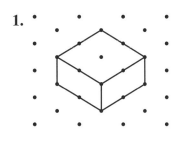   2. 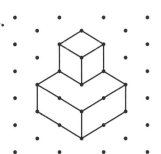   3.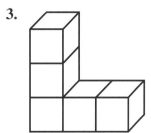

4. Here are two shapes made using four multilink cubes.

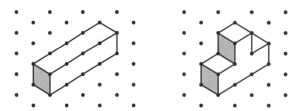

Make and then draw four more shapes, using four cubes, which are different from the two above.

5. You need 16 small cubes.
Make the two shapes shown and then arrange them into a 4 × 4 × 1 cuboid by adding a third shape, which you have to find. Draw the third shape on isometric paper. (There are two possible shapes.)

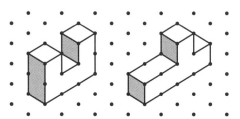

6. Three views of object A are shown.

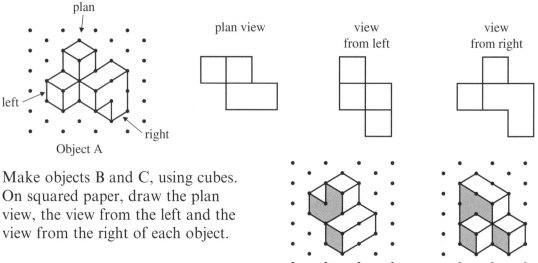

Make objects B and C, using cubes. On squared paper, draw the plan view, the view from the left and the view from the right of each object.

7. Make each of the objects whose views are given below. Draw an isometric picture of each one.

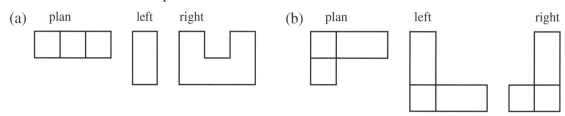

Nets

If the cube here was made of cardboard, and you cut along some of the edges and laid it out flat, you would have the **net** of the cube.

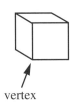
vertex

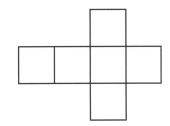

A cube has: 8 vertices;
            6 faces;
            12 edges.

Here is the net for a square-based pyramid.

This pyramid has: 5 vertices;
                   5 faces;
                   8 edges.

Exercise 4B

1. Which of the nets below can be used to make a cube?

(a) 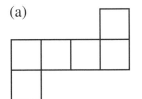   (b) 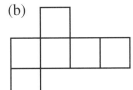   (c) 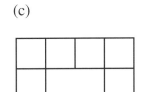   (d)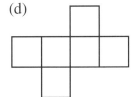

2. The numbers on opposite faces of a dice add up to 7. Take one of the possible nets for a cube from Question **1** and show the number of dots on each face.

3. Here we have started to draw the net of a cuboid (a closed rectangular box) measuring 4 cm × 3 cm × 1 cm.
Copy and then complete the net.

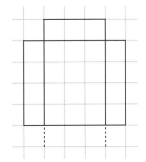

4. The diagram shown needs one more square to complete the net of a cube. Copy and cut out the shape and then draw the **four** possible nets which would make a cube.

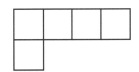

5. Describe the solid formed from each of these nets. State the number of vertices and faces for each object.

(a)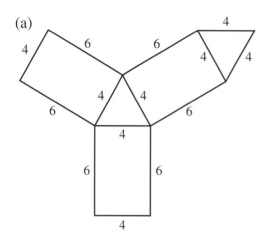

(b)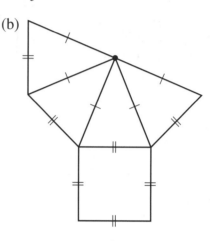

6. Sketch a possible net for each of the following:
   (a) a cuboid measuring 5 cm by 2 cm by 8 cm
   (b) a prism 10 cm long whose cross-section is a right-angled triangle with sides 3 cm, 4 cm and 5 cm.

7. The diagram shows the net of a pyramid. The base is shaded. The lengths are in cm.
   (a) How many edges will the pyramid have?
   (b) How many vertices will it have?
   (c) Find the lengths a, b, c, d.
   (d) Use the formula
   $$V = \frac{1}{3} \text{ base area} \times \text{height}$$
   to calculate the volume of the pyramid.

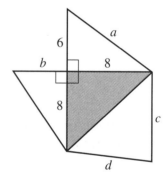

8. This is the net of a square-based pyramid. What are the lengths a, b, c, x, y?

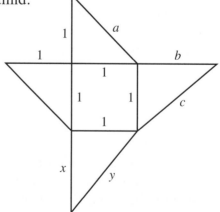

Symmetry

(a) Line symmetry

The letter M has one line of symmetry, shown dotted.

(b) Rotational symmetry

The shape may be turned about O into three identical positions. It has rotational symmetry of order three.

Exercise 4C

For each shape state:
(a) the number of lines of symmetry
(b) the order of rotational symmetry.

1.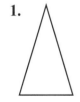
2.
3.
4.

5.
6.
7.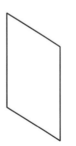
8.

9.
10.
11.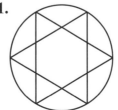
12.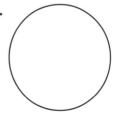

Planes of symmetry

- A plane of symmetry divides a 3-D shape into two congruent shapes.

Note: One shape must be a mirror image of the other shape.

Example

The shaded plane is a plane of symmetry of the cube.

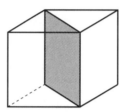

Exercise 4D

1. How many planes of symmetry does this cuboid have?

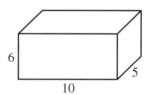

2. How many planes of symmetry do these shapes have?

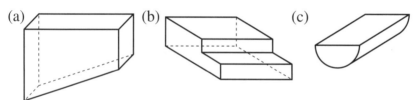

(a) (b) (c)

3. How many planes of symmetry does a cube have?

4. Draw a pyramid with a square base so that the vertex of the pyramid is vertically above the centre of the square base.
Show any planes of symmetry by shading.

5. The diagrams show the plan view and the side view of an object.

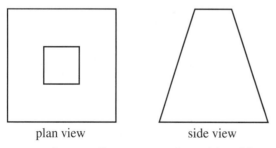

plan view   side view

How many planes of symmetry has this object?

4.2 Trigonometry

Trigonometry is used to calculate sides and angles in triangles.

Right-angled triangles

The side opposite the right angle is called
the **hypotenuse** (use hyp.).
It is the longest side.

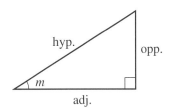

The side opposite the marked angle m is
called the **opposite** (use opp.).
The other side is called the **adjacent** (use adj.).

Consider two triangles, one of which is an enlargement of
the other.
It is clear that, for the angle 30°, the

$$\text{ratio} = \frac{\text{opposite}}{\text{hypotenuse}} = \frac{6}{12} = \frac{2}{4} = \frac{1}{2}$$

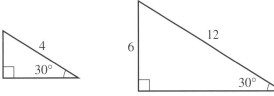

This is the same for both triangles.

Sine, cosine, tangent

Three important ratios are defined for angle x.

$$\sin x = \frac{\text{opp.}}{\text{hyp.}} \quad \cos x = \frac{\text{adj.}}{\text{hyp.}} \quad \tan x = \frac{\text{opp.}}{\text{adj.}}$$

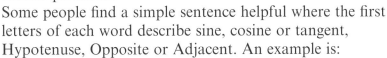

It is important to get the letters in the
correct positions.
Some people find a simple sentence helpful where the first
letters of each word describe sine, cosine or tangent,
Hypotenuse, Opposite or Adjacent. An example is:

Silly Old Harry Caught A Herring Trawling Off Afghanistan

e.g. S O H $\quad \sin = \dfrac{\text{opp.}}{\text{hyp.}}$

Finding the length of an unknown side

Example 1

Find the side marked x.

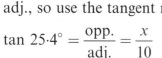

Label the sides of the triangle hyp., opp., adj.

In this example, we have only opp. and adj., so use the tangent ratio.

$$\tan 25 \cdot 4° = \frac{\text{opp.}}{\text{adj.}} = \frac{x}{10}$$

Find tan 25·4° on a calculator:

$$0 \cdot 4748 = \frac{x}{10}$$

Solve for x.

$$x = 10 \times 0 \cdot 4748 = 4 \cdot 748$$
$$x = 4 \cdot 75 \, \text{cm} \, (3 \, \text{s.f.})$$

Example 2

Find the side marked z.

Label hyp., opp., adj.

we have opp. and hyp. so use the sine ratio.

$$\sin 31 \cdot 3° = \frac{\text{opp.}}{\text{hyp.}} = \frac{7 \cdot 4}{z}$$

Multiply by z. $z \times (\sin 31 \cdot 3°) = 7 \cdot 4$

$$z = \frac{7 \cdot 4}{\sin 31 \cdot 3°}$$
$$z = 14 \cdot 2 \, \text{cm} \, (3 \, \text{s.f.})$$

Exercise 4E

Find the lengths of the sides marked with a letter.
Give your answers to three significant figures.

1.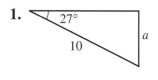

2.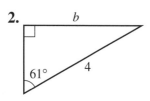

3.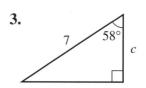

4.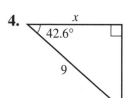

5.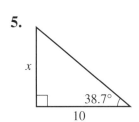

6.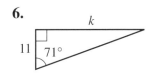

7.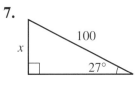

8.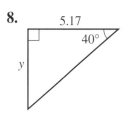

9.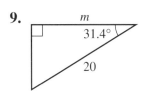

10.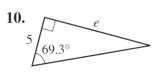

11.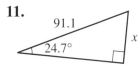

12.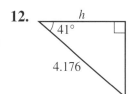

13.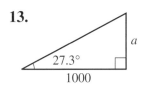

14.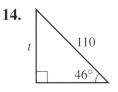

15.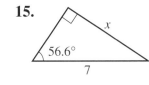

16.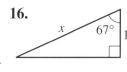

17.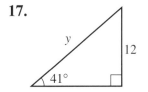

18.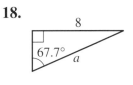

19.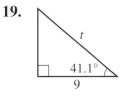

20.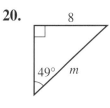

21.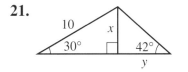

22.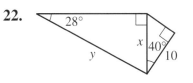

23.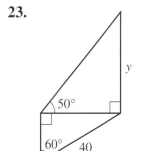

24.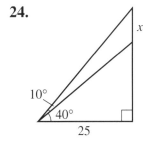

25.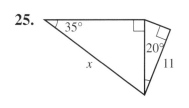

26.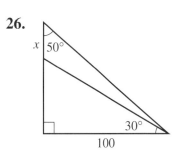

Finding an unknown angle

Example

Find the angle marked *m* to one decimal place.

Label the sides of the triangle hyp., opp. and adj. in relation to angle *m*.

Here only adj. and hyp. are known, so use cosine.

$$\cos m = \frac{\text{adj.}}{\text{hyp.}} = \frac{4}{5}$$

Using a calculator, the angle *m* can be found as follows.

(a) Press $\boxed{4}$ $\boxed{\div}$ $\boxed{5}$ $\boxed{=}$

(b) Press $\boxed{\text{INV}}$ and then $\boxed{\cos}$

This gives the angle as 36·869 898

$$m = 36.9° \text{ (1 d.p.)}$$

Exercise 4F

For Questions **1** to **15**, find the angle marked with a letter.
All lengths are in cm.

1.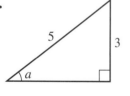

2.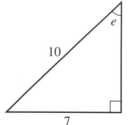

3.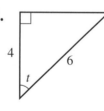

4.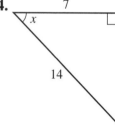

5.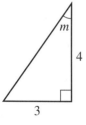

6.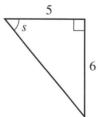

7.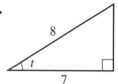

8.

9.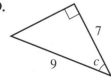

10.

11.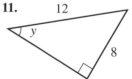

12.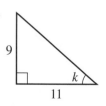

In Questions **13** to **17**, the triangle has a right angle at the middle letter.

13. In △ABC, BC = 4, AC = 7. Find $\hat{A}$.

14. In △DEF, EF = 5, DF = 10. Find $\hat{F}$.

15. In △GHI, GH = 9, HI = 10. Find $\hat{I}$.

16. In △JKL, JL = 5, KL = 3. Find $\hat{J}$.

17. In △MNO, MN = 4, NO = 5. Find $\hat{M}$.

In Questions **18** to **21**, find the angle x.

18. 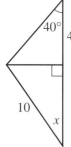   **19.** 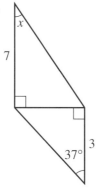   **20.** 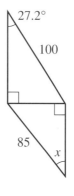   **21.**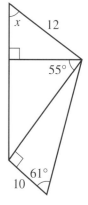

22. In △ABC
$\widehat{BAC} = 42°$ and AB = 12 cm.

M is the mid-point of CB. Find the angle x to 1 d.p.

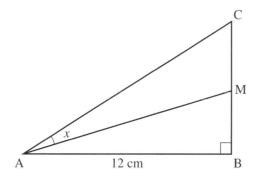

Bearings

Bearings are measured **clockwise** from North.

(a) 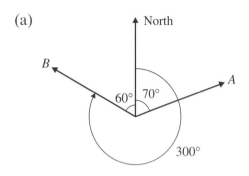   (b)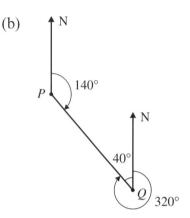

Ship *A* sails on a bearing 070°.   The bearing of *Q* from *P* is 140°.
Ship *B* sails on a bearing 300°.   The bearing of *P* from *Q* is 320°.

Angles of elevation and depression

(a) 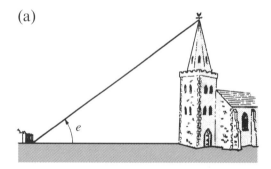   (b)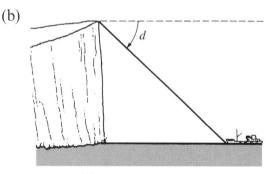

e is the angle of **elevation** of the steeple from the gate.

d is the angle of **depression** of the boat from the cliff top.

Exercise 4G

In this exercise, start each question by drawing a clear diagram.

1. A ladder of length 6 m leans against a vertical wall so that the base of the ladder is 2 m from the wall. Calculate the angle between the ladder and the wall.

2. A ladder of length 8 m rests against a wall so that the angle between the ladder and the wall is 31°. How far is the base of the ladder from the wall?

3. A point P is 90 m away from a vertical flagpole, which is 11 m high. What is the angle of elevation to the top of the flagpole from P?

4. A chord AB of length 10 cm is drawn in a circle of radius 7 cm, centre O. Work out the angle AOB.

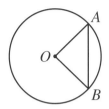

5. A ladder of length 5 m rests against a vertical wall. For safety reasons the angle between a ladder and the ground must be between 65° and 75°. Work out the possible safe distances from the foot of a ladder to the base of the wall.

6. An isosceles triangle has sides of length 8 cm, 8 cm and 5 cm. Find the angle between the two equal sides.

7. The angles of an isosceles triangle are 66°, 66° and 48°. If the shortest side of the triangle is 8·4 cm, find the length of one of the two equal sides.

8. An arctic explorer reached the North Pole and decided to erect his national flag and to sing his national anthem, which lasted a rather chilly four minutes.
His eye was 1·6 m from the ground and the angle of elevation to the top of the 3 m flagpole was 28°. How far from the base of the flagpole did he stand?

9. A ship sails 35 km on a bearing of 042°.
   (a) How far north has it travelled?
   (b) How far east has it travelled?

10. A ship sails 200 km on a bearing of 243·7°.
    (a) How far south has it travelled?
    (b) How far west has it travelled?

11. An aircraft flies 400 km from a point O on a bearing of 025° and then 700 km on a bearing of 080° to arrive at B.
    (a) How far north of O is B?
    (b) How far east of O is B?
    (c) Find the distance and bearing of B from O.

12. Find the acute angle between the diagonals of a rectangle whose sides are 5 cm and 7 cm.

13. A kite flying at a height of 55 m is attached to a string which makes an angle of 55° with the horizontal. What is the length of the string?

14. A rocket flies 10 km vertically, then 20 km at an angle of 15° to the vertical and finally 60 km at an angle of 26° to the vertical. Calculate the vertical height of the rocket at the end of the third stage.

15. Find x, given $AD = BC = 6$ m.

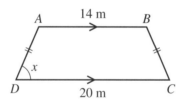

16. Ants can hear each other up to a range of 2 m. An ant A, 1 m from a wall sees her friend B about to be eaten by a spider. If the angle of elevation of B from A is 62°, will the spider have a meal or not? (Assume B escapes if he hears A calling.)

17. A regular pentagon is inscribed in a circle of radius 7 cm.
    Find the angle a and then the length of a side of the pentagon.

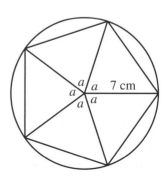

Exercise 4H

1. Find the length d.

   (a) 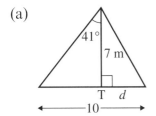   (b)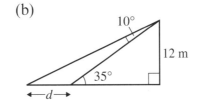

For Questions **2** to **5**, plot the points for each question on a sketch graph with x- and y-axes drawn to the same scale.

2. For the points $A(5, 0)$ and $B(7, 3)$, calculate the angle between AB and the x-axis.

3. For the points $C(0, 2)$ and $D(5, 9)$, calculate the angle between CD and the y-axis.

4. For the points $A(3, 0)$, $B(5, 2)$ and $C(7, {}^-2)$, calculate the angle BAC.

5. For the points $P(2, 5)$, $Q(5, 1)$ and $R(0, {}^-3)$, calculate the angle PQR.

6. From the top of a tower of height 75 m, a guard sees two prisoners, both due West of him.

   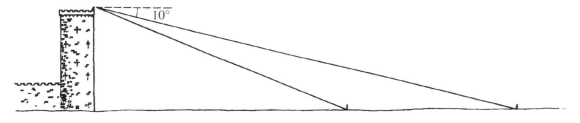

   If the angles of depression of the two prisoners are 10° and 17°, calculate the distance between them.

7. A hedgehog wishes to cross a road without being run over. He observes the angle of elevation of a lamp post on the other side of the road to be 27° from the edge of the road and 15° from a point 10 m back from the road. How wide is the road? If he can run at 1 m/s, how long will he take to cross?
   If cars are travelling at 20 m/s, how far apart must they be if he is to survive?

8. A symmetrical drawbridge is shown below. When lowered, the roads *AX* and *BY* just meet in the middle. Calculate the length *XY*.

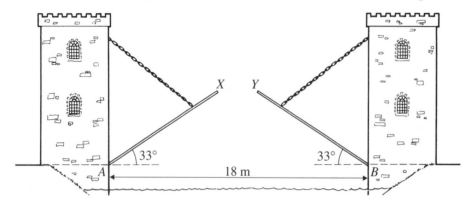

9. Here are three triangles.

 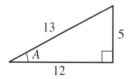 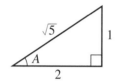

   (a) For each triangle write down the values of sin *A* and cos *A* and work out $(\sin A)^2 + (\cos A)^2$.
   (b) Write down what you notice.
   (c) Does the same result hold for a general triangle with sides *a*, *b*, *c*?

10. The diagram shows part of a polygon with *n* sides and centre *O*.

    (a) What is angle *AOB* in terms of *n*?
    (b) M is the mid-point of *AB*. What is angle *MOB* in terms of *n*?
    (c) Find the length *MB* and hence the length *AB* in terms of *n*.
    (d) Find an expression for the perimeter of the polygon.
    (e) Work out the perimeter for $n = 100$ and $n = 1000$. What do you notice?

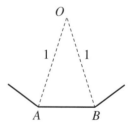 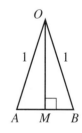

11. The diagram shows the cross section of a rectangular fish tank. When *AB* is inclined at 40°, the water just comes up to *A*.

    The tank is then lowered so that *BC* is horizontal. What is now the depth or water in the tank?

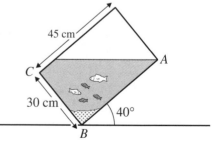

Three-dimensional problems

A projection is like a shadow on a surface or plane.

- The angle between a line and a plane is the angle between the line and its **projection** in the plane.

PA is a vertical pole standing at the vertex A of a horizontal rectangular field $ABCD$.

The angle between the line PC and the plane $ABCD$ is found as follows:

(a) The projection of PC in the plane $ABCD$ is AC.
(b) The angle between the line PC and the plane $ABCD$ is the angle PCA.

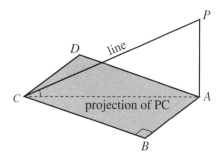

Example

A rectangular box with top $WXYZ$ and base $ABCD$ has $AB = 6$ cm, $BC = 8$ cm and $WA = 3$ cm. Calculate:

(a) the length of AC
(b) the angle between the line WC and the plane $ABCD$.

(a) Redraw triangle ABC.
$AC^2 = 6^2 + 8^2 = 100$
$AC = 10$ cm.

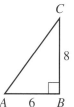

> Always draw a large, clear diagram. It is often helpful to redraw the particular triangle which contains the length or angle to be found.

(b) The projection of WC on the plane $ABCD$ is AC. The angle required is $W\hat{C}A$.

Redraw triangle WAC.
let $W\hat{C}A = \theta$.

$\tan \theta = \frac{3}{10}$

$\theta = 16{\cdot}7°$.

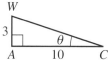

The angle between WC and the plane $ABCD$ is $16{\cdot}7°$.

Exercise 4I

1. In the rectangular box shown, find:
   (a) AC
   (b) AR
   (c) the angle between AC and AR.

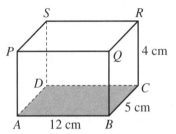

2. A vertical pole BP stands at one corner of a horizontal rectangular field as shown.
   If $AB = 10$ m, $AD = 5$ m and the angle of elevation of P from A is $22°$, calculate:
   (a) the height of the pole
   (b) the angle of elevation of P from C
   (c) the length of a diagonal of the rectangle $ABCD$
   (d) the angle of elevation of P from D.

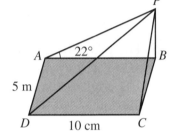

3. In the cube shown, find:
   (a) BD
   (b) AS
   (c) BS
   (d) the angle SBD.

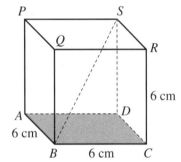

4. In the square-based pyramid shown, V is vertically above the middle of the base, $AB = 10$ cm and $VC = 20$ cm. Find:
   (a) AC
   (b) the height of the pyramid
   (c) the angle between VC and the base $ABCD$.

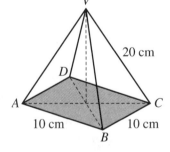

5. The figure shows a cuboid. Calculate:
   (a) the lengths of AC and AY
   (b) the angle between AY and the plane $ABCD$.

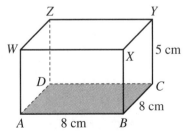

6. The figure shows a cuboid. Calculate:
   (a) the lengths ZX and KX
   (b) the angle between NX and the plane $WXYZ$
   (c) the angle between KY and the plane $KLWX$.

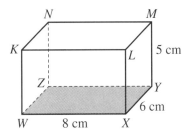

7. In the wedge shown, $PQRS$ is perpendicular to $ABRQ$; $PQRS$ and $ABRQ$ are rectangles with $AB = QR = 6$ m, $BR = 4$ m, $RS = 2$ m. Find:
   (a) BS (b) AS
   (c) the angle between AS and the plane $ABRQ$.

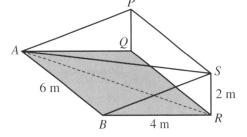

8. The pyramid $VPQRS$ has a square base PQRS. $VP = VQ = VR = VS = 12$ cm and $PQ = 9$ cm. Calculate the angle between VP and the plane $PQRS$.

9. The pyramid $VABCD$ has a rectangular base $ABCD$. $VA = VB = VC = VD = 15$ cm, $AB = 14$ cm and $BC = 8$ cm. Calculate:
   (a) the angle between VB and the plane $ABCD$
   (b) the angle between VX and the plane $ABCD$ where X is the mid-point of BC.

10. In the diagram A, B and O are points in a horizontal plane and P is vertically above O, where $OP = h$ m. A is due West of O, B is due South of O and $AB = 60$ m. The angle of elevation of P from A is $25°$ and the angle of elevation of P from B is $33°$.
    (a) Find the length AO in terms of h.
    (b) Find the length BO in terms of h.
    (c) Find the value of h.

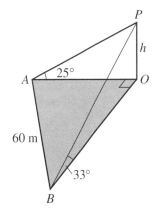

11. The angle of elevation of the top of a tower is $38°$ from a point A due South of it. The angle of elevation of the top of the tower from another point B, due East of the tower is $29°$. Find the height of the tower if the distance AB is 50 m.

4.3 Dimensions of formulae

Here are some formulae for finding volumes, areas and lengths met earlier in this book.

sphere: volume $= \frac{4}{3}\pi r^3$
surface area $= 4\pi r^2$

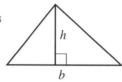
triangle: area $= \frac{1}{2}bh$

cylinder: volume $= \pi r^2 h$
curved surface area $= 2\pi rh$

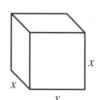
cube: volume $= x^3$
surface area $= 6x^2$

cone: volume $= \frac{1}{3}\pi r^2 h$
curved surface area $= \pi rl$

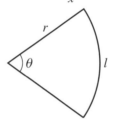
sector area $= \frac{\theta}{360} \times \pi r^2$
arc length $= \frac{\theta}{360} \times 2\pi r$

All the symbols in bold type are lengths. They have the **dimension** of length and are measured in cm, metres, km, etc.

(a) All the formulae for volume have **three** lengths multiplied together. They have three dimensions.
(b) All the formulae for area have **two** lengths multiplied together. They have two dimensions.
(c) Any formula for the length of an object will involve just **one** length (or one dimension).

Remember:
Numbers, like 3, ½, or π, have **no dimensions**.

A formula can have more than one term.
The formula $A = \pi r^2 + 3rd$ has two terms and each term has two dimensions.

It is **not** possible to have a mixture of terms some with, say, two dimensions and some with three dimensions.
So the formula $A = \pi r^2 + 3r^2 d$ could not possibly represent area. Nor could it represent volume.

We can use these facts to check that any formula we may be using has the correct number of dimensions.

The formula $z = \frac{2\pi r^2 h}{L}$ has three dimensions on the top line and one dimension on the bottom. The dimensions can be 'cancelled' so the expression for z has only two dimensions and can only represent an area.

Example

Here are four formulae where the letters c, d, r represent lengths:

(a) $t = 3c^2$
(b) $k = \dfrac{\pi}{3}r^3 + 4r^2 d$
(c) $m = \pi(c + d)$
(d) $f = 4c + 3cd$

State whether the formula gives:
    (i) a length
    (ii) an area
    (iii) a volume
    (iv) an impossible expression.

(a) $t = 3c^2 = 3c \times c$
This has **two** dimensions so t is an **area**.
[Notice that the number '3' has no dimensions.]

(b) $k = \dfrac{\pi}{3}r^3 + 4r^2 d$.

Both $\dfrac{\pi}{3}r^3$ and $4r^2 d$ have **three** dimensions so k is a **volume**.

$\left[\dfrac{\pi}{3} \text{ and '4' have no dimensions.} \right]$

(c) $m = \pi(c + d) = \pi c + \pi d$
πc is a length and πd is a length.
So m is a length plus a length.
$\therefore$ m is a length.

(d) $f = 4c + 3cd$
$4c$ is a length.
$3cd$ is a length multiplied by a length and is an area.
So f is a length plus an area which is an **impossible** expression.

Exercise 4J

The symbols a, b, d, h, l, r represent lengths.

1. For each expression, decide if it represents a length, an area or a volume:

   (a) $a + 3b$
   (b) $a^2 b$
   (c) ab
   (d) $5(b - a)$
   (e) $b^3 + a^3$
   (f) $7ab + a^2$

2. State the number of dimensions for each of the following:

   (a) πl^2
   (b) $3\pi lr$
   (c) $\dfrac{\pi}{2} b^2 h$
   (d) $\pi(a + b)$
   (e) $\dfrac{ab + h^2}{6}$
   (f) $abd \sin 30°$

3. Give the number of dimensions that a formula for each of the following should have.
   (a) Total area of windows in a room.
   (b) Volume of sand in a lorry.
   (c) Area of a sports field.
   (d) The diagonal of a rectangle.
   (e) The capacity of an oil can.
   (f) The perimeter of a trapezium.
   (g) The number of people in a cinema.
   (h) The surface area of the roof of a house.

4. One of the expressions gives the area of the shaded shape and another gives the perimeter.

   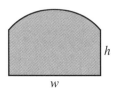

   $2h + w + \frac{5}{4}w$ $w^2h + h^2$

   $w + h + h + wh$ $wh + \frac{\pi}{6}wh$

   Write down the correct expression for:
   (a) the area
   (b) the perimeter.

5. From this list of expressions, choose the **two** that represent volume, the four that represent area and the **one** that represents a length. The other expression is impossible.
   (a) $\pi rh + \pi r^2$
   (b) $5a + 6c$
   (c) $3 \cdot 5\,abd$
   (d) $4hl + \pi rh$
   (e) $3r^2hl$
   (f) $2\pi r(r + h)$
   (g) $2(rb^2 + h^3)$
   (h) $\frac{\pi}{2}(l + d)^2$

6. State whether each expression represents length (L), area (A), volume (V) or is impossible (I).
   (a) $2a - b$
   (b) $a^2 + c$
   (c) $2a^3 + abc$
   (d) $\pi(ab + c^2)$
   (e) $a^2b + b^2a$
   (f) $(a - c)(b + a)$
   (g) $a(b^2 + c)$
   (h) $\pi r^2(a + b)$
   (i) $\pi^2(b + a)$
   (j) $4\pi a + b$
   (k) $\sqrt{\dfrac{a^2 + b^2}{c}}$
   (l) $\dfrac{\pi r^3}{4} + a^2$

7. In Sam's notes, the formula for the volume of a container was written with Tippex over the index for r. The formula was

$V = \dfrac{\pi}{3} r^{\boxed{}} h$. What was the missing index?

8. A physicist worked out a formula for the surface area of a complicated object and got:

$S = 3\pi (a+b)^2 \sin 20° + \dfrac{\pi}{2} a.$

Explain why the formula could not be correct.

4.4 Transformations

Reflection and rotation

- A reflection is specified by the choice of a mirror line.

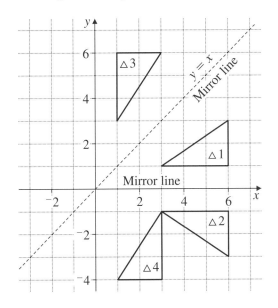

In the diagram:

(a) △2 is the image of △1 after reflection in the x-axis (the mirror line).

(b) △3 is the image of △1 after reflection in the mirror line $y = x$.
   - A rotation requires specification of **three** things: angle, direction and centre of rotation.

(c) △4 is the image of △2 after rotation through 90° clockwise about (3, ⁻1).

(d) △4 is the image of △3 after rotation through 180° in either direction about (2,1).
- By convention, an anticlockwise rotation is positive and a clockwise rotation is negative. So the rotation of △2 onto △4 is [⁻90°, centre (3, ⁻1).
  It is helpful to use tracing paper to obtain the result of a rotation.
- Under reflection and rotation (and translation) the object and image are always **congruent**.
  [The object and image are the same size.]

Exercise 4K

In Questions **1** and **2**, draw the object and its image after reflection in the broken line.

1.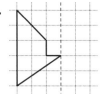
2.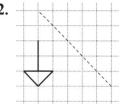

In Questions **3** to **6**, draw x- and y-axes with values from ⁻8 to ⁺8.

3. (a) Draw the triangle DEF at $D(^-6, 8)$, $E(^-2, 8)$, $F(^-2, 6)$. Draw the lines $x = 1$, $y = x$, $y = ^-x$.
   (b) Draw the image of $\triangle DEF$ after reflection in:
       (i) the line $x = 1$. Label it △1.
       (ii) the line $y = x$. Label it △2.
       (iii) the line $y = ^-x$. Label it △3.
   (c) Write down the coordinates of the image of point D in each case.

4. (a) Draw △1 at (3, 1), (7, 1), (7, 3).
   (b) Reflect △1 in the line $y = x$ onto △2.
   (c) Reflect △2 in the x-axis onto △3.
   (d) Reflect △3 in the line $y = ^-x$ onto △4.
   (e) Reflect △4 in the line $x = 2$ onto △5.
   (f) Write down the coordinates of △5.

5. (a) Draw a triangle PQR at $P(1, 2)$, $Q(3, 5)$, $R(6, 2)$.
   (b) Find the image of PQR under the following rotations:
      (i) 90° anticlockwise, centre (0, 0); label the image $P'Q'R'$
      (ii) 90° clockwise, centre ($^-2$, 2); label the image $P''Q''R''$
      (iii) 180°, centre, (1, 0); label the image $P*Q*R*$.
   (c) Write down the coordinates of P', P'', $P*$.

6. (a) Draw △1 at (1, 2), (1, 6), (3, 5).
   (b) Rotate △1 90° clockwise, centre (1, 2) onto △2.
   (c) Rotate △2 180°, centre (2, $^-1$) onto △3.
   (d) Rotate △3 90° clockwise, centre (2, 3) onto △4.
   (e) Write down the coordinates of △4.

Translation and enlargement

- A translation can be specified by a **column vector**.
  In the diagram:
  (a) △5 is the image of △4 after translation with the column vector $\begin{pmatrix} 4 \\ 2 \end{pmatrix}$.

  (b) △4 is the image of △1 after translation with the column vector $\begin{pmatrix} 2 \\ -3 \end{pmatrix}$.

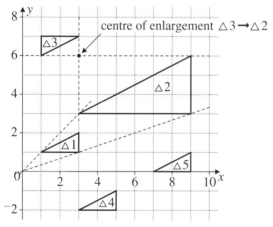

- An enlargement requires specification of two things:
  scale factor and **centre of enlargement**.
  (c) △2 is the image of △1 after enlargement by a scale factor 3 with centre of enlargement (0, 0).
  (d) △1 is the image of △2 after enlargement by a scale factor $\frac{1}{3}$ with centre of enlargement (0, 0). Note the construction lines.
  (e) △2 is the image of △3 after enlargement by a scale factor -3 with centre of enlargement (3, 6).

> Note that the negative scale factor causes the object to be inverted to form the image.

Exercise 4L

1. For the diagram below, write down the column vector for each of the following translations.

   (a) D onto A (b) B onto F
   (c) E onto A (d) A onto C
   (e) E onto C (f) C onto B
   (g) F onto E (h) B onto C.

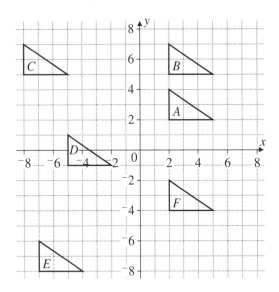

For Questions **2** to **5**, copy the diagram and draw an enlargement using the centre O and the scale factor given.

2. Scale factor 2

3. Scale factor 3

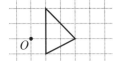

4. Scale factor 3

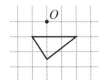

5. Scale factor $^-2$

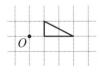

For Questions **6** to **11**, draw *x*- and *y*-axes with values from 0 to 15. Enlarge the object using the centre of enlargement and scale factor given.

	object			centre	scale factor
6.	(2, 4)	(4, 2)	(5, 5)	(1, 2)	$^+2$
7.	(1, 1)	(4, 2)	(2, 3)	(1, 1)	$^+3$
8.	(1, 2)	(13, 2)	(1, 10)	(0, 0)	$^+\tfrac{1}{2}$
9.	(5, 10)	(5, 7)	(11, 7)	(2, 1)	$^+\tfrac{1}{3}$
10.	(1, 1)	(3, 1)	(3, 2)	(4, 3)	$^-2$
11.	(9, 2)	(14, 2)	(14, 6)	(7, 4)	$-\tfrac{1}{2}$

Questions **12** and **13** involve rotation, reflection, enlargement and translation.

12. Copy the diagram opposite. Describe fully the following transformations:

(a) △1 → △2   (b) △1 → △3
(c) △4 → △1   (d) △1 → △5
(e) △3 → △6   (f) △6 → △4

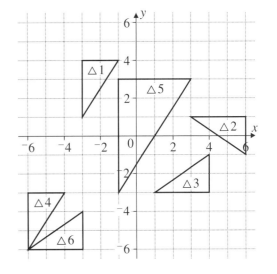

13. Draw *x*- and *y*-axes from $^-8$ to $^+8$. Plot and label the following triangles:

△1 : ($^-5$, $^-5$), ($^-1$, $^-5$), ($^-1$, $^-3$)
△2 : (1, 7), (1, 3), (3, 3)
△3 : (3, $^-3$), (7, $^-3$), (7, $^-1$)
△4 : ($^-5$, $^-5$), ($^-5$, $^-1$), ($^-3$, $^-1$)
△5 : (1, $^-6$), (3, $^-6$), (3, $^-5$)
△6 : ($^-3$, 3), ($^-3$, 7), ($^-5$, 7)

Describe fully the following transformations:

(a) △1 → △2   (b) △1 → △3
(c) △1 → △4   (d) △1 → △5
(e) △1 → △6   (f) △5 → △3
(g) △2 → △3

Successive transformations

In the diagram △2 is the image of △1 after rotation 90° clockwise about (1, 3).

△3 is the image of △2 after translation $\begin{pmatrix} 3 \\ 3 \end{pmatrix}$.

So △1 has been moved onto △3 by successive transformations.
The equivalent single transformation is rotation 90° clockwise about (3, 3). [Check this.]

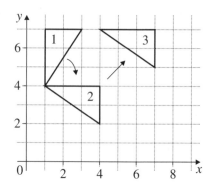

Exercise 4M

1. Copy the diagram shown.

   (a) Rotate △1 180° about (4, 2). Label the image △2.
   (b) Reflect △2 in the line $y = 2$. Label the image △3.
   (c) Describe the **single** transformation which maps △1 onto △3.

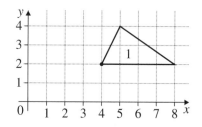

2. Copy the diagram.

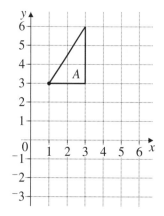

   (a) Reflect △A in the line $y = x$. Label the image △B.
   (b) Reflect △B in the x-axis. Label the image △C.
   (c) Describe fully the single transformation which maps △A onto △C.

In Questions **3** to **8** transformations **A**, **B**, ... **H**, are as follows:

A denotes reflection in $x = 2$
B denotes 180° rotation, centre (1, 1)
C denotes translation $\begin{pmatrix} -6 \\ 2 \end{pmatrix}$
D denotes reflection in $y = x$
E denotes reflection in $y = 0$
F denotes translation $\begin{pmatrix} 4 \\ 3 \end{pmatrix}$
G denotes 90° rotation clockwise, centre (0, 0)
H denotes enlargement, scale factor $+\frac{1}{2}$, centre (0, 0).

Draw x- and y-axes with values from $^-8$ to $^+8$.

3. Draw triangle LMN at $L(2, 2)$, $M(6, 2)$, $N(6, 4)$. Find the image of LMN under the following transformations:
(a) **A** followed by **C**
(b) **D** followed by **E**
(c) **B** followed by **D**
(d) **E** followed by **B**.
Write down the coordinates of the image of point L in each case.

4. Draw triangle PQR at $P(2, 2)$, $Q(6, 2)$, $R(6, 4)$. Find the image of PQR under the following transformations:
(a) **F** followed by **A**
(b) **G** followed by **C**
(c) **G** followed by **A**
(d) **E** followed by **H**.
Write down the coordinates of the image of point P in each case.

5. Draw triangle XYZ at $X(^-2, 4)$, $Y(^-2, 1)$, $Z(4, 1)$. Find the image of XYZ under the following combinations of transformations and state the equivalent single transformation in each case.
(a) **E**, then **G**, then **G**
(b) **B**, then **C**
(c) **A**, then **D**.

6. Draw triangle OPQ at $O(0, 0)$, $P(0, 2)$, $Q(3, 2)$. Find the image of OPQ under the following combinations of transformations and state the equivalent single transformation in each case.
(a) **E**, then **D**
(b) **C**, then **F**
(c) **C**, then **E**, then **D**
(d) **E**, then **F**, then **D**.

7. Draw triangle RST at R($^-$4, $^-$1), S($^-$2$\frac{1}{2}$, $^-$2), T($^-$4, $^-$4).
   Find the image of RST under the following combinations of transformations and state the equivalent single transformation in each case.
   (a) **G**, then **A**, then **E**
   (b) **H**, then **F**
   (c) **F**, then **G**.

8. This inverse of a transformation is the transformation which takes the image back to the object.
   Write down the inverses of the transformations **A, B, ... H**.

4.5 Vectors

A vector quantity has both magnitude and direction. A translation is described by a vector and a vector can also represent physical quantities such as velocity, force, acceleration, etc. The symbol for a vector is a bold letter, e.g. **a, x**. On a coordinate grid, the magnitude and direction of the vector can be shown by a column vector, e.g. $\begin{pmatrix} 2 \\ 1 \end{pmatrix}, \begin{pmatrix} 1 \\ -3 \end{pmatrix}$ where the upper number shows the distance across the page and the lower number shows the distance up the page.

Example
Describe each of the vectors shown on the grid.

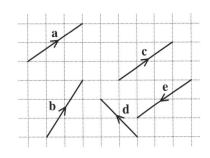

$\mathbf{a} = \begin{pmatrix} 3 \\ 2 \end{pmatrix}$ $\quad$ $\mathbf{d} = \begin{pmatrix} -2 \\ 2 \end{pmatrix}$

$\mathbf{b} = \begin{pmatrix} 2 \\ 3 \end{pmatrix}$ $\quad$ $\mathbf{e} = \begin{pmatrix} -3 \\ -2 \end{pmatrix}$

$\mathbf{c} = \begin{pmatrix} 3 \\ 2 \end{pmatrix}$

Equal vectors
Two vectors are equal if they have the same length **and** the same direction. The actual position of the vector on the diagram or in space is of no consequence.

Thus in the example, vectors **a** and **c** are equal because they have the same magnitude and direction. Even though vector **b** also has the same length (magnitude) as **a** and **c**, it is not equal to **a** or **c** because it acts in a different direction. Likewise, vector **e** has the same length as vector **c** but acts in the reverse direction and so cannot equal **c**.
Equal vectors have identical column vectors.

Addition of vectors

Vectors **a** and **b** are represented by the line segments shown below, they can be added by using the 'nose-to-tail' method to give a single equivalent vector.

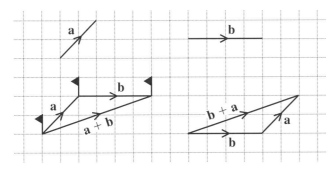

The 'tail' of vector **b** is joined to the 'nose' of vector **a**.

Alternatively the tail of **a** can be joined to the 'nose' of vector **b**.

In both cases the vector $\overrightarrow{XY}$ has the same length and direction and therefore **a** + **b** = **b** + **a**.

In the first diagram the flag is moved by translation **a** and then translation **b**. The translation **a** + **b** is the equivalent or **resultant** translation.

Multiplication by a scalar

A scalar quantity has magnitude but no direction (e.g. mass, volume, temperature). Ordinary numbers are scalars.

When vector **x** is multiplied by 2, the result is 2**x**.

When **x** is multiplied by ¯3 the result is ¯3**x**.

Note:
(1) The negative sign reverses the direction of the vector.
(2) The result $\mathbf{a} - \mathbf{b}$ is $\mathbf{a} +^- \mathbf{b}$.
   So, subtracting $\mathbf{b}$ is equivalent to adding the negative of $\mathbf{b}$.

Example

The diagram on the right shows vectors $\mathbf{a}$ and $\mathbf{b}$. Draw a diagram to show $\overrightarrow{OP}$ and $\overrightarrow{OQ}$ such that $\overrightarrow{OP} = 3\mathbf{a} + \mathbf{b}$ $\overrightarrow{OQ} = ^-2\mathbf{a} - 3\mathbf{b}$

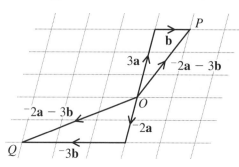

Exercise 4N

In Questions **1** to **15**, use the diagram below to describe the vectors given in terms of $\mathbf{c}$ and $\mathbf{d}$ where $\mathbf{c} = \overrightarrow{QN}$ and $\mathbf{d} = \overrightarrow{QR}$. For example, $\overrightarrow{QS} = 2\mathbf{d}$, $\overrightarrow{TD} = \mathbf{c} + \mathbf{d}$.

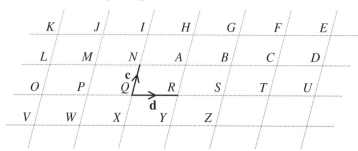

1. $\overrightarrow{AB}$ 2. $\overrightarrow{SG}$ 3. $\overrightarrow{VK}$ 4. $\overrightarrow{KH}$ 5. $\overrightarrow{OT}$
6. $\overrightarrow{WJ}$ 7. $\overrightarrow{FH}$ 8. $\overrightarrow{FT}$ 9. $\overrightarrow{KV}$ 10. $\overrightarrow{NQ}$
11. $\overrightarrow{OM}$ 12. $\overrightarrow{SD}$ 13. $\overrightarrow{PI}$ 14. $\overrightarrow{YG}$ 15. $\overrightarrow{OI}$

In Questions **16** to **21**, use the same diagram above to find vectors for the following in terms of the capital letters, starting from Q each time.

For example, $3\mathbf{d} = \overrightarrow{QT}$, $\mathbf{c} + \mathbf{d} = \overrightarrow{QA}$.

16. $2\mathbf{c}$ 17. $4\mathbf{d}$ 18. $2\mathbf{c} + \mathbf{d}$
19. $2\mathbf{d} + \mathbf{c}$ 20. $3\mathbf{d} + 2\mathbf{c}$ 21. $2\mathbf{c} - \mathbf{d}$

In Questions **22** and **23**, use the diagram below. $\overrightarrow{LM} = \mathbf{a}$, $\overrightarrow{LQ} = \mathbf{b}$.

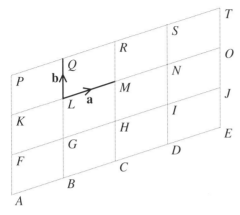

22. Write these vectors in terms of **a** and **b**.
   (a) $\overrightarrow{GN}$
   (b) $\overrightarrow{CO}$
   (c) $\overrightarrow{TN}$
   (d) $\overrightarrow{FT}$
   (e) $\overrightarrow{KC}$
   (f) $\overrightarrow{CJ}$

23. From your answers to Question **22**, find the vector which is:
   (a) parallel to $\overrightarrow{LR}$
   (b) 'opposite' to $\overrightarrow{LR}$
   (c) parallel to $\overrightarrow{CJ}$ with twice the magnitude
   (d) parallel to the vector $(\mathbf{a} - \mathbf{b})$.

In Questions **24** to **27**, write each vector in terms of **a**, **b**, or **a** and **b**.

24. (a) $\overrightarrow{BA}$
   (b) $\overrightarrow{AC}$
   (c) $\overrightarrow{DB}$
   (d) $\overrightarrow{AD}$

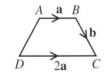

25. (a) $\overrightarrow{ZX}$
   (b) $\overrightarrow{YW}$
   (c) $\overrightarrow{XY}$
   (d) $\overrightarrow{XZ}$

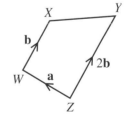

26. (a) $\overrightarrow{MK}$
   (b) $\overrightarrow{NL}$
   (c) $\overrightarrow{NK}$
   (d) $\overrightarrow{KN}$

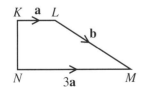

27. (a) $\overrightarrow{FE}$
   (b) $\overrightarrow{BC}$
   (c) $\overrightarrow{FC}$
   (d) $\overrightarrow{DA}$

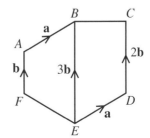

Vector geometry

Example

In the diagram, $OA = AP$ and $BQ = 3OB$.
N is the mid-point of PQ.
$\overrightarrow{OA} = \mathbf{a}$ and $\overrightarrow{OB} = \mathbf{b}$.

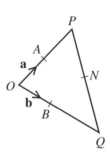

Express each of the following vectors in terms of **a**, **b**, or **a** and **b**.

(a) $\overrightarrow{AP}$ (b) $\overrightarrow{AB}$ (c) $\overrightarrow{OQ}$ (d) $\overrightarrow{PO}$ (e) $\overrightarrow{PQ}$ (f) $\overrightarrow{PN}$ (g) $\overrightarrow{ON}$ (h) $\overrightarrow{AN}$

(a) $\overrightarrow{AP} = \mathbf{a}$ (b) $\overrightarrow{AB} = {}^-\mathbf{a} + \mathbf{b}$ (c) $\overrightarrow{OQ} = 4\mathbf{b}$ (d) $\overrightarrow{PO} = {}^-2\mathbf{a}$

(e) $\overrightarrow{PQ} = \overrightarrow{PO} + \overrightarrow{OQ}$ (f) $\overrightarrow{PN} = \tfrac{1}{2}\overrightarrow{PQ}$ (g) $\overrightarrow{ON} = \overrightarrow{OP} + \overrightarrow{PN}$

     $= 2\mathbf{a} + 4\mathbf{b}$ $= {}^-\mathbf{a} + 2\mathbf{b}$ $= 2\mathbf{a} + ({}^-\mathbf{a} + 2\mathbf{b})$

                                                                                                    $= \mathbf{a} + 2\mathbf{b}$

(h) $\overrightarrow{AN} = \overrightarrow{AP} + \overrightarrow{PN}$

     $= \mathbf{a} + ({}^-\mathbf{a} + 2\mathbf{b})$

     $= 2\mathbf{b}$

Exercise 40

In Questions **1** to **4**, $\overrightarrow{OA} = \mathbf{a}$ and $\overrightarrow{OB} = \mathbf{b}$. Copy each diagram and use the information given to express the following vectors in terms of **a**, **b** or **a** and **b**.

(a) $\overrightarrow{AP}$ (b) $\overrightarrow{AB}$ (c) $\overrightarrow{OQ}$ (d) $\overrightarrow{PO}$ (e) $\overrightarrow{PQ}$
(f) $\overrightarrow{PN}$ (g) $\overrightarrow{ON}$ (h) $\overrightarrow{AN}$ (i) $\overrightarrow{BP}$ (j) $\overrightarrow{QA}$

1. A, B and N are mid-points of OP, OB and PQ respectively.

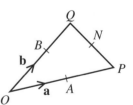

2. A and N are mid-points of OP and PQ; $BQ = 2OB$.

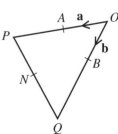

3. $AP = 2OA$, $BQ = OB$, $PN = NQ$

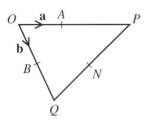

4. $OA = 2AP$, $BQ = 3OB$, $PN = 2NQ$

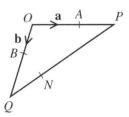

5. In $\triangle XYZ$, the mid-point of YZ is M. If $\overrightarrow{XY} = \mathbf{s}$ and $\overrightarrow{ZX} = \mathbf{t}$, find $\overrightarrow{XM}$ in terms of $\mathbf{s}$ and $\mathbf{t}$.

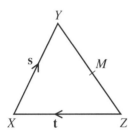

6. In $\triangle AOB$, $AM : MB = 2 : 1$. If $\overrightarrow{OA} = \mathbf{a}$, and $\overrightarrow{OB} = \mathbf{b}$ find $\overrightarrow{OM}$ in terms of $\mathbf{a}$ and $\mathbf{b}$.

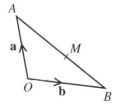

7. $ABCDEF$ is a regular hexagon with $\overrightarrow{AB}$, representing the vector $\mathbf{m}$ and $\overrightarrow{AF}$, representing the vector $\mathbf{n}$. Find the vector representing $\overrightarrow{AD}$.

8. *ABCDEF* is a regular hexagon with centre *O*.
   $\vec{FA} = \mathbf{a}$ and $\vec{FB} = \mathbf{b}$.

   Express the following vectors in terms of **a** and/or **b**.

   (a) $\vec{AB}$ (b) $\vec{FO}$ (c) $\vec{FC}$
   (d) $\vec{BC}$ (e) $\vec{AO}$ (f) $\vec{FD}$

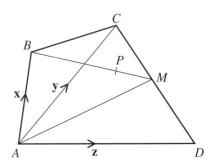

9. In the diagram, *M* is the mid-point of *CD*, $BP:PM = 2:1$, $\vec{AB} = \mathbf{x}$, and $\vec{AC} = \mathbf{y}$ and $\vec{AD} = \mathbf{z}$.

   Express the following vectors in terms of **x**, **y** and **z**.

   (a) $\vec{DC}$ (b) $\vec{DM}$ (c) $\vec{AM}$
   (d) $\vec{BM}$ (e) $\vec{BP}$ (f) $\vec{AP}$

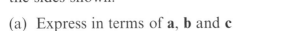

10. In the quadrilateral shown $\vec{OA} = 2\mathbf{a}$, $\vec{OB} = 2\mathbf{b}$, $\vec{OC} = 2\mathbf{c}$. Points *P*, *Q*, *R* and *S* are the mid-points of the sides shown.

    (a) Express in terms of **a**, **b** and **c**
       (i) $\vec{AB}$
       (ii) $\vec{BC}$
       (iii) $\vec{PQ}$
       (iv) $\vec{QR}$
       (v) $\vec{PS}$
    (b) Describe the relationship between *QR* and *PS*.
    (c) What sort of quadrilateral is *PQRS*?

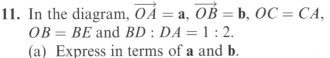

11. In the diagram, $\vec{OA} = \mathbf{a}$, $\vec{OB} = \mathbf{b}$, $OC = CA$, $OB = BE$ and $BD:DA = 1:2$.
    (a) Express in terms of **a** and **b**.
       (i) $\vec{BA}$
       (ii) $\vec{BD}$
       (iii) $\vec{CD}$
       (iv) $\vec{CE}$
    (b) Explain why points *C*, *D* and *E* lie on a straight line.

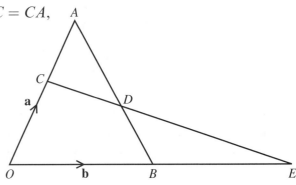

4.6 Sine, cosine, tangent for any angle

So far we have used sine, cosine and tangent only in right-angled triangles. For angles greater than 90°, there is a close connection between trigonometric ratios and circles.

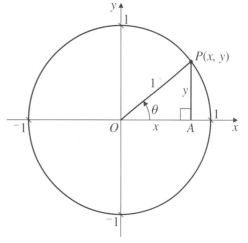

The circle on the right is of radius 1 unit with centre (0, 0). A point P with coordinates (x, y) moves round the circumference of the circle. The angle that OP makes with the positive x-axis as it turns in an anticlockwise direction is θ.

In triangle OAP, $\cos \theta = \dfrac{x}{1}$ and $\sin \theta = \dfrac{y}{1}$

The x-coordinate of P is $\cos \theta$.
The y-coordinate of P is $\sin \theta$.

This idea is used to define the cosine and the sine of any angle, including angles greater than 90°. Here are two angles that are greater than 90°.

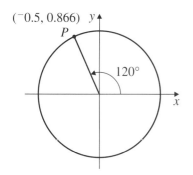
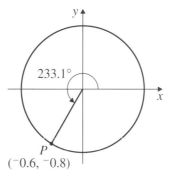

$\cos 120° = {}^-0.5$ $\cos 233.1° = {}^-0.6$
$\sin 120° = 0.866$ $\sin 233.1° = {}^-0.8$

A graphics calculator can be used to show the graph of $y = \sin x$ for any range of angles. The graph on page 382 shows $y = \sin x$ for x from 0° to 360°. The curve above the x-axis has line symmetry about $x = 90°$ and that below the x-axis has line symmetry about $x = 270°$.

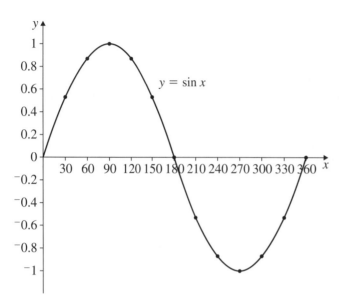

Note that:
sin 150° = sin 30° and cos 150° = ⁻cos 30°
sin 110° = sin 70°   cos 110° = ⁻cos 70°
sin 163° = sin 17°   cos 163° = ⁻cos 17°

or sin x = sin (180° − x)
and cos x = ⁻cos (180° − x)

These two results are particularly important for use with obtuse angles (90° < x < 180°).

Exercise 4P

1. (a) Use a calculator to find the cosine of all the angles 0°, 30°, 60°, 90°, 120°, … 330°, 360°.
   (b) Draw a graph of $y = \cos x$ for $0 \leqslant x \leqslant 360°$. Use a scale of 1 cm to 30° on the x-axis and 5 cm to 1 unit on the y-axis.

2. Draw the graph of $y = \sin x$, using the same angles and scales as in Question **1**.

3. Find the tangent of the angles 0°, 20°, 40°, 60°, … 320°, 340°, 360°.
   (a) Notice that we have deliberately omitted 90° and 270°. Why has this been done?
   (b) Draw a graph of $y = \tan x$. Use a scale of 1 cm to 20° on the x-axis and 1 cm to 1 unit on the y-axis.
   (c) Draw a vertical dotted line at $x = 90°$ and $x = 270°$.

These lines are **asymptotes** to the curve. As the value of x approaches 90° from either side, the curve gets nearer and nearer to the asymptote but it **never** quite reaches it.
Your graph should look something like this:

> Use a calculator to find tan 89°, tan 89·9°, tan 89·99°, tan 89·999° etc.

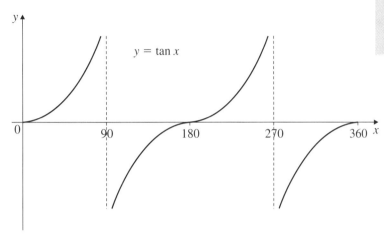

4. Sort the following into pairs of equal value.

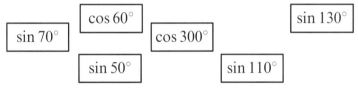

sin 70°   cos 60°   cos 300°   sin 130°   sin 50°   sin 110°

5. Sort the following into pairs of equal value.

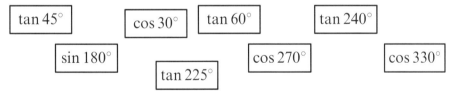

tan 45°   cos 30°   tan 60°   tan 240°   sin 180°   cos 270°   cos 330°   tan 225°

In Questions **6** to **17** do not use a calculator. Use the symmetry of the graphs $y = \sin x$, $y = \cos x$, and $y = \tan x$.
Angles are from 0° to 360° and answers should be given to the nearest degree.

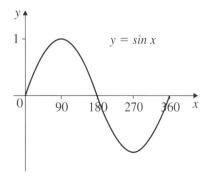

6. If $\sin 18° = 0·309$, give another angle whose sine is 0·309.

7. If $\sin 27° = 0·454$, give another angle whose sine is 0·454.

8. Give another angle which has the same sine as
   (a) 40°   (b) 70°   (c) 130°

9. If cos 70° = 0·342, give another angle whose cosine is 0·342.

10. If cos 45° = 0·707, give another angle whose cosine is 0·707.

11. Give another angle which has the same cosine as
    (a) 10°   (b) 56°   (c) 300°

12. If tan 40° = 0·839, give another angle whose tangent is 0·839.

13. If sin 20° = 0·342, what other angle has a sine of 0·342?

14. If sin 98° = 0·990, give another angle whose sine is 0·990.

15. If tan 135° = ⁻1, give another angle whose tangent is ⁻1.

16. If cos 120° = ⁻0·5, give another angle whose cosine is ⁻0·5.

17. If tan 70° = 2·75, give another angle whose tangent is 2·75.

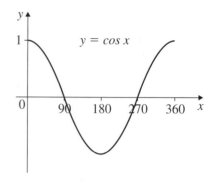

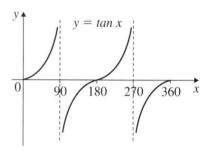

Exercise 4Q

In Question **1** to **8** give angles correct to 1 decimal place.

1. Find two values for x, between 0° and 360°, if sin x = 0·848.

2. If sin x = 0·35, find two solutions for x between 0° and 360°.

3. If cos x = 0·6, find two solutions for x between 0° and 360°.

4. If tan x = 1, find two solutions for x between 0° and 360°.

5. Find two solutions between 0° and 360°.
   (a) sin x = 0·339   (b) sin x = 0·951
   (c) sin x = $\frac{1}{2}$ (d) sin x = $\frac{\sqrt{3}}{2}$

6. Find two solutions between 0° and 360°.
   (a) $\sin x = 0.72$
   (b) $\cos x = 0.3$
   (c) $\tan x = 5$
   (d) $\sin x = {}^-0.65$

7. Find **four** solutions of the equation $(\sin x)^2 = \frac{1}{4}$, for x between 0° and 360°.

8. Find four solutions of the equation $(\tan x)^2 = 1$, for x between 0° and 360°.

9. The triangle is drawn to find sin, cos and tan of 30° and 60°. For example, $\sin 30° = \frac{1}{2}$ and $\tan 60° = \sqrt{3}$.

   Copy and complete this table.

	30°	60°	120°	150°
sin	$\frac{1}{2}$			
cos				
tan		$\sqrt{3}$		

   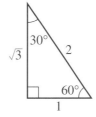

10. (a) Between 0° and 360°, for what values of x is $\sin x > 0$?
    (b) Between 0° and 360°, for what values of x is $\cos x < 0$?

11.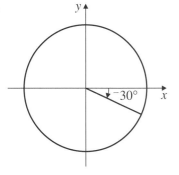

    A negative angle is measured *clockwise* from the x-axis.
    $\sin({}^-30°) = {}^-\frac{1}{2}$
    $\cos({}^-30°) = -\frac{\sqrt{3}}{2}$

    An angle greater than 360° is shown.
    $\sin 390° = \sin 30°$
    $\cos 400° = \cos 40°$ etc.

    Sketch the graphs of $\sin \theta$, $\cos \theta$ and $\tan \theta$ for θ between ${}^-360°$ and 360°.

12. (a) Find three values of θ for which $\sin \theta = 1$.
    (b) Find four values of θ for which $\cos \theta = 0$.

13. (a) Write down the value of $\sin(360n°)$, where n is an integer.
    (b) Write down the value of $\cos(360n°)$, where n is an integer.

14. (a) Copy and complete the following.
    $$\sin\theta = \frac{a}{c}, \quad \cos\theta = \boxed{\frac{-}{-}}, \quad \tan\theta = \boxed{\frac{-}{-}}$$
    $$\sin\theta \div \cos\theta = \frac{a}{c} \div \boxed{\frac{-}{-}} = \boxed{\frac{-}{-}}$$
    $$\therefore \frac{\sin\theta}{\cos\theta} = \tan\theta$$

    (b) Check if this relationship holds for angles greater than 90°. Work out:
    (i) $\dfrac{\sin 135°}{\cos 135°}$, $\tan 135°$
    (ii) $\dfrac{\sin 240°}{\cos 240°}$, $\tan 240°$
    (iii) $\dfrac{\sin 320°}{\cos 320°}$, $\tan 320°$.

15. Suppose the cos button on your calculator was broken, but the sin button was working. Explain how you could work out $\cos 40°$.

16. Find all possible solutions between 0° and 360°.
    (a) $\sin x = \cos x$
    (b) $\sin 2x = \sin 60°$
    (c) $\tan 3x = 1$ [Hint: There are six solutions.]

17. Draw the graph of $y = 2\sin x + 1$ for $0 \leqslant x \leqslant 180°$, taking 1 cm to 10° for x and 5 cm to 1 unit for y. Find approximate solutions to the equations:
    (a) $2\sin x + 1 = 2\cdot3$
    (b) $\dfrac{1}{(2\sin x + 1)} = 0\cdot5$

18. Draw the graph of $y = 2\sin x + \cos x$ for $0 \leqslant x \leqslant 180°$, taking 1 cm to 10° for x and 5 cm to 1 unit for y.
    (a) Solve approximately the equations:
        (i) $2\sin x + \cos x = 1\cdot5$
        (ii) $2\sin x + \cos x = 0$
    (b) Estimate the maximum value of y.
    (c) Find the value of x at which the maximum occurs.

19. Draw the graph of $y = 3\cos x - 4\sin x$ for $0° \leqslant x \leqslant 220°$, taking 1 cm to 10° for x and 2 cm to 1 unit for y.
    Solve approximately the equations:
    (a) $3\cos x - 4\sin x + 1 = 0$
    (b) $3\cos x = 4\sin x$

4.7 Sine and cosine rules

The sine rule enables us to calculate sides and angles in some triangles where there is not a right angle.

In triangle ABC, we use the convention that

a is the side opposite $\widehat{A}$
b is the side opposite $\widehat{B}$

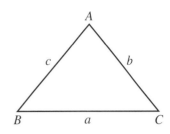

Sine rule

either $\quad \dfrac{a}{\sin A} = \dfrac{b}{\sin B} = \dfrac{c}{\sin C} \quad \ldots [1]$

or $\quad \dfrac{\sin A}{a} = \dfrac{\sin B}{b} = \dfrac{\sin C}{c} \quad \ldots [2]$

Use [1] when finding a **side**,
and [2] when finding an **angle**.

Remember:
To find a **side**, have the **sides** on **top**.
To find an **angle**, have the **angles** on **top**.

Example

Find c.

$\dfrac{c}{\sin C} = \dfrac{b}{\sin B}$

$\dfrac{c}{\sin 50°} = \dfrac{7}{\sin 60°}$

$c = \dfrac{7 \times \sin 50°}{\sin 60°} = 6.19 \text{ cm (3 s.f.)}$

Find $\widehat{B}$.

$\dfrac{\sin B}{b} = \dfrac{\sin A}{a}$

$\dfrac{\sin B}{6} = \dfrac{\sin 120°}{15}$

$\sin B = \dfrac{6 \times \sin 120°}{15}$

$\sin B = 0.346$

$\widehat{B} = 20.3°$ [1 d.p.]

Exercise 4R

For Questions **1** to **6**, find each side marked with a letter.

1.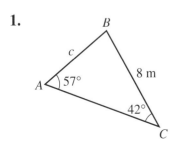

2.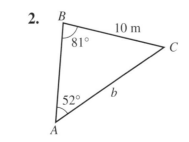

3.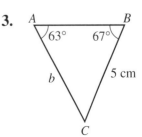

4. 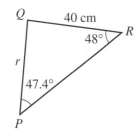   **5.** 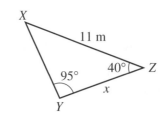   **6.**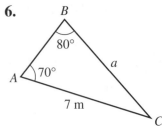

7. In $\triangle ABC$, $\hat{A} = 61°$, $\hat{B} = 47°$, $AC = 7\cdot 2$ cm. Find BC.

8. In $\triangle XYZ$, $\hat{Z} = 32°$, $\hat{Y} = 78°$, $XY = 5\cdot 4$ cm. Find XZ.

9. In $\triangle PQR$, $\hat{} = 100°$, $\hat{R} = 21°$, $PQ = 3\cdot 1$ cm. Find PR.

10. In $\triangle LMN$, $\hat{L} = 21°$, $\hat{N} = 30°$, $MN = 7$ cm. Find LN.

In Questions **11** to **16**, find each angle marked *.
All lengths are in centimetres.

11. 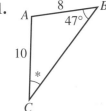   **12.** 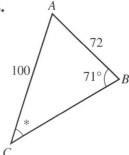   **13.**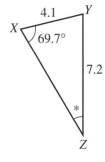

14. 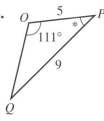   **15.** 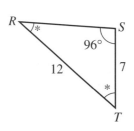   **16.**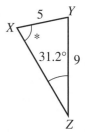

17. In $\triangle ABC$, $\hat{A} = 62°$, $BC = 8$, $AB = 7$. Find $\hat{C}$.

18. In $\triangle XYZ$, $\hat{Y} = 97\cdot 3°$, $XZ = 22$, $XY = 14$. Find $\hat{Z}$.

19. The sine rule can be ambiguous.

The diagram shows two possible triangles with $AB = 5\cdot 4$, $\hat{A} = 32°$ and $BC = 3$.
Find the two possible values of angle C.

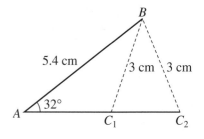

Cosine rule

We use the cosine rule when we know either
(a) two sides and the included angle or
(b) all three sides.

There are two forms.

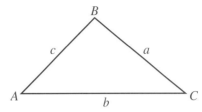

1. To find the length of a side.
$$a^2 = b^2 + c^2 - (2bc \cos A)$$
or $\quad b^2 = c^2 + a^2 - (2ac \cos B)$
or $\quad c^2 = a^2 + b^2 - (2ab \cos C)$

2. To find an angle when given all three sides.
$$\cos A = \frac{b^2 + c^2 - a^2}{2bc}$$
or $\quad \cos B = \dfrac{a^2 + c^2 - b^2}{2ac}$

or $\quad \cos C = \dfrac{a^2 + b^2 - c^2}{2ab}$

For an obtuse angle x we have $\cos x = {}^-\cos(180° - x)$.

For example, $\quad \cos 120° = {}^-\cos 60°$

$\cos 142° = {}^-\cos 38°$

Example
(a) Find b.

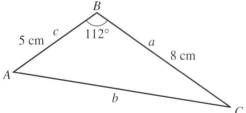

$b^2 = a^2 + c^2 - (2ac \cos B)$
$b^2 = 8^2 + 5^2 - (2 \times 8 \times 5 \times \cos 112°)$
$b^2 = 64 + 25 - [80 \times ({}^-0.3746)]$
$b^2 = 64 + 25 + 29.968$
$b = \sqrt{(118.968)} = 10.9$ cm (to 3 s.f.)

(b) Find angle C.

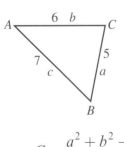

$\cos C = \dfrac{a^2 + b^2 - c^2}{2ab}$

$\cos C = \dfrac{5^2 + 6^2 - 7^2}{2 \times 5 \times 6} = \dfrac{12}{60} = 0.200$

$\widehat{C} = 78.5°$ (to 1 d.p.)

Exercise 4S

Find the sides marked *.

1.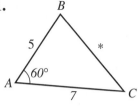

2.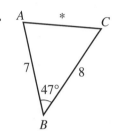

3.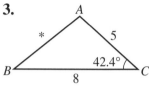

4.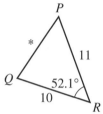

5.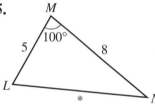

6.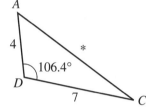

7. In $\triangle ABC$, $AB = 4$ cm, $AC = 7$ cm, $\hat{A} = 57°$. Find BC.

8. In $\triangle XYZ$, $XY = 3$ cm, $YZ = 3$ cm, $\hat{Y} = 90°$. Find XZ.

9. In $\triangle LMN$, $LM = 5.3$ cm, $MN = 7.9$ cm, $\hat{M} = 127°$. Find LN.

10. In $\triangle PQR$, $\hat{Q} = 117°$, $PQ = 80$ cm, $QR = 100$ cm. Find PR.

In Questions **11** to **16**, find the angles marked *.

11.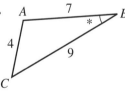

12.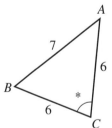

13.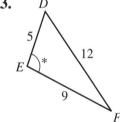

14.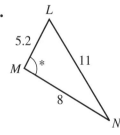

15.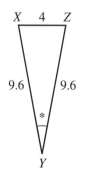

16.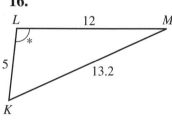

17. In $\triangle ABC$, $a = 4\cdot 3$, $b = 7\cdot 2$, $c = 9$. Find $\hat{C}$.

18. In $\triangle DEF$, $d = 30$, $e = 50$, $f = 70$. Find $\hat{E}$.

19. In $\triangle PQR$, $p = 8$, $q = 14$, $r = 7$. Find $\hat{Q}$.

20. In $\triangle LMN$, $l = 7$, $m = 5$, $n = 4$. Find $\hat{N}$.

Example

A ship sails from a port P a distance of $7\,\text{km}$ on a bearing of $306°$ and then a further $11\,\text{km}$ on a bearing of $070°$ to arrive at X. Calculate the distance from P to X.

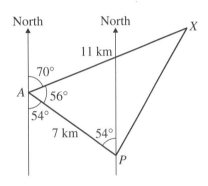

We know two sides and the included angle. Using the cosine rule gives:

$PX^2 = 7^2 + 11^2 - (2 \times 7 \times 11 \times \cos 56°)$

$ = 49 + 121 - (86\cdot 12)$

$PX^2 = 83\cdot 88$

$PX = 9\cdot 16\,\text{km}$ (to 3 s.f.)

The distance from P to X is $9\cdot 16\,\text{km}$.

Exercise 4T

1. Ship B is $58\,\text{km}$ south-east of Ship A. Ship C is $70\,\text{km}$ due south of Ship A.

   (a) How far is Ship B from Ship C?
   (b) What is the bearing of Ship B from Ship C?

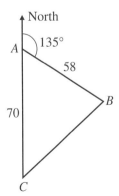

2. A destroyer D and a cruiser C leave port P at the same time. The destroyer sails $25\,\text{km}$ on a bearing $040°$ and the cruiser sails $30\,\text{km}$ on a bearing of $320°$. How far apart are the ships?

3. Two honeybees *A* and *B* leave the hive *H* at the same time; *A* flies 27 m due south and *B* flies 9 m on a bearing of 111°. How far apart are they?

4. Find the sides and angles marked with letters. All lengths are in cm.

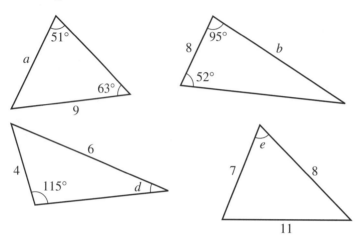
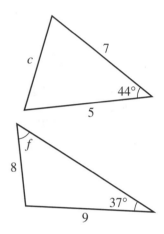

5. Find the largest angle in a triangle in which the sides are in the ratio 5 : 6 : 8.

6. A rhombus has sides of length 8 cm and angles of 50° and 130°. Find the length of the longer diagonal of the rhombus.

7. From *A*, *B* lies 11 km away on a bearing of 041° and *C* lies 8 km away on a bearing of 341°. Find:
   (a) the distance between *B* and *C*
   (b) the bearing of *B* from *C*.

8. From a lighthouse *L* an aircraft carrier *A* is 15 km away on a bearing of 112° and a submarine *S* is 26 km away on a bearing of 200°. Find:
   (a) the distance between *A* and *S*
   (b) the bearing of *A* from *S*.

9. The diagram show three towns *A*, *B*, *C*.
   The bearing of *B* from *A* is 110°.
   The bearing of *C* from *A* is 160°.
   The bearing of *C* from *B* is 240°.
   The distance from *C* to *B* is 110 km.
   Find the distance from *A* to *B*.

10. An aircraft flies from its base 200 km on a bearing 162°, then 350 km on a bearing 260°, and then returns directly to base. Calculate the length and bearing of the return journey.

11. The diagram shows a point A which lies 10 km due south of a point B. A straight road AD is such that the bearing of D from A is 043°.
P and Q are two points on this road which are 8 km from B.
Calculate the bearing of P from B.
(Hint: Find angle BPA first.)

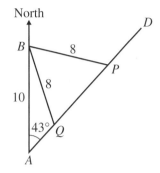

12. The diagram shows wires attached to a communications antenna. Find the length h, correct to the nearest metre.

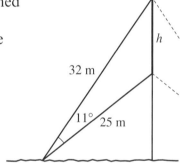

13. Here is a proof of the cosine rule.
    (a) Using $\triangle ADC$ show that $x = b\cos C$.
    (b) Using Pythagoras' theorem in $\triangle ADC$, find h^2 in terms of b and x.

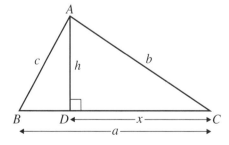

    (c) Using $\triangle ABD$ find c^2 in terms of h, a and x.
    (d) Use your results to prove that
    $c^2 = a^2 + b^2 - 2ab\cos C$.

14. Calculate WX, given $YZ = 15$ m.

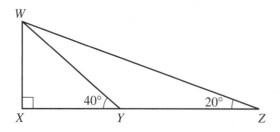

15. Find angle DCB. Check your answer by making a scale drawing.

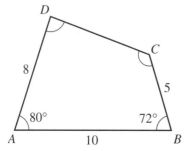

16. Find (a) WX (b) WZ (c) WY.

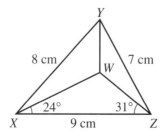

17. The diagram shows a cube of side 10 cm from which one corner has been cut.
Calculate the angle PQR.

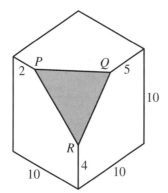

4.8 Circle theorems

Theorem 1
The angle subtended at the centre of a circle is twice the angle subtended at the circumference.

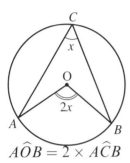

$A\hat{O}B = 2 \times A\hat{C}B$

The proof of this theorem is given on page 402.

Theorem 2
Angles subtended by an arc in the same segment of a circle are equal.

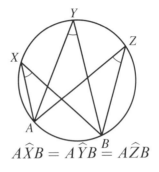

$A\hat{X}B = A\hat{Y}B = A\hat{Z}B$

Example
(a) Given $A\hat{B}O = 50°$, find $B\hat{C}A$.

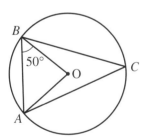

Triangle OBA is isosceles ($OA = OB$).

∴ $O\hat{A}B = 50°$
∴ $B\hat{O}A = 80°$ (angle sum of a triangle)
∴ $B\hat{C}A = 40°$ (angle at the centre)

(b) Given $B\hat{D}C = 62°$ and $D\hat{C}A = 44°$ find $B\hat{A}C$ and $A\hat{B}D$.

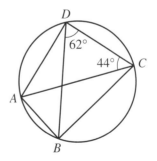

$B\hat{D}C = B\hat{A}C$
(both subtended by arc BC)
∴ $B\hat{A}C = 62°$

$D\hat{C}A = A\hat{B}D$
(both subtended by arc DA)
∴ $A\hat{B}D = 44°$

Exercise 4U

Find the angles marked with letters. A line passes through the centre only when point O is shown.

1.

2.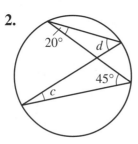

3.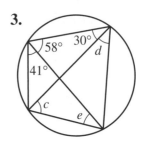

4.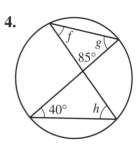

5.

6.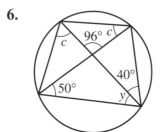

7.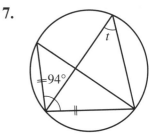

8.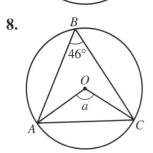

9.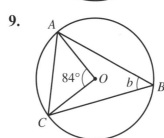

10.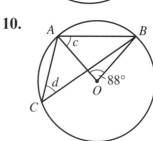

11.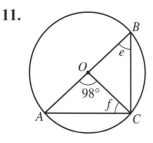

12.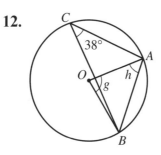

13.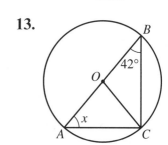

14.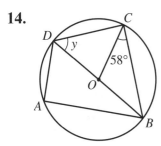

15.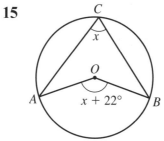

16.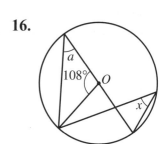

Circle theorems

Theorem 3
The opposite angles in a cyclic quadrilateral add up to 180° (the angles are supplementary).

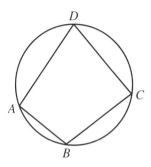

$\hat{A} + \hat{C} = 180°$
$\hat{B} + \hat{D} = 180°$

Find a and x.
$a = 180° - 81°$

(opposite angles of a cyclic quadrilateral)
$\therefore\ a = 99°$

$x + 2x = 180°$
(opposite angles of a cyclic quadrilateral)
$3x = 180°$
$\therefore\ x = 60°$

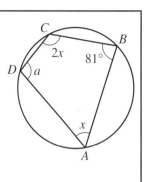

Theorem 4
The angle in a semicircle is a right angle.

In the diagram, AB is a diameter.
$A\hat{C}B = 90°$

Example
Find b given that AOB is a diameter.

$A\hat{C}B = 90°$ (angle in a semicircle)
$\therefore\ b = 180° - (90 + 37)°$
$= 53°$

Exercise 4V
Find the angles marked with a letter.

1.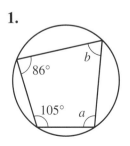
2.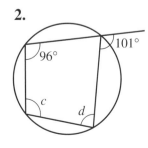
3.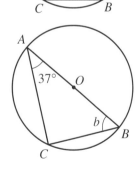
4.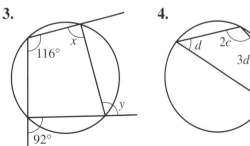

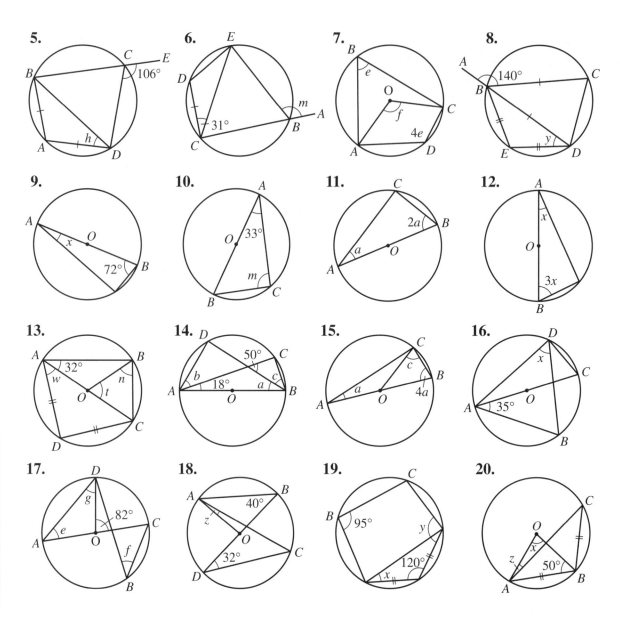

Tangents to circles

Theorem 1
The angle between a tangent and the radius drawn to the point of contact is 90°.

$T\hat{A}O = 90°$
$TA = TB$

Theorem 2
From any point outside a circle just two tangents to the circle may be drawn and they are of equal length.

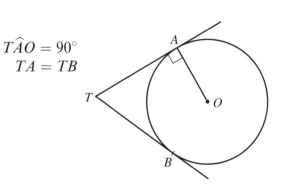

Theorem 3 – Alternate segment theorem

The angle between a tangent and a chord through the point of contact is equal to the angle subtended by the chord in the alternate segment. The theorem is proved in Question **16** of the next exercise.

$$T\widehat{A}B = B\widehat{C}A$$
$$S\widehat{A}C = C\widehat{B}A$$

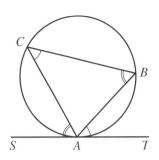

Example

TA and *TB* are tangents to the circle, centre *O*.
Given $A\widehat{T}B = 50°$, find:

(a) $A\widehat{B}T$ (b) $O\widehat{B}A$ (c) $A\widehat{C}B$.

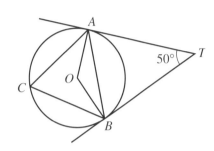

(a) △*TBA* is isosceles (*TA* = *TB*)
 ∴ $A\widehat{B}T = \frac{1}{2}(180 - 50) = 65°$

(b) $O\widehat{B}T = 90°$ (tangent and radius)
 ∴ $O\widehat{B}A = 90 - 65 = 25°$

(c) $A\widehat{C}B = A\widehat{B}T$ (alternate segment theorem)
    $A\widehat{C}B = 65°$

Exercise 4W

Find the angles marked with letters.

1.
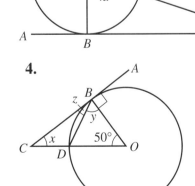

2.
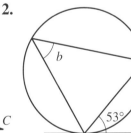

3.
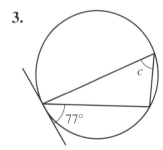

4.
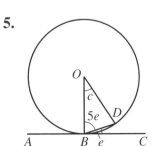

5.

6.
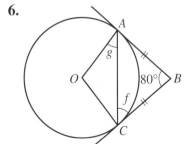

7.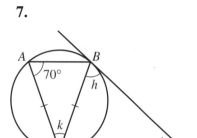

8.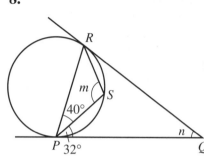

9.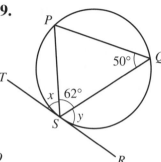

10.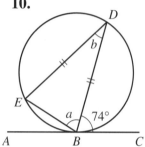

11.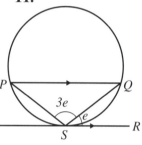

12.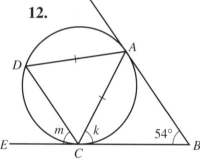

13.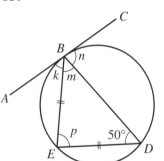

14. Find, in terms of p
(a) $B\hat{A}C$ (b) $X\hat{C}A$
(c) $A\hat{C}O$

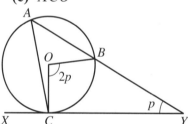

15. Find x, y and z.

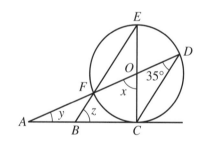

16. Copy and complete the following proof of the alternate segment theorem [i.e. to prove $a = b$].

Draw a diameter RP.
$R\hat{P}A = 90°$ [angle between tangent and radius]
$\therefore\ R\hat{P}Q = 90 - a$

$R\hat{Q}P = \square$ [angle in a semicircle]

$\therefore\ P\hat{R}Q = \square$ [angles in a triangle]

$P\hat{R}Q = P\hat{T}Q$ [angles in the same segment]

$\therefore\ Q\hat{P}A = Q\hat{T}P$ as required.

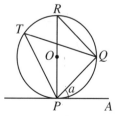

Exercise 4X Mixed questions

Find the angles marked with letters. The centre of the circle is O.

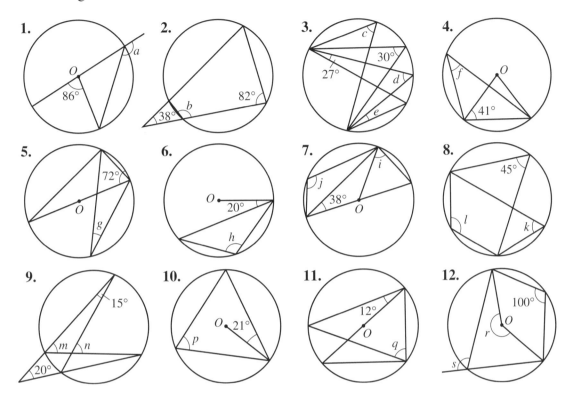

Questions **13** to **18** involve tangents to the circles.

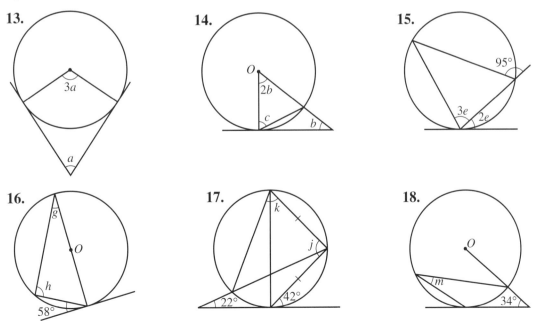

4.9 Geometric proof

When you are writing a geometric proof you should state clearly after each line what you have used.
For example 'angles in a triangle add up to 180°' or 'opposite angles are equal'.

Example 1

Prove that the angle at the centre of a circle is twice the angle at the circumference.

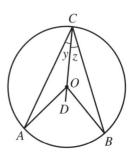

Draw the straight line COD.
Let $A\hat{C}O = y$ and $B\hat{C}O = z$.

In triangle AOC:

$$AO = OC \quad \text{(radii)}$$

$\therefore \quad O\hat{C}A = O\hat{A}C \quad$ (isosceles triangle)

$\therefore \quad C\hat{O}A = 180 - 2y \quad$ (angle sum of triangle)

$\therefore \quad A\hat{O}D = 2y \quad$ (angles on a straight line)

Similarly from triangle COB, we find:

$$D\hat{O}B = 2z$$

Now $\quad A\hat{C}B = y + z$
and $\quad A\hat{O}B = 2y + 2z$

$\therefore \quad A\hat{O}B = 2 \times A\hat{C}B$ as required.

Example 2

Prove that opposite angles in a cyclic quadrilateral add up to 180°.

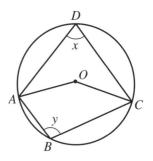

Draw radii OA and OC.
Let $A\hat{D}C = x$ and $A\hat{B}C = y$.

$\quad A\hat{O}C$ obtuse $= 2x$ (angle at the centre)

$\quad A\hat{O}C$ reflex $= 2y$ (angle at the centre)

$\therefore \quad 2x + 2y = 360°$ (angles at a point)

$\therefore \quad x + y = 180°$ as required.

Exercise 4Y

1. Triangle ABC is isosceles.

   Prove that $C\hat{B}D = 2 \times C\hat{A}B$.

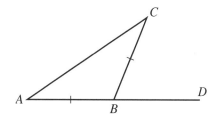

2. Triangle PQR is isosceles and $P\hat{S}Q = 90°$.

   Prove that PS bisects $Q\hat{P}R$.

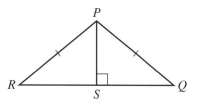

3. Two chords of a circle AB and CD intersect at X.

   Use similar triangles to prove that $AX \times BX = CX \times DX$. This result is called the Intersecting Chords Theorem.

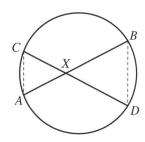

4. Line ATB touches a circle at T and TC is a diameter. AC and BC cut the circle at D and E respectively. Prove that the quadrilateral $ADEB$ is cyclic.

5. Prove that the angle in a semicircle is a right angle.

6. TC is a tangent to a circle at C and BA produced meets this tangent at T.

   Show that triangles TCA and TBC are similar and hence prove that $TC^2 = TA \times TB$.

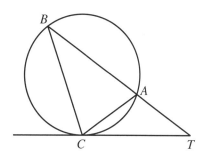

7. Given that BOC is a diameter and that $A\hat{D}C = 90°$, prove that AC bisects $B\hat{C}D$.

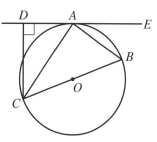

Summary

1. You can draw a net.

2. You can recognise lines and planes of symmetry.

3. You know and can use trigonometric ratios in a right-angled triangle.

4. You can consider dimensions in a formula.

5. You can reflect, rotate, translate and enlarge a shape.

Check out AS4

1. Draw a possible net of a cuboid of sides 2 cm, 5 cm and 3 cm.

2. State:
   (a) the number of lines of symmetry and
   (b) the order of rotational symmetry of the figure below:

3. Find angle PRQ.

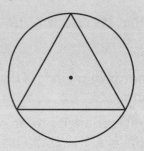

4. Which of the following could be an area?
   (a) $2\pi rh$ (b) $\pi r^2 h$ (c) $3 + 7r^2$

5.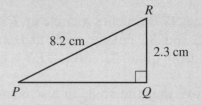

   (a) Reflect the shape in the line $x = 2$.
   (b) Rotate the shape about the point $(3, 0)$ through 90° anticlockwise.
   (c) Translate the shape by $\begin{pmatrix} 5 \\ 1 \end{pmatrix}$.
   (d) Enlarge the shape, by a scale factor of 2, about $(0, 0)$.

6. You can use vectors.

6. D is the mid-point of AB and $AE = \frac{1}{3}AC$.

$\overrightarrow{AD} = \mathbf{a}$, $\overrightarrow{AE} = \mathbf{b}$.

Find $\overrightarrow{BC}$.

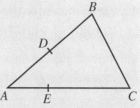

7. You can use the sine, cosine and tangent of any angle.

7. (a) Find $\sin 290°$.
   (b) If $\cos \theta = 0.5$, write down two possible angles for θ.

8. You can use the sine rule.

8. (a) Find AC.
   (b) Find angle BAC and hence find the side BC.

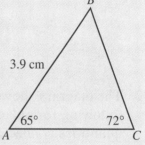

9. You can use the cosine rule.

9. Find the smallest angle in a triangle with lengths 5 cm, 6 cm and 7 cm.

10. You know and can use the circle theorems.

10.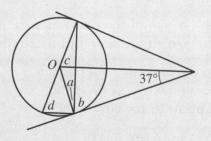

Find the angles a, b, c, d.

11. You can prove the circle theorems.

11. (a) Prove the angle at the centre of a circle equals twice the angle at the circumference.
    (b) Prove that opposite angles in a cyclic quadrilateral add up to $180°$.

Revision exercise AS4

1. Describe fully the **single** transformation which maps triangle A onto triangle B. [SEG]

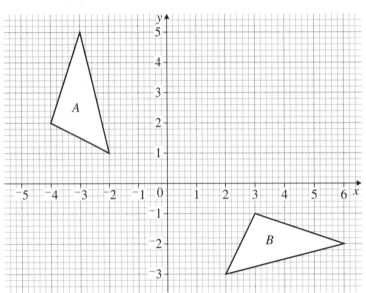

2. The diagram shows a prism. The following formulae represent certain quantities connected with the prism.

   $wx + wy$ $\frac{1}{2}z(x+y)w$

   $\frac{z(x+y)}{2}$ $2(v + 2w + x + y + z)$

   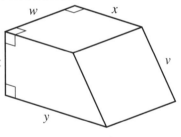

   Not to scale

   (a) Which of these formulae represents length?
   (b) Which of these formulae represents volume? [SEG]

3. In the diagram, P, Q and R lie on a circle. NPT is the tangent to the circle at P.

   Work out the size of angle QPR.
   Give a reason for each step of your working. [SEG]

   Not to scale

   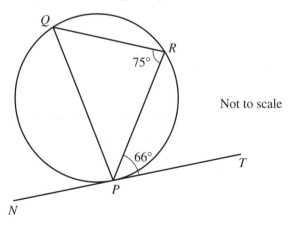

4. $OPQR$ is a parallelogram.
   A and B are points on OP and OR respectively.

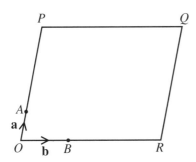

$\overrightarrow{OA} = \mathbf{a}$ and $\overrightarrow{OB} = \mathbf{b}$.
$\overrightarrow{OP} = 4\overrightarrow{OA}$ and $\overrightarrow{OR} = 3\overrightarrow{OB}$.

(a) Find in terms of $\mathbf{a}$ and $\mathbf{b}$.
  (i) $\overrightarrow{AB}$ (ii) $\overrightarrow{RA}$ (iii) $\overrightarrow{BQ}$
(b) S is a point on PQ such that BQ is parallel to AS.
  Find $\overrightarrow{AS}$ in terms of $\mathbf{a}$ and $\mathbf{b}$. [SEG]

5. A and B are two similar cylinders.
   The height of cylinder A is 10 cm and its volume is 625 cm^3.

   Calculate the height of cylinder B. [SEG]

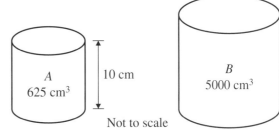

6. A ladder is 2.5 m long and is leaning against a wall, as shown in the diagram. The angle between the ladder and the ground is 61°.

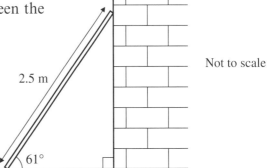

Calculate the height of the top of the ladder above the ground. [AQA]

7. (a) The diagram shows a sketch of the graph of $y = \sin x$ for $^-180° \leqslant x \leqslant 180°$.

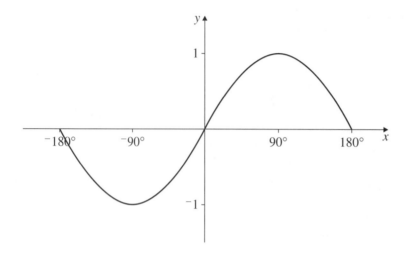

   (i) One solution of the equation $\sin x = 0.9$ is $64°$ to the nearest degree.
   Find the other solution of the equation $\sin x = 0.9$ for $^-180° \leqslant x \leqslant 180°$.
   (ii) Find all the solutions of the equation $\sin x = {}^-0.9$ for $^-180° \leqslant x \leqslant 180°$.
   (b) Sketch the graph of $y = \cos x$ for $^-180° \leqslant x \leqslant 180°$.
   [SEG]

8. A, B and C are three points which lie in a straight line on horizontal ground. BT is a vertical tower.

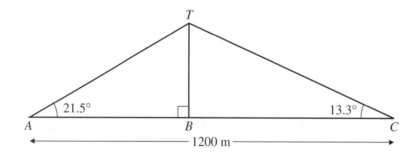

The angle of elevation of T from A is $21.5°$.
The angle of elevation of T from C is $13.3°$.
$AC = 1200\,\text{m}$.

Calculate the height of the tower. [SEG]

AS5 Algebra 3

This unit will show you how to:

- Find and use gradients of a curve
- Find areas under a curve
- Draw and use distance–time graphs
- Factorise quadratic expressions
- Solve quadratic equations
- Manipulate algebraic fractions
- Transform curves
- Solve simultaneous equations where one equation is a quadratic.

Before you start:

You should know how to...	Check in AS5
1. Draw graphs. For example, draw the graph of $y = x^3 - 4x$ for values of x from $^-4$ to $^+4$. 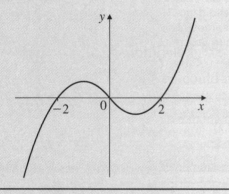	**1.** (a) Draw the graph of $y = 2^x$ for values of x from $^-2$ to $^+3$. (b) Draw the graph of $y = x^2 - 3x + 2$ for values of x from $^-1$ to $^+4$.
2. Solve simultaneous equations by substitution. For example, solve: $y + 5 = 2x$ ① and $2y + 3x = 7$ ② From ①: $y = 2x - 5$ Substituting into ②: $2(2x - 5) + 3x = 7$ $7x = 17$ $x = \dfrac{17}{7} \Rightarrow y = -\dfrac{1}{7}$	**2.** Using the substitution method, solve: $y + 3x = 9$ and $3y + 5x = 11$

5.1 Gradient of a curve

Mathematicians have been interested in finding the gradient of a curve for hundreds of years. When we say 'gradient of the curve', we really mean the 'gradient of the tangent to the curve' at that point.

In the sketch, the gradient of the curve at B is greater than the gradient of the curve at A.

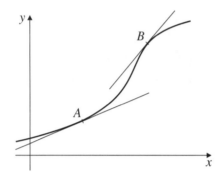

The gradient of a curve is used to find speed or acceleration and such questions appear later in this unit.

Example

The graph of $y = \dfrac{12}{x} + x - 6$ is drawn on the right.

Find the gradient of the tangent to the curve drawn at the point where $x = 5$.

The tangent AB is drawn to touch the curve at $x = 5$.

The gradient of $AB = \dfrac{BC}{AC}$.

$$\text{gradient} = \dfrac{3}{8 - 2 \cdot 4} = \dfrac{3}{5 \cdot 6} \approx 0 \cdot 54$$

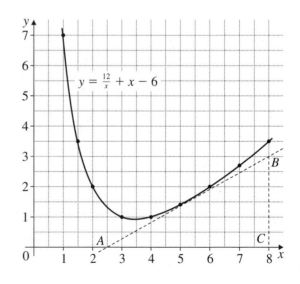

It is difficult to obtain an accurate value for the gradient of a tangent so the result is more realistically 'approximately $0 \cdot 5$'.

Exercise 5A

1. The graph shows $y = x^2 - 4x + 3$ with two tangents drawn.
   (a) Find the gradient of the tangent to the curve at $x = 3$.
   (b) Find the gradient of the tangent to the curve at $x = 0$.
   [Remember to count **units**, not just squares!]

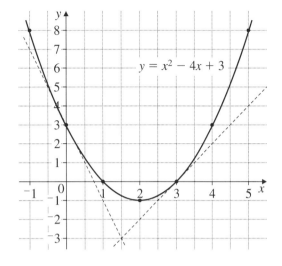

2. Draw the graph of $y = x^2 - 3x$, for $^-2 \leqslant x \leqslant 5$.
   (Scales: 2 cm = 1 unit for x; 1 cm = 1 unit for y).
   Find:
   (a) the gradient of the tangent to the curve at $x = 3$,
   (b) the gradient of the tangent to the curve at $x = {}^-1$,
   (c) the value of x where the gradient of the curve is zero.

3. Draw the graph of $y = 3^x$, for $^-3 \leqslant x \leqslant 3$.
   (Scales: 2 cm = 1 unit for x; 1 cm = 2 units for y).
   Find the gradient of the curve at $x = 1$.

4. Two tangents are drawn to the curve $y = \dfrac{10}{x}$.
   (a) Find the gradient of the curve at $x = 2$.
   (b) Find the gradient of the curve at $x = 5$.
   (c) Copy and complete this sentence:

   'As $x \to \infty$, the gradient of the curve $y = \dfrac{10}{x} \to \Box$.'

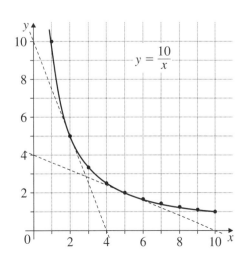

5.2 Area under a curve

It is sometimes useful to know the area under a curve. In the next section we look at the area under a speed-time graph to calculate distance travelled.
The method below, using trapeziums, gives only an approximate value for the area. When a computer is used, the answer can be found to a high degree of accuracy.

Example
Find an approximate value for the area under the curve $y = \dfrac{12}{x}$ between $x = 2$ and $x = 5$.

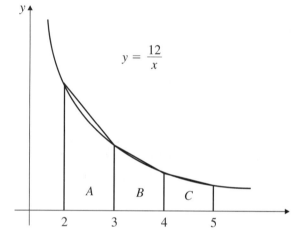

(a) Draw lines parallel to the y-axis at $x = 2, 3, 4, 5$ to form three trapeziums. The area under the curve is approximately equal to the area of the three trapeziums A, B and C.

(b) Calculate the values of y at $x = 2, 3, 4, 5$

at $x = 2$, $y = \frac{12}{2} = 6$

at $x = 3$, $y = \frac{12}{3} = 4$

at $x = 4$, $y = \frac{12}{4} = 3$

at $x = 5$, $y = \frac{12}{5} = 2\cdot 4$

(c) Calculate the area of each of the trapeziums using the formula area $= \frac{1}{2}(a+b)h$, where a and b are the parallel sides and h is the distance in between.

Area $A = \frac{1}{2}(6+4) \times 1 = 5$ sq. units

Area $B = \frac{1}{2}(4+3) \times 1 = 3\cdot 5$ sq. units

Area $C = \frac{1}{2}(3+2\cdot 4) \times 1 = 2\cdot 7$ sq. units

Total area under the curve $\approx (5 + 3\cdot 5 + 2\cdot 7)$

$\approx 11\cdot 2$ sq. units

Exercise 5B

1. Find an approximate value for the area under the curve $y = 3 + 4x - x^2$ between $x = 0$ and $x = 4$. Divide the area into four trapeziums as shown.

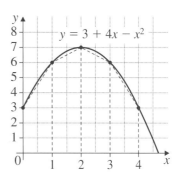

2. Estimate the area under each curve below.

(a)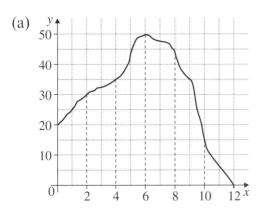

(b)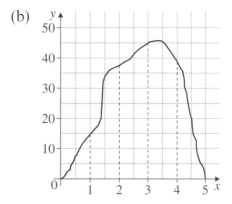

3. Find an approximate value for the area under the curve $y = \dfrac{8}{x}$ between $x = 2$ and $x = 5$.

Divide the area into three trapeziums as shown.

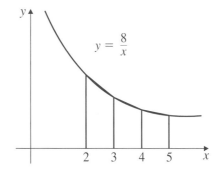

4. Find an approximate value for the area under the curve $y = \dfrac{12}{(x+2)}$ between $x = 0$ and $x = 4$.

Divide the area into four trapeziums as shown.

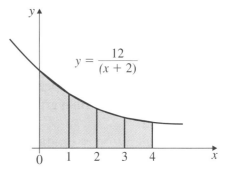

5. (a) Find an approximate value for the area under the curve $y = x^2 + 2x + 5$ between $x = 0$ and $x = 4$.
   Divide the area into four trapeziums.
   (b) State whether your approximate value is greater than or less than the actual value for the area.

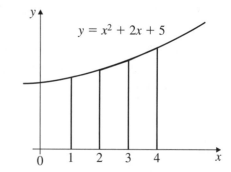

6. The sketch shows the curve $y = x^2 - 5x + 8$ and the line $y = 4$.
   (a) Calculate the x-values at the points A and B.
   (b) Find an approximate value for the area shaded.

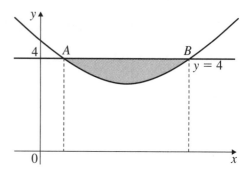

5.3 Distance, velocity, acceleration

- When a **distance–time** graph is drawn, the gradient of the graph gives the speed of the object.

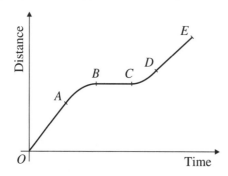

From O to A : constant speed
   A to B : speed goes down to zero
   B to C : at rest
   C to D : speed increases
   D to E : constant speed (not as fast as O to A)
- When a **velocity–time** graph is drawn two quantities can be found:
   (i) acceleration = gradient of graph,
   (ii) distance travelled = area under graph.

Example 1

The diagram is the speed–time graph of the first 30 seconds of a car journey.

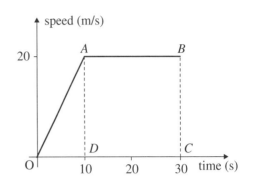

(a) The gradient of line $OA = \frac{20}{10} = 2$.

∴ The acceleration in the first 10 seconds is $2 \, \text{m/s}^2$.

(b) The distance travelled in the first 30 seconds is given by the area of OAD plus the area of $ABCD$.

Distance $= (\frac{1}{2} \times 10 \times 20) + (20 \times 20) = 500 \, \text{m}$

Example 2

The sketch shows the velocity–time graph of a rocket as it takes off. The equation of the curve is $v = 11t^2$.

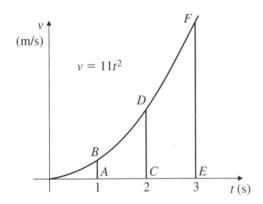

Find an approximate value for the distance travelled by the rocket in the first 3 seconds.

We use 'distance travelled = area under graph'.

Divide the area into 3 trapeziums (one of which is a triangle).

$AB = 11$, $CD = 44$, $EF = 99$ (from $v = 11t^2$)

Area $\approx \frac{1}{2}(0 + 11) \times 1 + \frac{1}{2}(11 + 44) \times 1 + \frac{1}{2}(44 + 99) \times 1$

$\approx 104 \cdot 75$

Distance travelled in the first 3 seconds is about $105 \, \text{m}$.

Note:
To find the acceleration of the rocket at $t = 2$, we could draw a tangent to the curve at this point and so find its gradient.

Exercise 5C

The graphs show speed v in m/s and time t in seconds.

1. Find:
   (a) the acceleration when $t = 4$
   (b) the total distance travelled.

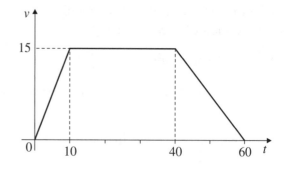

2. Find:
   (a) the total distance travelled
   (b) the distance travelled in the first 10 seconds
   (c) the acceleration when $t = 20$.

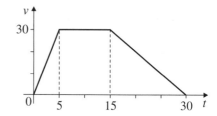

3. Find:
   (a) the total distance travelled
   (b) the distance travelled in the first 40 seconds
   (c) the acceleration when $t = 15$.

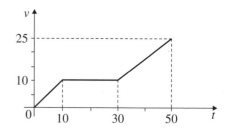

4. Find:
   (a) V if the total distance travelled is 900 m
   (b) the distance travelled in the first 60 seconds.

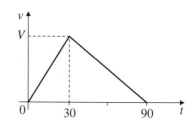

5. Find:
   (a) T if the initial acceleration is $2 \, \text{m/s}^2$
   (b) the total distance travelled
   (c) the average speed for the whole journey.

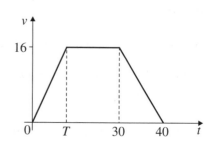

6. Each graph represents one of the following situations.
    A A car moving in congested traffic
    B A ball rolling along a lane at a bowling alley
    C An apple thrown vertically in the air
    D A parachutist after jumping from a stationary hot-air balloon
    E A lift moving from level 2 to level 3
    F A table tennis ball during a game.

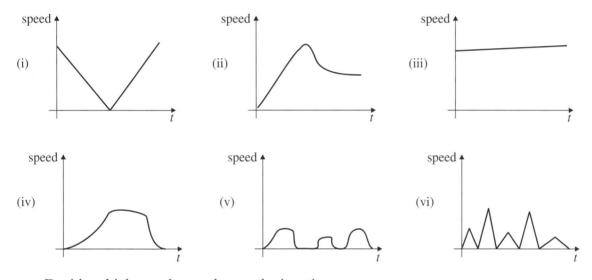

Decide which graph matches each situation.

7. The speed of a train is measured at regular intervals of time from $t = 0$ to $t = 60$ s, as shown in the table.

t s	0	10	20	30	40	50	60
v m/s	0	10	16	19·7	22·2	23·8	24·7

Draw a speed–time graph to illustrate the motion. Plot t on the horizontal axis with a scale of 1 cm to 5 s and plot v on the vertical axis with a scale of 2 cm to 5 m/s. Use the graph to estimate:
(a) the acceleration at $t = 10$
(b) the distance travelled by the train from $t = 30$ to $t = 60$.

8. The speed of a car is measured at regular intervals of time from $t = 0$ to $t = 60$ s, as shown below.

t s	0	10	20	30	40	50	60
v m/s	0	1·3	3·2	6	10·1	16·5	30

Draw a speed–time graph using the same scales as in Question **6**. Use the graph to estimate:
(a) the acceleration at $t = 30$
(b) the distance travelled by the car from $t = 20$ to $t = 50$.

9. The sketch graph shows the distance–time graph of an object in an experiment.
[N.B. Not a velocity–time graph.]

The graph passes through (0, 0), (0·2, 1·2), (0·5, 1·6), (1, 2), (2, 2·5), (3, 2·9).

Draw your own accurate graph and then draw a suitable tangent to find the speed of the object at $t = 0·5$ seconds.
Use a scale of 5 cm to 1 second across the page and 5 cm to 1 unit up the page.

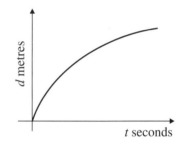

10. Given that the average speed for the whole journey is 37·5 m/s and that the deceleration between T and $2T$ is 2·5 m/s², find:
(a) the value of V (b) the value of T.

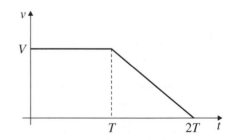

11. Given that the total distance travelled is 4 km and that the initial deceleration is 4 m/s², find:
(a) the value of V (b) the value of T.

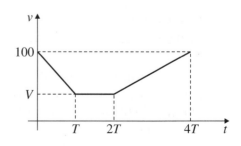

5.4 Quadratic equations

Factorising quadratic expressions

Earlier in the book a pair of brackets like $(x + 4)(x - 3)$ were multiplied to give $x^2 + x - 12$.

The reverse of this process is called **factorising**.

Example 1

Factorise $x^2 + 6x + 8$.

(a) Find two numbers which multiply to give 8 and add up to 6.

(b) Put these numbers into brackets.
So $x^2 + 6x + 8 = (x + 4)(x + 2)$

Example 2

Factorise: (a) $x^2 + 2x - 15$
(b) $x^2 - 6x + 8$

(a) Two numbers which multiply to give $^-15$ and add up to $^+2$ are $^-3$ and 5.
$\therefore \quad x^2 + 2x - 15 = (x - 3)(x + 5)$

(b) Two numbers which multiply to give $^+8$ and add up to $^-6$ are $^-2$ and $^-4$.
$\therefore \quad x^2 - 6x + 8 = (x - 2)(x - 4)$

Exercise 5D

Factorise the following:

1. $x^2 + 7x + 10$
2. $x^2 + 7x + 12$
3. $x^2 + 8x + 15$
4. $x^2 + 10x + 21$
5. $x^2 + 8x + 12$
6. $y^2 + 12y + 35$
7. $y^2 + 11y + 24$
8. $y^2 + 10y + 25$
9. $y^2 + 15y + 36$
10. $a^2 - 3a - 10$
11. $a^2 - a - 12$
12. $z^2 + z - 6$
13. $x^2 - 2x - 35$
14. $x^2 - 5x - 24$
15. $x^2 - 6x + 8$
16. $y^2 - 5y + 6$
17. $x^2 - 8x + 15$
18. $a^2 - a - 6$
19. $a^2 + 14a + 45$
20. $b^2 - 4b - 21$
21. $x^2 - 8x + 16$
22. $y^2 + 2y + 1$
23. $y^2 - 3y - 28$
24. $x^2 - x - 20$
25. $x^2 - 8x - 240$
26. $x^2 - 26x + 165$
27. $y^2 + 3y - 108$
28. $x^2 - 49$
29. $x^2 - 9$
30. $x^2 - 16$

31. Factorise:
   (a) $2x^2 + 4x - 30$
   (b) $3x^2 + 21x + 30$
   (c) $3x^2 + 24x + 45$
   (d) $2n^2 - 6n - 20$
   (e) $5a^2 + 5a - 30$
   (f) $4x^2 - 64$

> **Hint for 31a:**
> $2x^2 + 4x - 30 = 2(x^2 + 2x - 15)$
> Complete the factorisation.

Example
Factorise $3x^2 + 13x + 4$.

(a) Find two numbers which multiply to give (3×4), i.e. 12, and add up to 13. In this case the numbers are 1 and 12.

(b) Split the '$13x$' term, $3x^2 + x + 12x + 4$

(c) Factorise in pairs, $x(3x + 1) + 4(3x + 1)$

(d) $(3x + 1)$ is common, $(3x + 1)(x + 4)$

Exercise 5E
Factorise the following:

1. $2x^2 + 5x + 3$
2. $2x^2 + 7x + 3$
3. $3x^2 + 7x + 2$
4. $2x^2 + 11x + 12$
5. $3x^2 + 8x + 4$
6. $2x^2 + 7x + 5$
7. $3x^2 - 5x - 2$
8. $2x^2 - x - 15$
9. $2x^2 + x - 21$
10. $3x^2 - 17x - 28$
11. $6x^2 + 7x + 2$
12. $3x^2 - 11x + 6$
13. $3y^2 - 11y + 10$
14. $6y^2 + 7y - 3$
15. $10x^2 + 9x + 2$
16. $6x^2 - 19x + 3$
17. $8x^2 - 10x - 3$
18. $12x^2 + 23x + 10$

The difference of two squares

- $x^2 - y^2 = (x - y)(x + y)$
Remember this result.

Example
Factorise: (a) $y^2 - 16$ (b) $4a^2 - b^2$

(a) $y^2 - 16 = (y - 4)(y + 4)$
(b) $4a^2 - b^2 = (2a - b)(2a + b)$

Exercise 5F
Factorise the following:

1. $y^2 - a^2$
2. $m^2 - n^2$
3. $x^2 - t^2$
4. $y^2 - 1$
5. $x^2 - 9$
6. $a^2 - 25$
7. $x^2 - \dfrac{1}{4}$
8. $x^2 - \dfrac{1}{9}$
9. $4x^2 - y^2$
10. $a^2 - 4b^2$
11. $25x^2 - 4y^2$
12. $9x^2 - 16y^2$
13. $4x^2 - \dfrac{z^2}{100}$
14. $x^3 - x$
15. $a^3 - ab^2$
16. $4x^3 - x$
17. $8x^3 - 2xy^2$
18. $y^3 - 9y$

19. Find the exact value of $100\,003^2 - 99\,997^2$.

20. Find the exact value of $1\,500\,002^2 - 1\,499\,998^2$.

21. Rewrite 9991 as the difference of two squares. Use your answer to find the prime factors of 9991.

Methods of solving quadratic equations

The last three exercises have been about factorising expressions. These contain no 'equals' sign, but equations do. Quadratic equations always have an x^2 term, and often an x term as well as a number term. They generally have two different solutions.

Here three different methods of solution are considered.

(a) Solution by factors

Consider the equation $a \times b = 0$, where a and b are numbers. The product $a \times b$ can only be zero if either a or b (or both) is equal to zero. Can you think of other possible pairs of numbers which multiply together to give zero?

Example

(a) Solve the equation $x^2 + x - 12 = 0$.

Factorising, $(x - 3)(x + 4) = 0$

either $x - 3 = 0$ or $x + 4 = 0$
$x = 3$ $x = {}^-4$

(b) Solve the equation $6x^2 + x - 2 = 0$.

Factorising, $(2x - 1)(3x + 2) = 0$

either $2x - 1 = 0$ or $3x + 2 = 0$
$2x = 1$ $3x = {}^-2$
$x = \tfrac{1}{2}$ $x = {}^-\tfrac{2}{3}$

Exercise 5G

Solve the following equations.

1. $x^2 + 7x + 12 = 0$
2. $x^2 + 7x + 10 = 0$
3. $x^2 + 2x - 15 = 0$
4. $x^2 + x - 6 = 0$
5. $x^2 - 8x + 12 = 0$
6. $x^2 + 10x + 21 = 0$
7. $x^2 - 5x + 6 = 0$
8. $x^2 - 4x - 5 = 0$
9. $x^2 + 5x - 14 = 0$
10. $2x^2 - 3x - 2 = 0$
11. $3x^2 + 10x - 8 = 0$
12. $2x^2 + 7x - 15 = 0$
13. $6x^2 - 13x + 6 = 0$
14. $4x^2 - 29x + 7 = 0$
15. $10x^2 - x - 3 = 0$

16. $y^2 - 15y + 56 = 0$
17. $12y^2 - 16y + 5 = 0$
18. $y^2 + 2y - 63 = 0$
19. $x^2 + 2x + 1 = 0$
20. $x^2 - 6x + 9 = 0$
21. $x^2 + 10x + 25 = 0$
22. $x^2 - 14x + 49 = 0$
23. $6a^2 - a - 1 = 0$
24. $4a^2 - 3a - 10 = 0$
25. $z^2 - 8z - 65 = 0$
26. $6x^2 + 17x - 3 = 0$
27. $10k^2 + 19k - 2 = 0$
28. $y^2 - 2y + 1 = 0$
29. $36x^2 + x - 2 = 0$
30. $20x^2 - 7x - 3 = 0$
31. $x^4 - 5x^2 + 4 = 0$
32. $x^4 - 13x^2 + 36 = 0$

[Hint: Put $y = x^2$]

33. $4x^4 - 17x^2 + 4 = 0$
34. $x^6 - 9x^3 + 8 = 0$

Sometimes a quadratic expression can have only two terms.

Example

(a) Solve the equation $x^2 - 7x = 0$.

Factorising, $x(x - 7) = 0$

either $x = 0$ or $x - 7 = 0$

$x = 7$

The solutions are $x = 0$ and $x = 7$.

(b) Solve the equation $x^2 - 100 = 0$.

Rearranging, $x^2 = 100$

Take square root, $x = 10$ or $^-10$

[Alternatively, use the difference of squares.]

Exercise 5H

Solve the following equations.

1. $x^2 - 3x = 0$
2. $x^2 + 7x = 0$
3. $2x^2 - 2x = 0$
4. $3x^2 - x = 0$
5. $x^2 - 16 = 0$
6. $x^2 - 49 = 0$
7. $4x^2 - 1 = 0$
8. $9x^2 - 4 = 0$
9. $6y^2 + 9y = 0$
10. $6a^2 - 9a = 0$
11. $10x^2 - 55x = 0$
12. $16x^2 - 1 = 0$
13. $y^2 - \frac{1}{4} = 0$
14. $56x^2 - 35x = 0$
15. $36x^2 - 3x = 0$
16. $x^2 = 6x$
17. $x^2 = 11x$
18. $2x^2 = 3x$
19. $x^2 = x$
20. $4x = x^2$

(b) Solution by formula

- The solutions of the quadratic equation $ax^2 + bx + c = 0$

are given by the formula $x = \dfrac{^-b \pm \sqrt{(b^2 - 4ac)}}{2a}$.

Use this formula only after trying (and failing) to factorise.

Example

Solve the equation $2x^2 - 3x - 4 = 0$.

In this case $a = 2$, $b = ^-3$, $c = -4$.

$$x = \frac{^-(^-3) \pm \sqrt{[(^-3)^2 - (4 \times 2 \times ^-4)]}}{2 \times 2}$$

$$x = \frac{3 \pm \sqrt{[9+32]}}{4} = \frac{3 \pm \sqrt{41}}{4}$$

$$x = \frac{3 \pm 6\cdot403}{4}$$

either $\quad x = \dfrac{3 + 6\cdot403}{4} = 2\cdot35$ (2 decimal places)

or $\quad x = \dfrac{3 - 6\cdot403}{4} = \dfrac{^-3\cdot403}{4} = {^-}0\cdot85$ (2 decimal places)

Exercise 5I

Solve the following, giving answers to 2 decimal places where necessary.

1. $2x^2 + 11x + 5 = 0$
2. $3x^2 + 11x + 6 = 0$
3. $6x^2 + 7x + 2 = 0$
4. $3x^2 - 10x + 3 = 0$
5. $5x^2 - 7x + 2 = 0$
6. $6x^2 - 11x + 3 = 0$
7. $2x^2 + 6x + 3 = 0$
8. $x^2 + 4x + 1 = 0$
9. $5x^2 - 5x + 1 = 0$
10. $x^2 - 7x + 2 = 0$
11. $2x^2 + 5x - 1 = 0$
12. $3x^2 + x - 3 = 0$
13. $3x^2 + 8x - 6 = 0$
14. $3x^2 - 7x - 20 = 0$
15. $2x^2 - 7x - 15 = 0$
16. $x^2 - 3x - 2 = 0$
17. $2x^2 + 6x - 1 = 0$
18. $6x^2 - 11x - 7 = 0$
19. $3x^2 + 25x + 8 = 0$
20. $3y^2 - 2y - 5 = 0$
21. $2y^2 - 5y + 1 = 0$
22. $\frac{1}{2}y^2 + 3y + 1 = 0$
23. $2 - x - 6x^2 = 0$
24. $3 + 4x - 2x^2 = 0$

25. (a) To find the x-coordinate of A and B, solve the equation
$$x^2 - 5x + 1 = 0.$$

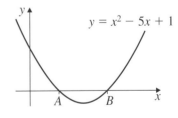

(b) Explain why the equation $x^2 - 5x + 10 = 0$ has no solutions.

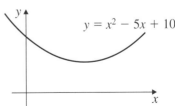

26. Copy and complete the sentence: 'The equation $ax^2 + bx + c = 0$ has solutions if $(b^2 - 4ac)$ is _____.'

The solution to a problem can involve an equation which does not at first appear to be quadratic. The terms in the equation may need to be rearranged as shown below.

Example
Solve:
$$2x(x-1) = (x+1)^2 - 5$$
$$2x^2 - 2x = x^2 + 2x + 1 - 5$$
$$2x^2 - 2x - x^2 - 2x - 1 + 5 = 0$$
$$x^2 - 4x + 4 = 0$$
$$(x-2)(x-2) = 0$$
$$x = 2$$

In this example the quadratic has a repeated root of $x = 2$.

Exercise 5J
Solve the following, giving answers to 2 decimal places where necessary.

1. $x^2 = 6 - x$
2. $x(x+10) = -21$
3. $3x + 2 = 2x^2$
4. $x^2 + 4 = 5x$
5. $6x(x+1) = 5 - x$
6. $(2x)^2 = x(x-14) - 5$
7. $(x-3)^2 = 10$
8. $(x+1)^2 - 10 = 2x(x-2)$
9. $(2x-1)^2 = (x-1)^2 + 8$
10. $3x(x+2) - x(x-2) + 6 = 0$
11. $x = \dfrac{15}{x} - 22$
12. $x + 5 = \dfrac{14}{x}$
13. $4x + \dfrac{7}{x} = 29$
14. $10x = 1 + \dfrac{3}{x}$
15. $2x^2 = 7x$
16. $16 = \dfrac{1}{x^2}$
17. $2x + 2 = \dfrac{7}{x} - 1$
18. $\dfrac{2}{x} + \dfrac{2}{x+1} = 3$
19. $\dfrac{3}{x-1} + \dfrac{3}{x+1} = 4$
20. $\dfrac{2}{x-2} + \dfrac{4}{x+1} = 3$

21. Hassan says you can make it easier to solve some equations by dividing through by a common factor.
So, by dividing through by 2, the equation $2x^2 + 10x - 48 = 0$ can be written as $x^2 + 5x - 24 = 0$.
Is this OK?
Check by solving the two equations above.
Are the solutions the same?

22. Solve the equations:
   (a) $4x^2 - 10x - 6 = 0$ (b) $3x^2 - 6x - 72 = 0$

23. One of the solutions published by Cardan in 1545 for the solution of cubic equations is given below. For an equation in the form $x^3 + px = q$

$$x = \sqrt[3]{\left[\sqrt{\left(\frac{p}{3}\right)^3 + \left(\frac{q}{2}\right)^2} + \frac{q}{2}\right]} - \sqrt[3]{\left[\sqrt{\left(\frac{p}{3}\right)^3 + \left(\frac{q}{2}\right)^2} - \frac{q}{2}\right]}$$

Use the formula to solve the following equations, giving answers to 4 sig. fig. where necessary.
   (a) $x^3 + 7x = {}^-8$ (b) $x^3 + 6x = 4$
   (c) $x^3 + 3x = 2$ (d) $x^3 + 9x - 2 = 0$

Girolamo Cardan (1501–1576) was a colourful character who became Professor of Mathematics at Milan. As well as being a distinguished academic, he was an astrologer, a physician, a gambler and a heretic, yet he received a pension from the Pope. His mathematical genius enabled him to open up the general theory of cubic and quartic equations, although a method for solving cubic equations which he claimed as his own was pirated from Niccolo Tartaglia.

(c) Solution by completing the square

Look at the function $f(x) = x^2 + 6x$

Completing the square, this becomes $f(x) = (x + 3)^2 - 9$

This is done as follows.
1. 3 is half of 6 and gives $6x$.
2. Having added 3 to the square term, 9 needs to be subtracted from the expression to cancel the ${}^+9$ obtained.

Here are some more examples.

(a) $x^2 - 12x = (x - 6)^2 - 36$

(b) $x^2 + 3x = (x + \frac{3}{2})^2 - \frac{9}{4}$

(c) $x^2 + 6x + 1 = (x + 3)^2 - 9 + 1$
$ = (x + 3)^2 - 8$

(d) $x^2 - 10x - 17 = (x - 5)^2 - 25 - 17$
$ = (x - 5)^2 - 42$

(e) $2x^2 - 12x + 7 = 2[x^2 - 6x + \frac{7}{2}]$
$ = 2[(x - 3)^2 - 9 + \frac{7}{2}]$
$ = 2[(x - 3)^2 - \frac{11}{2}]$

Example 1

Solve the quadratic equation $x^2 - 6x + 7 = 0$ by completing the square.

$(x - 3)^2 - 9 + 7 = 0$
$(x - 3)^2 = 2$
$\therefore \quad x - 3 = {}^+\sqrt{2}$ or ${}^-\sqrt{2}$
$ = 3 + \sqrt{2}$ or $3 - \sqrt{2}$

So, $x = 4\cdot41$ or $1\cdot59$ to 2 d.p.

Example 2

Given $f(x) = x^2 - 8x + 18$, show that $f(x) \geqslant 2$ for all values of x.

Completing the square, $f(x) = (x-4)^2 - 16 + 18$
$$f(x) = (x-4)^2 + 2.$$

Now $(x-4)^2$ is always greater than or equal to zero because it is 'something squared'.

$$\therefore \quad f(x) \geqslant 2$$

Exercise 5K

In Questions **1** to **10**, complete the square for each expression by writing each one in the form $(x+a)^2 + b$ where a and b can be positive or negative.

1. $x^2 + 8x$
2. $x^2 - 12x$
3. $x^2 + x$
4. $x^2 + 4x + 1$
5. $x^2 - 6x + 9$
6. $x^2 + 2x - 15$
7. $x^2 + 16x + 5$
8. $x^2 - 10x$
9. $x^2 + 3x$

10. Solve the equations by completing the square.
    (a) $x^2 - 8x + 12 = 0$
    (b) $x^2 + 10x + 21 = 0$
    (c) $x^2 - 4x - 5 = 0$

11. Solve these equations by completing the square.
    (a) $x^2 + 4x - 3 = 0$
    (b) $x^2 - 3x - 2 = 0$
    (c) $x^2 + 12x = 1$

12. Try to solve the equation $x^2 + 6x + 10 = 0$, by completing the square. Explain why you can find no solutions.

13. Given $f(x) = x^2 + 6x + 12$, show that $f(x) \geqslant 3$ for all values of x.

14. Given $g(x) = x^2 - 7x + \frac{1}{4}$, show that the least possible value of $g(x)$ is $^-12$.

15. If $f(x) = x^2 + 4x + 7$ find:
    (a) the smallest possible value of $f(x)$,
    (b) the value of x for which this smallest value occurs,
    (c) the greatest possible value of $\dfrac{1}{(x^2 + 4x + 7)}$.

Using quadratic equations to solve problems

Example

The perimeter of a rectangle is 42 cm. If the diagonal is 15 cm, find the width of the rectangle.

Let the width of the rectangle be x cm.
Since the perimeter is 42 cm, the sum of the length and the width is 21 cm.
∴ length of rectangle = $(21 - x)$ cm

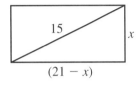

By Pythagoras' theorem
$$x^2 + (21 - x)^2 = 15^2$$
$$x^2 + (21 - x)(21 - x) = 15^2$$
$$x^2 + 441 - 42x + x^2 = 225$$
$$2x^2 - 42x + 216 = 0$$
$$x^2 - 21x + 108 = 0$$
$$(x - 12)(x - 9) = 0$$
$$x = 12 \quad \text{or} \quad x = 9$$

Note that the dimensions of the rectangle are 9 cm and 12 cm, whichever value of x is taken.
∴ The width of the rectangle is 9 cm.

Exercise 5L

Solve by forming a quadratic equation.

1. Two numbers, which differ by 3, have a product of 88. Find them.
   [Call the numbers x and $x + 3$.]

2. The product of two consecutive odd numbers is 143. Find the numbers. [Hint: If the first odd number is x, what is the next odd number?]

3. The height of a photo exceeds the width by 7 cm. If the area is 60 cm², find the height of the photo.

4. The area of the rectangle exceeds the area of the square by 24 m². Find x.

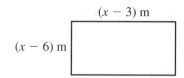
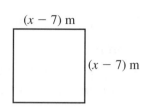

5. Three consecutive integers are written as x, $x+1$, $x+2$. The square of the largest number is 45 less than the sum of the squares of the other numbers. Find the three numbers.

6. $(x-1)$, x and $(x+1)$ represent three positive integers. The product of the three numbers is five times their sum.
   (a) Write an equation in x.
   (b) Show that your equation simplifies to $x^3 - 16x = 0$.
   (c) Factorise $x^3 - 16x$ completely.
   (d) Hence find the three positive integers.

7. An aircraft flies a certain distance on a bearing of 045° and then twice the distance on a bearing of 135°. Its distance from the starting point is then 350 km. Find the length of the first part of the journey.

8. The area of rectangle A is twice the area of rectangle B. Find x.

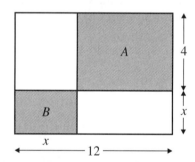

9. The perimeter of a rectangle is 68 cm. If the diagonal is 26 cm, find the dimensions of the rectangle.

10. A stone is thrown in the air. After t seconds its height, h, above sea level is given by the formula $h = 80 + 3t - 5t^2$. Find the value of t when the stone falls into the sea.

11. The total surface area of a cylinder, A, is given by the formula $A = 2\pi r^2 + 2\pi rh$.
    Given that $A = 200\,\text{cm}^2$ and $h = 10\,\text{cm}$, find the value of r, correct to 1 d.p.

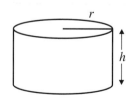

12. A rectangular pond, 6 m × 4 m, is surrounded by a uniform path of width x. The area of the path is equal to the area of the pond.
Find x.

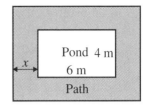

13. The perimeters of a square and a rectangle are equal. The length of the rectangle is 11 cm and the area of the square is 4 cm² more than the area of the rectangle. Find the side of the square.

14. The sequence 3, 6, 15, 24, … can be written (1 × 3), (2 × 4), (3 × 5), (4 × 6)… Write down an expression for the nth term of the sequence. One term in the sequence is 255. Form an equation and hence find what number term it is.

15. When each edge of a cube is decreased by 1 cm, its volume is decreased by 91 cm³. Find the length of a side of the original cube.

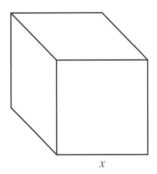

16. A cyclist travels 40 km at a speed x km/h. Find the time taken in terms of x. Find the time taken when his speed is reduced by 2 km/h. If the difference between the times is 1 hour, find the original speed.

17. An increase of speed of 4 km/h on a journey of 32 km reduces the time taken by 4 hours. Find the original speed.

18. A train normally travels 60 miles at a certain speed. One day, due to bad weather, the train's speed is reduced by 10 mph so that the journey takes 3 hours longer. Find the normal speed.

19. A number exceeds four times its reciprocal by 3. Find the number.

20. Paper often comes in standard sizes A0, A1, A2, A3 etc. The large sheet shown is size A0; A0 can be cut in half to give two sheets of A1; A1 can be cut in half to give two sheets of A2 and so on.

Call the sides of A1 x and y, as shown.
   (a) What are the sides of A0 in terms of x and y?
   (b) All the sizes of paper are similar. Form an equation involving x and y.
   (c) Size A0 has an area of 1 m². Calculate the values of x and y correct to the nearest mm.
   (d) Hence calculate the long side of a sheet of A4. Check your answer by measuring a sheet of A4.

5.5 Algebraic fractions

Example

(a) (i) $\dfrac{3a}{5a^2} = \dfrac{3 \times \cancel{a}}{5 \times a \times \cancel{a}} = \dfrac{3}{5a}$ (ii) $\dfrac{3y + y^2}{6y} = \dfrac{y(3+y)}{6y}$

$\phantom{(a) (i) \dfrac{3a}{5a^2} = \dfrac{3 \times \cancel{a}}{5 \times a \times \cancel{a}} = \dfrac{3}{5a} (ii) \dfrac{3y + y^2}{6y}} = \dfrac{3+y}{6}$

(b) Write as a single fraction: (i) $\dfrac{4\cancel{x}}{3} \times \dfrac{2}{\cancel{x}^{\,2}}$ (ii) $\dfrac{5}{n} \div \dfrac{3}{n}$

$ = \dfrac{8}{3x}$ $ = \dfrac{5}{\cancel{n}} \times \dfrac{\cancel{n}}{3} = \dfrac{5}{3}$

Exercise 5M

1. Simplify as far as possible.

   (a) $\dfrac{25}{35}$ (b) $\dfrac{5y^2}{y}$ (c) $\dfrac{y}{2y}$ (d) $\dfrac{8x^2}{2x^2}$

   (e) $\dfrac{2x}{4y}$ (f) $\dfrac{6y}{3y}$ (g) $\dfrac{5ab}{10b}$ (h) $\dfrac{8ab^2}{12ab}$

2. Sort these into four pairs of equivalent expressions.

   A $\dfrac{x^2}{2x}$ B $\dfrac{x(x+1)}{x^2}$ C $\dfrac{x^2+x}{x^2-x}$ D $\dfrac{3x+6}{3x}$

   E $\dfrac{x+1}{x}$ F $\dfrac{x+2}{x}$ G $\dfrac{x}{2}$ H $\dfrac{x+1}{x-1}$

3. Write as a single fraction.

   (a) $\dfrac{3x}{2} \times \dfrac{2a}{3x}$ (b) $\dfrac{5mn}{3} \times \dfrac{2}{n}$ (c) $\dfrac{3y^2}{3} \times \dfrac{2x}{9y}$ (d) $\dfrac{2}{q} \div \dfrac{a}{2}$

   (e) $\dfrac{4x}{3} \div \dfrac{x}{2}$ (f) $\dfrac{x}{5} \times \dfrac{y^2}{x^2}$ (g) $\dfrac{a^2}{5} \div \dfrac{a}{10}$ (h) $\dfrac{x^2}{x^2 + 2x} \div \dfrac{x}{x+2}$

4. Simplify as far as possible.

   (a) $\dfrac{7a^2b}{35ab^2}$ (b) $\dfrac{(2a)^2}{4a}$ (c) $\dfrac{7yx}{8xy}$ (d) $\dfrac{3x}{4x - x^2}$

   (e) $\dfrac{5x + 2x^2}{3x}$ (f) $\dfrac{9x + 3}{3x}$ (g) $\dfrac{4a + 5a^2}{5a}$ (h) $\dfrac{5ab}{15a + 10a^2}$

5. Copy and complete.

   (a) $\dfrac{x^2}{3} \times \dfrac{\square}{x} = 2x$ (b) $\dfrac{8}{x} \div \dfrac{2}{x} = \square$ (c) $\dfrac{a}{3} + \dfrac{a}{3} = \dfrac{\square}{3}$

6. Sort these into four pairs of equivalent expressions.

   A $\dfrac{x^2}{3x}$ B $\dfrac{x}{2} \times \dfrac{x}{2}$ C $\dfrac{12x + 6}{6}$ D $\dfrac{x}{5} - \dfrac{2}{5}$

   E $\dfrac{2x^2 + x}{x}$ F $\dfrac{x(x+1)}{3x+3}$ G $\dfrac{x-2}{5}$ H $\dfrac{ax^2}{4a}$

7. Write in a more simple form.

   (a) $\dfrac{5x + 10y}{15xy}$ (b) $\dfrac{18a - 3ab}{6a^2}$ (c) $\dfrac{4ab + 8a^2}{2ab}$ (d) $\dfrac{(2x)^2 - 8x}{4x}$

8. Simplify as far as possible.

   (a) $\dfrac{x^2 + 2x}{x^2 - 3x}$ (b) $\dfrac{x^2 - 3x}{x^2 - 2x - 3}$ (c) $\dfrac{x^2 + 4x}{2x^2 - 10x}$

   (d) $\dfrac{x^2 + 6x + 5}{x^2 - x - 2}$ (e) $\dfrac{x^2 - 4x - 21}{x^2 - 5x - 14}$ (f) $\dfrac{x^2 + 7x + 10}{x^2 - 4}$

Addition and subtraction of algebraic fractions

Example
(a) Write as a single fraction: $\dfrac{2}{x} + \dfrac{3}{y}$

The LCM of x and y is xy.

$\therefore \quad \dfrac{2}{x} + \dfrac{3}{y} = \dfrac{2y}{xy} + \dfrac{3x}{xy} = \dfrac{2y + 3x}{xy}$

(b) Write as a single fraction: $\dfrac{4}{x} + \dfrac{5}{x-1}$

The LCM of x and $(x-1)$ is $x(x-1)$.

$$\therefore \dfrac{4}{x} + \dfrac{5}{x-1} = \dfrac{4(x-1) + 5x}{x(x-1)}$$

$$= \dfrac{9x - 4}{x(x-1)}$$

Exercise 5N

1. Write as a single fraction.

   (a) $\dfrac{2x}{5} + \dfrac{x}{5}$ (b) $\dfrac{2}{x} + \dfrac{1}{x}$ (c) $\dfrac{x}{7} + \dfrac{3x}{7}$

   (d) $\dfrac{1}{7x} + \dfrac{3}{7x}$ (e) $\dfrac{5x}{8} + \dfrac{x}{4}$ (f) $\dfrac{5}{8x} + \dfrac{1}{4x}$

   (g) $\dfrac{2x}{3} + \dfrac{x}{6}$ (h) $\dfrac{2}{3x} + \dfrac{1}{6x}$

2. Sort into four pairs of equivalent fractions.

   A $\dfrac{x}{2} + \dfrac{x}{4}$ B $\dfrac{2}{x} + \dfrac{2}{x}$ C $\dfrac{9x}{8} - \dfrac{x}{4}$ D $\dfrac{7x}{8}$

   E $\dfrac{5x}{x^2} - \dfrac{2}{x}$ F $\dfrac{3x}{4}$ G $\dfrac{3}{x}$ H $\dfrac{4}{x}$

3. Simplify.

   (a) $\dfrac{3x}{4} + \dfrac{2x}{5}$ (b) $\dfrac{3}{4x} + \dfrac{2}{5x}$ (c) $\dfrac{3x}{4} - \dfrac{2x}{3}$

   (d) $\dfrac{3}{4x} - \dfrac{2}{3x}$ (e) $\dfrac{x}{2} + \dfrac{x+1}{3}$ (f) $\dfrac{x-1}{3} + \dfrac{x+2}{4}$

4. Work out these subtractions.

   (a) $\dfrac{x+1}{3} - \dfrac{(2x+1)}{4}$ (b) $\dfrac{x-3}{3} - \dfrac{(x-2)}{5}$ (c) $\dfrac{(x+2)}{2} - \dfrac{(2x+1)}{7}$

5. Copy and complete.

   (a) $\dfrac{x}{5} - \dfrac{\square}{\square} = \dfrac{x}{10}$ (b) $\dfrac{\square}{2} + \dfrac{x}{4} = \dfrac{7x}{4}$ (c) $\dfrac{3}{2x} - \dfrac{1}{8x} = \dfrac{\square}{8x}$

6. Write as a single fraction. [Hint: Use brackets freely.]

(a) $\dfrac{1}{x} + \dfrac{2}{x+1}$ 　　(b) $\dfrac{3}{x-2} + \dfrac{4}{x}$

(c) $\dfrac{5}{x-2} + \dfrac{3}{x+3}$ 　　(d) $\dfrac{7}{x+1} - \dfrac{3}{x+2}$

(e) $\dfrac{2}{x+3} - \dfrac{5}{x-1}$ 　　(f) $\dfrac{3}{x-2} - \dfrac{4}{x+1}$

7.
The perimeter of the rectangle shown is 24 units. Form an equation and solve it to find x.

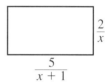

5.6 Transformation of curves

The notation f(x) means 'function of x'. A function of x is an expression which (usually) varies, depending on the value of x. Examples of functions are:

$f(x) = x^2 + 3; \quad f(x) = \dfrac{1}{x} + 7; \quad f(x) = \sin x.$

We can imagine a box which performs the function f on any input.

If $f(x) = x^2 + 7x + 2, \quad f(3) = 3^2 + 7 \times 3 + 2 = 32.$

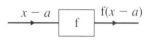

Here are **four** ways in which any function f(x) can be transformed:

$$f(x) + a; \quad f(x - a); \quad af(x); \quad f(ax).$$

(a) $y = f(x) + a$

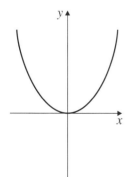
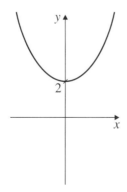
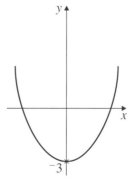

① $y = f(x)$ 　　② $y = f(x) + 2$ 　　③ $y = f(x) - 3$

① to ②: translate the curve 2 units up.
① to ③: translate the curve 3 units down.

(b) $y = f(x - a)$

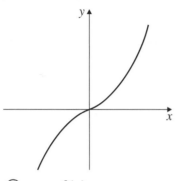 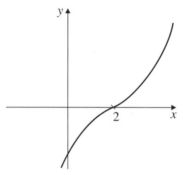 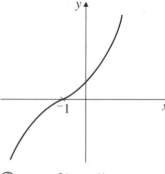

① $y = f(x)$ ② $y = f(x - 2)$ ③ $y = f(x + 1)$

① to ②: translate the curve 2 units to the right.
① to ③: translate the curve 2 units to the left.

(c) $y = af(x)$

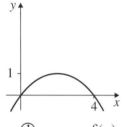 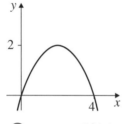 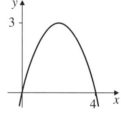 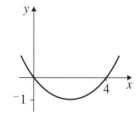

① $y = f(x)$ ② $y = 2f(x)$ ③ $y = 3f(x)$ ④ $y = {}^-f(x)$

① to ②: stretch vertically by a factor 2.
① to ③: stretch vertically by a factor 3.
① to ④: stretch vertically by a factor $^-1$.

(d) $y = f(ax)$

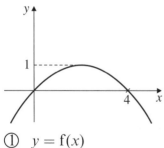 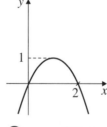 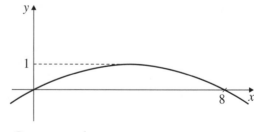

① $y = f(x)$ ② $y = f(2x)$ ③ $y = f(\tfrac{1}{2}x)$

① to ②: stretch parallel to the x-axis (horizontally) by a scale factor $\tfrac{1}{2}$.
① to ③: stretch parallel to the x-axis by a scale factor 2.
Similarly $f(3x)$ would be a stretch with scale factor $\tfrac{1}{3}$ and $f(\tfrac{1}{5}x)$ would be a stretch with scale factor 5.

Summary of rules for curve transformations

(a) $y = f(x) + a$: Translation by a units parallel to the y-axis.
(b) $y = f(x - a)$: Translation by a units parallel to the x-axis.
(note the negative sign).
(c) $y = af(x)$: Stretch parallel to the y-axis by a scale factor a.
(d) $y = f(ax)$: Stretch parallel to the x-axis by a scale factor $\frac{1}{a}$
(note the inverse of a).

Example

From the sketch of $y = f(x)$, draw
(a) $y = f(x + 3)$,   (b) $y = f(3x)$.

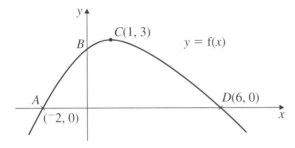

(a) This is a translation of ⁻3 units parallel to the x-axis.

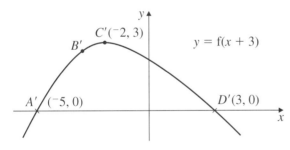

(b) This is a stretch parallel to the x-axis by a scale factor $\frac{1}{3}$. Notice that B remains in the same place and that the y-coordinate of C remains unchanged.

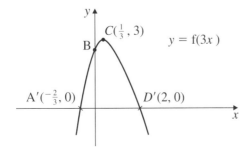

Exercise 50

A computer or calculator which sketches curves can be used effectively in this exercise, although it is not essential.

1. (a) Draw an accurate graph of $y = x^2$ for values of x from $^-3$ to 3. Label the graph $y = f(x)$
   (b) On the same axes draw an accurate graph of $y = x^2 + 4$ and label the graph $y = f(x) + 4$.
   (c) On the same axes draw the graph of $y = (x - 1)^2$ and label the graph $y = f(x - 1)$.
   Scales: x from $^-3$ to $^+3$, 2 cm = 1 unit
   y from 0 to 14, 1 cm = 1 unit.

2. This is the sketch graph of $y = f(x)$.
   (a) Sketch the graph of $y = f(x) + 3$.
   (b) Sketch the graph of $y = f(x + 1)$.
   (c) Sketch the graph of $y = {}^-f(x)$.

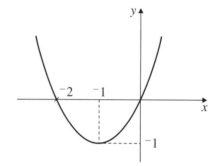

3. This is the sketch graph of $y = f(x)$.
   (a) Sketch $y = f(x) - 2$.
   (b) Sketch $y = f(x - 7)$.
   Give the new coordinates of the point A on the two sketches.

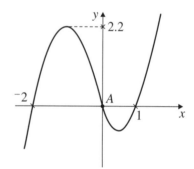

4. This is the sketch of $y = f(x)$ which passes through A, B, C.

   Sketch the following curves, giving the new coordinates of A, B, C in each case.

   (a) $y = {}^-f(x)$
   (b) $y = f(x - 2)$
   (c) $y = f(2x)$

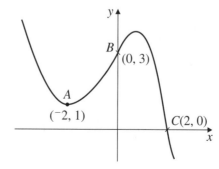

In Questions **5** to **10** each graph shows a different function f(x).
On squared paper draw a sketch to show the given transformation.
The scales are 1 square = 1 unit on both axes.

5.

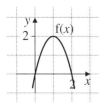

Sketch f(x) + 3.

6.

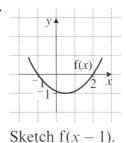

Sketch f(x − 1).

7.

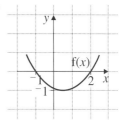

Sketch 2f(x).

8.

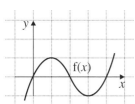

Sketch f(x + 2).

9.

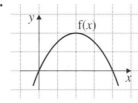

Sketch f(2x).

10.

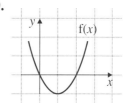

Sketch $f\left(\dfrac{x}{2}\right)$.

Exercise 5P

1. On the same axes, sketch and label the graphs of:
   (a) $y = x^2$ (b) $y = x^2 - 4$ (c) $y = x^2 + 2$

2. On the same axes, sketch and label:
   (a) $y = x^2$ (b) $y = (x - 3)^2$ (c) $y = (x + 2)^2$

3. On the same axes, sketch and label:
   (a) $y = x^2$ (b) $y = 4x^2$ (c) $y = \tfrac{1}{2}x^2$

4. On the same axes, sketch and label:
   (a) $y = x^3$ (b) $y = (x - 1)^3$ (c) $y = (x - 1)^3 + 4$

5. On the same axes, sketch and label:
   (a) $y = x(x - 1)$ (b) $y = 2x(2x - 1)$

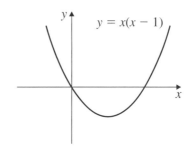

6. On the same axes, sketch and label:
   (a) $y = \dfrac{1}{x}$
   (b) $y = \dfrac{1}{x-2}$

7. Copy the sketch of $y = \cos x$.
   Using different colours, sketch:
   (a) $y = \cos(x + 90°)$
   (b) $y = 2\cos x$
   (c) $y = \cos 2x$

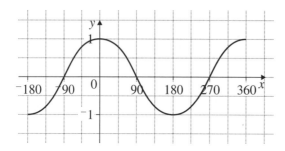

8. On squared paper, copy the sketch graphs of $y = \sin x$ and $y = \tan x$.

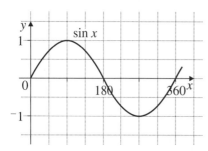
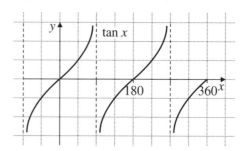

   Using different colours, sketch the graphs of:
   (a) $y = \sin\dfrac{x}{2}$
   (b) $y = \tan(x - 90°)$
   (c) $y = 2\sin x$
   (d) $y = \tan 2x$

9. (a) Draw an accurate graph of $y = (x+2)^2$ for values of x from $^-5$ to 1. Label the graph $y = f(x)$.
   (b) Draw an accurate graph of $y = (\tfrac{1}{2}x + 2)^2$ for values of x from $^-9$ to 1. Label the graph $y = f(\tfrac{1}{2}x)$.
   Scales: x from $^-10$ to 1, $1\text{ cm} = 1$ unit
   $\,y$ from $\,0$ to 10, $1\text{ cm} = 1$ unit.
   Describe the transformation from $y = (x+2)^2$ onto $y = (\tfrac{1}{2}x + 2)^2$.

10. It is possible to perform two or more successive transformations on the same curve. This is a sketch of $y = f(x)$.

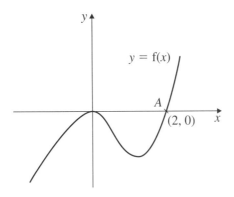

(a) Sketch $y = f(x+1) + 5$.
(b) Sketch $y = f(x-3) - 4$.

Show the new coordinates of the point A on each sketch.

11. Find the equation of the curve obtained when the graph of $y = x^2 + 3x$ is:
    (a) translated 5 units in the direction ↑
    (b) translated 2 units in the direction →
    (c) reflected in the x-axis.

12. $f(x) = x^2$ and $g(x) = x^2 - 4x + 7$.
    (a) If $g(x) = f(x-a) + b$, find the values of a and b.
    (b) Hence sketch the graphs of $y = f(x)$ and $y = g(x)$ showing the transformation from f to g.

13. Here are the sketches of $y = f(x)$ and $y = g(x)$.

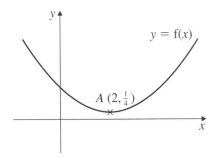 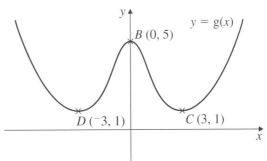

(a) Sketch $y = \dfrac{1}{f(x)}$, showing the new coordinates of A.

(b) Sketch $y = \dfrac{1}{g(x)}$, showing the new coordinates of B, C, D.

5.7 Simultaneous equations: one linear, one quadratic

To simultaneously solve a linear equation such as $y + 2x = 9$, and a quadratic equation such as $y^2 + x^2 = 18$, you could sketch their graphs and see whether they intersect.

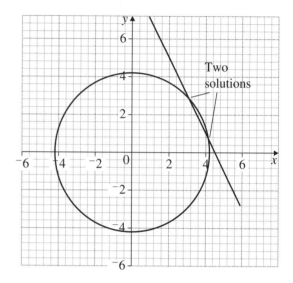

Two solutions

An algebraic method of finding a solution is by the substitution method.

To do this, make x or y the subject of the linear equation, and substitute this into the quadratic equation. Then solve the resulting equation.

For example, to solve $y + 2x = 9$ and $y^2 + x^2 = 18$, find y from the linear equation.

$y + 2x = 9 \Rightarrow y = 9 - 2x$

Substituting into $y^2 + x^2 = 18$, $(9 - 2x)^2 + x^2 = 18$

$$81 - 36x + 4x^2 + x^2 = 18$$
$$5x^2 - 36x + 63 = 0$$
$$(5x - 21)(x - 3) = 0$$
$$x = 3, \frac{21}{5}$$

If you found x you would introduce fractions which would not be helpful.

Substituting these values in turn into the **linear** equation we obtain the solutions $x = 3$, $y = 3$; $x = \frac{21}{5}$, $y = \frac{3}{5}$.

Exercise 5Q

1. Find the solution of $2x + y = 4$ and $x^2 + y^2 = 16$.
2. Find the solution of $3x + y = 5$ and $x^2 + y^2 = 25$.
3. Find the solution of $2x^2 + y^2 = 9$ and $x + 3y = 5$.
4. Find the solution of $x^2 - xy + y^2 = 7$ and $y = x + 1$.

Summary

1. You can find the gradient of a curve.

2. You can use trapeziums to find an approximate value for the area under a curve.

3. You can factorise quadratic expressions.

4. You can solve quadratic equations by factorisation, formula and by completion of the square.

5. You can solve equations using algebraic fractions.

6. You can transform curves.

7. You can solve algebraically a linear equation and a quadratic equation.

Check out AS5

1. Draw the curve $y = x^3 - 4x$.
   Find the gradient of the curve at:
   (a) $(2, 0)$ (b) $(1, {}^-3)$

2. Sketch the graph of $y = 4x - x^2$ for values of x from 0 to 5.
   Using five trapeziums, find an approximate value for the area bounded by the curve and the x-axis.

3. Factorise:
   (a) $x^2 - 5x + 5$ (b) $2x^2 - 3x - 2$
   (c) $6x^2 + 11x + 3$ (d) $x^3 - 4x$

4. (a) Solve by factorisation:
   $x^2 - 5x + 6 = 0$
   (b) Solve by use of the formula:
   $2x^2 - 3x - 4 = 0$
   (c) Solve by completion of the square:
   $x^2 - 4x + 3.1 = 0$

5. Solve:
   $$\frac{x}{5} + \frac{4}{x+1} = \frac{15 - 2x}{3}$$

6. The graph of $y = f(x)$ is shown below.

   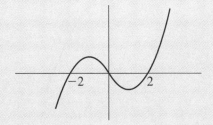

   Sketch: (a) $y = f(x + 3)$ (b) $y = f(2x)$
            (c) $y = f(x) + 4$

7. Solve: $x^2 + y^2 = 10$ and $y = 4x - 1$.

Revision exercise AS5

1. Simplify the expression $\dfrac{x^2 - 2x}{x^2 - 4}$. [SEG]

2. (a) The diagrams, **which are not drawn to scale**, show two transformations of the graph $y = x^2$. Write down the equations of the transformed graphs.

   (i)   (ii)

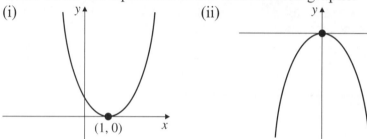

   (b) This is the graph of $y = \cos x$.

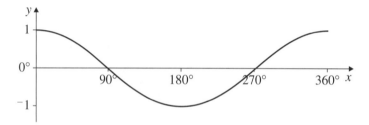

   Below and on the next page are shown two transformations of the graph $y = \cos x$. Write down the equation of each transformed graph.

   (i)

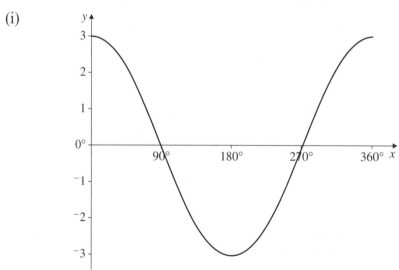

(ii)

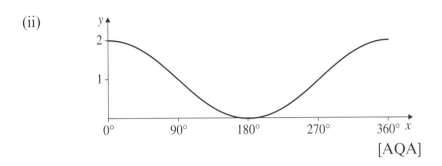

[AQA]

3. (a) Express the following as a single fraction in its simplest form.
$$\frac{x}{3} + \frac{x}{4} - \frac{x}{12}$$

(b) Solve the following equation.
$$\frac{1}{x} + \frac{1}{x+3} = \frac{2}{x+1}$$ [SEG]

4. The graph of $y = 2x^2 - 7x + 4$ is sketched below.

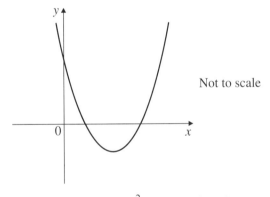

Not to scale

(a) Solve the equation $2x^2 - 7x + 4 = 0$.
Give your answers correct to 2 decimal places.
(b) Use your answers to (a) to find the equation of the line of symmetry of this graph. [SEG]

5. (a) Express the following as a single fraction in its simplest form.
$$\frac{3}{x} - \frac{5}{2x}$$

(b) Simplify this expression.
$$\frac{y^2 - 2y - 15}{2y^2 - 11y + 5}$$ [SEG]

Module 5 Practice Calculator Test

1. Use trial and improvement to solve the equation
   $x^3 + x^2 = 500$.
   One trial has been completed for you.
   Show all your trials.
   Give your answer correct to 1 decimal place.

x	$x^3 + x^2$	
7	$7^3 + 7^2 = 392$	Too small

   (4 marks)   [NEAB]

2. (a) Find the area of a semi-circle, diameter 8 cm.
   (b) Find the total perimeter of a semi-circle, diameter 8 cm. **(5 marks)**

3. Solve the equations:
   (a) $5x + 17 = 4(x - 2)$ **(3 marks)**
   (b) $\dfrac{7 - 2x}{5} = 3$ **(3 marks)**

4. (a) Solve the inequality
   $$3x + 4 \leqslant 7.$$
   (b) (i) Solve the inequality
   $$3x + 11 \geqslant 4 - 2x.$$
   (ii) If x is an integer what is the smallest possible value of x?   **(5 marks)**   [NEAB]

5. The diagram shows a kite $ABCD$
   Calculate the length of AC.

   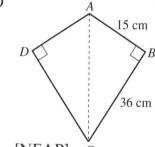

   Not drawn to scale

   (3 marks)   [NEAB]

6. The diagram shows part of a framework for a roof.

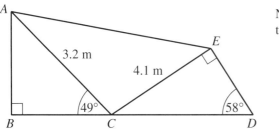

Not drawn to scale

Triangles ABC and CED are right-angled.
$AC = 3.2\,\text{m}$ $CE = 4.1\,\text{m}$
Angle ACB is $49°$.
Angle EDC is $58°$.

(a) Calculate the length BC.

(b) Calculate the length CD.

(c) BCD is a straight line.
    Calculate the length AE.   **(10 marks)**   [NEAB]

7. Cone A has a height of 10 cm and a volume of $1280\,\text{cm}^3$.

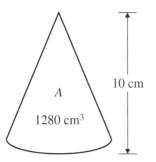 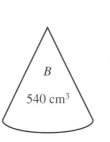

Not to scale

(a) Calculate the radius of cone A.

Cone B is similar to cone A.
Cone B has a volume of $540\,\text{cm}^3$.

(b) Calculate the height of cone B.   **(6 marks)**   [SEG]

8. OAB is a sector of radius 8.5 centimetres.
The length of the arc, AB, is 6.3 centimetres.

Calculate angle AOB.

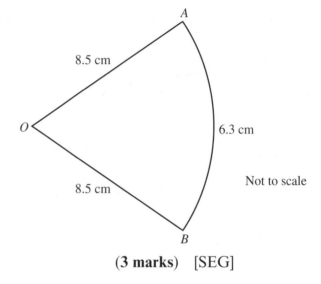

Not to scale

(3 marks) **[SEG]**

9. Solve the equation
$$3x^2 - 5x + 1 = 0.$$

Give your answers correct to 2 decimal places.

(3 marks) **[NEAB]**

10. Vector **a**, vector **b** and the point X are shown on this grid.

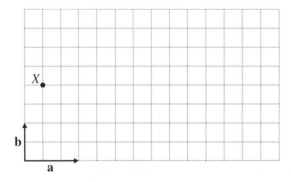

(a) Vector $\overrightarrow{XY} = 3\mathbf{a} + 1\tfrac{1}{2}\mathbf{b}$ and vector $\overrightarrow{XZ} = 2\mathbf{a} - \tfrac{1}{2}\mathbf{b}$.
Copy the grid and mark Y and Z on the diagram.

(b) P is the midpoint of YZ.
Express $\overrightarrow{XP}$ in terms of **a** and **b**.
You **must** show all your working.   **(6 marks)** **[SEG]**

11. (a) Factorise $2x^2 - 50y^2$.
    (b) Solve $y = 3x - 2$ and $x^2 + y^2 = 58$ **(10 marks)**

12. The angles in triangles ABC are 45°, 53° and 82°, as shown in the diagram. $BC = 11$ cm.

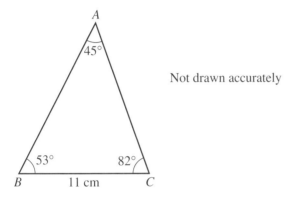

Not drawn accurately

(a) Calculate the length of AB, correct to 1 decimal place.
(b) Hence, or otherwise, find the area of triangle ABC.

A circle, centre O, is drawn through the points A, B and C of the **same** triangle.

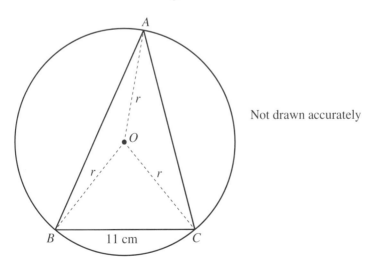

Not drawn accurately

(c) Calculate the radius, r, of the circle.

(9 marks)   [NEAB]

Module 5 Practice Non-Calculator Test

1. (a) A number sequence begins:

   3, 5, 7, 9, 11, ...

   Write an expression in terms of n for the nth term of the sequence.

   (b) Factorise completely:
   $6a^2 - 2ab$

   (c) Multiply out and simplify:

   $(x + 3)(4x - 1)$ **(6 marks)**

2. The exterior angle of a regular polygon is 40°.
   How many sides has the polygon? **(2 marks)**

3. Simplify:

   (a) $p^2 \times p^5$
   (b) $t^{10} \div t^2$
   (c) $(^-y)^2$ **(3 marks)** [NEAB]

4. The diagram shows a sketch of the line $2y + x = 10$.

   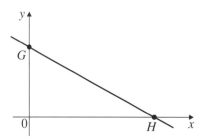

   (a) Find the coordinates of points G and H.

   (b) (i) On a copy of the same diagram, sketch the graph of $y = 2x$.
   (ii) Solve the simultaneous equations:
   $$2y + x = 10$$
   $$y = 2x$$

   (c) The equation of a straight line is $y = ax + b$.
   Rearrange the equation to give x in terms of y, a and b. **(8 marks)** [SEG]

5. One of these formulae gives the volume of an egg of height H cm and width w cm.

Formula A: $\frac{1}{6}\pi H^2 w^2$

Formula B: $\frac{1}{6}\pi H w^2$

Formula C: $\frac{1}{6}\pi H w$

(a) Which is the correct formula?
(b) Explain how you can tell that this is the correct formula.
(c) What are the units of the volume of the egg? **(3 marks)** [NEAB]

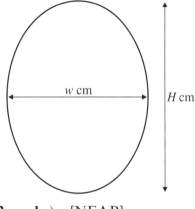

6. Find the area of trapezium $ABCD$.

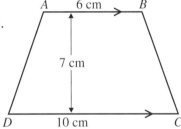

(3 marks)

7. Simplify $(4x^5 y^3) \times (x^3 y^2)$. **(2 marks)**

8. In the diagram, O is the centre of the circle.
SAT is a tangent to the circle at A.
Angle $BAC = 80°$ and angle $SAB = $ angle TAC.
(a) Calculate angle:
 (i) BOC (ii) OBC
 (iii) ABO (iv) ACO
(b) Explain why triangles OAB and OAC are congruent.

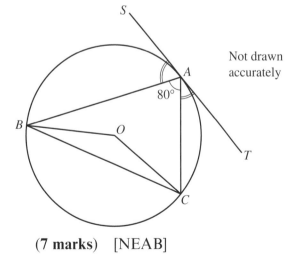

Not drawn accurately

(7 marks) [NEAB]

9. Solve the simultaneous equations:
$$5x - y = 9$$
$$2x + y = 5$$
(2 marks)

10. The diagram below represents a flag which measures 80 cm by 60 cm.
    The flag has two axes of symmetry.

    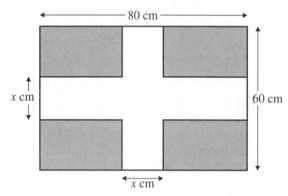

    The flag has two colours, grey and white.
    The bands of white are of equal width, x cm.
    The white area forms half the area of the flag.
    (a) Show that x satisfies the equation
    $x^2 - 140x + 2400 = 0$.
    (b) Solve the equation $x^2 - 140x + 2400 = 0$ to find the width of the white bands. **(6 marks)** [NEAB]

11. Find the value of the letter in:
    (a) $2^w = 1$ (b) $2^x = \frac{1}{4}$
    (c) $32^y = 2$ (d) $16^{z+3} = 2^z$ **(5 marks)** [SEG]

12. (a) Find the equation of the straight line through (0, 5) which is parallel to $y = 4x$.
    (b) Find p when $3(p - 7) = q(7 + 2p)$. **(6 marks)**

13. Here is a sketch of the graph of $y = \sin x$ for $0° \leqslant x \leqslant 360°$.

    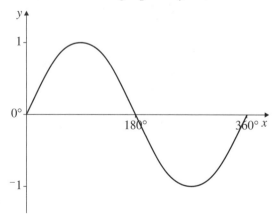

(a) Copy the sketch and show the locations of the two solutions of the equation
$$\sin x = \tfrac{1}{2}.$$

(b) Work out accurately the two solutions of the equation
$$4 \sin x = {}^{-}3$$
in the range $0° \leqslant x \leqslant 360°$. **(4 marks)** [NEAB]

14. The sketch below is of $y = x^2$.

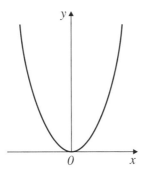

Sketch the following graphs on a separate copy of the graph $y = x^2$.
In each case p is a positive integer greater than 1.

(a) Sketch $y = px^2$.
(b) (i) Sketch $y = x^2 + p$.
    (ii) Write down the coordinates of the point where the graph crosses the y-axis.
(c) (i) Sketch $y = (x + p)^2$.
    (ii) Write down the coordinates of the point where the graph touches the x-axis. **(6 marks)** [NEAB]

15. Solve: $\dfrac{5}{2x + 1} - \dfrac{2}{x - 3} = 3$ **(7 marks)**

COURSEWORK GUIDANCE

For your GCSE Maths you will need to produce two pieces of coursework:

1. Investigative – module 4
2. Statistical – module 2

Each piece is worth 10% of your exam mark.

This unit gives you guidance on approaching your coursework and will tell you how to get good marks for each piece.

Module 4: Investigative coursework

In your investigative coursework you will need to:

- say how you are going to carry out the task and provide a plan of action
- collect results for the task and consider an appropriate way to represent your results
- write down observations you make from your diagrams or calculations
- look at your results and write down any observations, rules or patterns – try to make a general statement
- check out your general statements by testing further data – say if your test works or not
- develop the task by posing your own questions and provide a conclusion to the questions posed
- extend the task by using techniques and calculations from the content of the higher tier
- make your conclusions clear and link them to the original task giving comments on your methods.

Assessing the task

Before you start, it is helpful for you to know how your work will be marked. This way you can make sure you are familiar with the sort of things which examiners are looking for.

Investigative work is marked under three headings:

1. **Making and monitoring decisions to solve problems**
2. **Communicating mathematically**
3. **Developing skills of mathematical reasoning**

Each strand assesses a different aspect of your coursework. The criteria are:

1. Making and monitoring decisions to solve problems

This strand is about deciding what needs to be done then doing it. The strand requires you to select an appropriate approach, obtain information and introduce your own questions which develop the task further.

For the higher marks you need to analyse alternative mathematical approaches and develop the task using work from the higher GCSE syllabus content.

For this strand you need to:

- solve the task by collecting information
- break down the task by solving it in a systematic way
- extend the task by introducing your own **relevant** questions
- extend the task by following through alternative approaches
- develop the task by including a number of mathematical features
- explore the task extensively using higher level mathematics.

2. Communicating mathematically

This strand is about communicating what you are doing using words, tables, diagrams and symbols. You should make sure your chosen presentation is accurate and relevant.

For the higher marks you need to use mathematical symbols accurately, concisely and efficiently in presenting a reasoned argument.

For this strand you need to:

- illustrate your information using diagrams and calculations
- interpret and explain your diagrams and calculations
- begin to use mathematical symbols to explain your work
- use mathematical symbols consistently to explain your work
- use mathematical symbols accurately to argue your case
- use mathematical symbols concisely and efficiently to argue your case.

3. Developing skills of mathematical reasoning

This strand is about testing, explaining and justifying what you have done. It requires you to search for patterns and provide generalisations. You should test, justify and explain any generalisations.

For the higher marks you will need to provide a sophisticated and rigorous justification, argument or proof which demonstrates a mathematical insight into the problem.

For this strand you need to:

- make a general statement from your results
- confirm your general statement by further testing
- make a justification for your general statement
- develop your justification further
- provide a sophisticated justification
- provide a rigorous justification, argument or proof.

This unit uses a series of investigative tasks to demonstrate how each of these strands can be achieved.

Task 1

TRIANGLES

Patterns of triangles are made as shown in the following diagrams:

  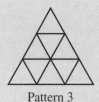

Pattern 1   Pattern 2   Pattern 3

What do you notice about the pattern number and the number of small triangles?
Investigate further.

Note:
You are reminded that any coursework submitted for your GCSE examination must be your own. If you copy from someone else then you may be disqualified from the examination.

Planning your work

A straightforward approach to this task is to continue the pattern further. You can record your results in a table and then see if you can make any generalisations.

Collecting information

A good starting point for any investigation is to collect information about the task.
You can see that:

- Pattern 1 shows 1 small triangle
- Pattern 2 shows 4 small triangles
- Pattern 3 shows 9 small triangles.

You can then extend this by considering further patterns.

Pattern 4

16 triangles

Pattern 5

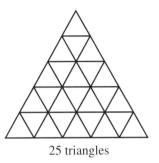

25 triangles

Moderator comment:
It is a good idea not to collect too much data as this is time consuming. Aim to collect 4 or 5 items of data to start with.

Drawing up tables

The information is not easy to follow so it is **always** a good idea to illustrate it in a table.

A table to show the relationship between
the pattern number and the number of triangles

pattern number	1	2	3	4	5
number of triangles	1	4	9	16	25

It is important to give your table a title and to make it quite clear what the table is showing.

Now you try ...

Using the table:
- Can you see anything special about the number of triangles?
- Ask yourself: are they odd numbers, even numbers, square numbers, triangular numbers, are they multiples, do they get bigger, do they get smaller ...?
- What other relationships might you look for?

Being systematic

Another important aspect to your work is to be systematic. This task shows you what that means.

Task 2

ARRANGE A LETTER

How many different ways can you arrange the letters ABCD? Investigate further.

Note:
You are reminded that any coursework submitted for your GCSE examination must be your own. If you copy from someone then you may be disqualified from the examination

You could haphazardly try out some different possibilities:

ABCD
BCDA
CDAB ...

You may or may not find all the arrangements this way. To be sure of finding them all you should list the arrangements systematically.
To be systematic you group letters starting with A:

ABCD	ACBD	ADBC
ABDC	ACDB	ADCB

Notice the system here:
A followed by B,
then A followed by C,
then A followed by D
and so on.

> **Now you try ...**
> Now see if you can systematically produce all of the arrangements starting with B (you should find six of them), then complete the other arrangements.

To approach the whole task systematically, you should start by finding the arrangements for:

- 1 letter: A
- 2 letters: A and B
- 3 letters: A, B and C.

This will give you more information about the task.
For different numbers of letters you should find:

Arrangement of different numbers of letters

number of letters	1	2	3	4
number of arrangements	1	2	6	24

You should now try to explain what your table tells you:

From my table I can ...

Finding a relationship – making a generalisation

To make a generalisation, you need to find a relationship from your table of results. A useful method is to look at the differences between terms in the table.

Here is the table of results from Task 1:

A table to show the relationship between the pattern number and the number of triangles.

pattern number	1	2	3	4	5
number of triangles	1	4	9	16	25

+3 +5 +7 +9
 +2 +2 +2

In this table the 'first differences' are going up in 2s.

The 'second differences' are all the same ... this tells you that the relationship is quadratic, so you should compare the numbers with the square numbers.

Here are some generalisations you could make:

From my table I notice that the number of triangles are all square numbers.

From my table I notice that the number of triangles are the pattern numbers squared.

A graph may help you see the relationship:

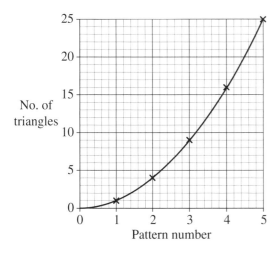

Moderator comment:
Remember that the idea is to identify some relationship from your table or graph so that there is some point in you using it in your work ... a table or graph without any comment on the findings is of little use.

From my graph I notice that there is a quadratic relationship between the number of triangles and the pattern number.

Using algebra

You will gain marks if you can express your generalisation or rule using algebra.

The rule:

> *I notice that the number of triangles are the pattern numbers squared.*

can be written in algebra:

> $t = p^2$ *where t is the number of triangles and p is the pattern number.*

Remember that you must explain what *t* and *p* stand for.

Testing generalisations

You need to confirm your generalisation by further testing.

> *From my table I notice that the number of triangles are the pattern numbers squared so that the sixth pattern will have $6^2 = 36$ triangles.*

You can confirm this with a diagram:

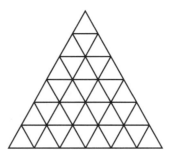

> *My diagram confirms the generalisation for the sixth pattern.*

Always confirm your testing by providing some comment, even if your test has not worked!

Now try to use all of these ideas by following Task 3.

Now you try ...
for Task 3 (Perfect Tiles) given next:
- Collect the information for different arrangements.
- Make sure you are systematic.
- Draw up a table of your results.

Task 3

PERFECT TILES

Floor spacers are used to give a perfect finish when laying tiles on a kitchen floor.

Three different spacers are used including:

    L spacer
    T spacer
    + spacer

Here is a 3 × 3 arrangement of tiles:
This uses   4 **L** spacers
                 8 **T** spacers and
                 4 **+** spacers

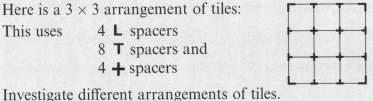

Investigate different arrangements of tiles.

Note:
You are reminded that any coursework submitted for your GCSE examination must be your own. If you copy from someone else then you may be disqualified from the examination.

You should be able to produce this table:

Number of spacers for different arrangements of tiles

size of square	1 × 1	2 × 2	3 × 3	4 × 4
number of **L** spacers	4	4	4	4
number of **T** spacers	0	4	8	12
number of **+** spacers	0	1	4	9

Now you try ...
What patterns do you notice from the table?
What general statements can you make?
Now test your general statements.

Generalisations for the 'Perfect Tiles' task might include:
$L = 4$ where L is the number of **L** spacers
$T = 4(n - 1)$ where T is the number of **T** spacers and
                    n is the size of the arrangement ($n \times n$)

The formula $T = 4(n - 1)$ can also be written as $T = 4n - 4$.

Test:
For a 5 × 5 arrangement (i.e. $n = 5$)

$T = 4(n - 1)$
$T = 4(5 - 1)$
$T = 4 \times 4$
$T = 16$

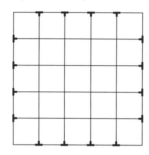

The number of **T** spacers is 16 so the formula works for a 5 × 5 arrangement.

$P = (n - 1)^2$ where p is the number of **+** spacers and n is the size of the arrangement ($n \times n$)

> It is sensible to replace **+** by P so that P stands for the number of **+** spacers.

Test:
For a 6 × 6 arrangement (i.e. $n = 6$)

$P = (n - 1)^2$
$P = (6 - 1)^2$
$P = (5)^2$
$P = 25$

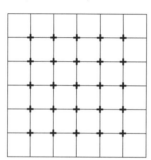

The number of **+** spacers is 25 so the formula works for a 6 × 6 arrangement.

Making justifications

Once you have found the general statement and tested it then you need to justify it. You justify the statement by explaining **why** it works.

For example, here are possible justifications for the rules in Task 3:

Why is the number of **L** spacers always 4?

> *The number of L spacers is always 4 because there are 4 corners to each arrangement.*

Why are all of the **T** spacers multiples of 4?

> The number of T spacers are multiples of 4 because each time you add another tile to the side then you add another T spacer.

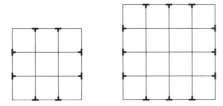

> As there are 4 sides to the arrangement then the number of T spacers will go up by 4.

Why are all of the ✚ spacers square numbers?

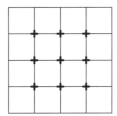

Now you try ...
Explain why the number of ✚ spacers are **always** square numbers.

Extending the problem – investigating further

Once you have understood and explained the basic task, you should extend your work to get better marks.

To extend a task you need to pose your own questions. This means that you must think of different ways to extend the original task.

Extending Task 1: 'Triangles'

You could ask and investigate:

What about different patterns?

For example, triangles:

What about patterns of different shapes?

For example, squares:

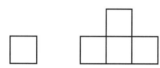 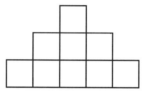

What about three-dimensional patterns?

For example, cubes:

 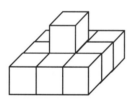 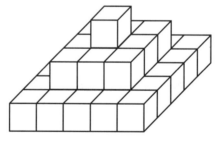

Extending Task 2: 'Arrange a letter'

What about different numbers of letters? For example, ABCDE

What happens when:
- *a letter is repeated?   For example AABCD*
- *a letter is repeated more than once?   For example, AAABC*
- *more than one letter is repeated?   For example, AAABB*

Of course, you will have to do more than just pose a question – you will have to carry out the investigation and come to a conclusion.

Extending Task 3: 'Perfect Tiles'
You could ask:

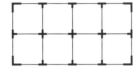

What about rectangular arrangements of tiles?

What about different shaped tiles?

For example, triangles:

Consider rectangular arrangements of tiles:

Remember to work systematically.

Number of spacers for different arrangements of tiles

size of arrangement	2 × 1	2 × 2	2 × 3	2 × 4	2 × 5
number of **L** spacers	4	4	4	4	4
number of **T** spacers	2	4	6	8	10
number of **+** spacers	0	1	2	3	4
number of spacers	6	9	12	15	18

Number of spacers for different arrangements of tiles

size of arrangement	3 × 1	3 × 2	3 × 3	3 × 4	3 × 5
number of **L** spacers	4	4	4	4	4
number of **T** spacers	4	6	8	10	12
number of **+** spacers	0	2	4	6	8
number of spacers	8	12	16	20	24

Rules for the 'Perfect Tiles' task extension might include:

```
L = 4                              where L is the number of L spacers.
T = 2(x − 1) + 2(y − 1)            where T is the number of T spacers
                                   and x is the length and y is the width
                                   of the arrangement.
P = (x − 1)(y − 1)                 where P is the number of + spacers
                                   and x is the length and y is the width
                                   of the arrangement.
N = (x + 1)(y + 1)                 where N is the number of + spacers
                                   and x is the length and y is the width
                                   of the arrangement.
```

> Note that the formula $T = 2(x − 1) + 2(y − 1)$ can also be written
> $$T = 2(x − 1) + 2(y − 1) = 2x + 2y − 4$$

```
The total number of spacers should equal the number of L spacers
plus the number of T spacers plus the number of + spacers.
Proof:   N = L + T + P
           = 4 + [2(x − 1) + 2(y − 1)] + (x − 1)(y − 1)
           = 4 + 2x − 2 + 2y − 2 + xy − y − x + 1
           = xy + x + y + 1
           = (x + 1)(y + 1)

Since N = (x + 1)(y + 1) is true, then my theory is proved.
```

Moderator comment:
The use of higher order algebra can result in high marks awarded on the middle strand as well as an opportunity to provide a sophisticated justification.

Task 4

SQUARE SEA SHELLS

Square sea shells are formed from squares whose pattern of growth is shown as follows:

Day 1

Day 2

The length of the new square is half that of the previous day.

Day 3

The length of the new square is half that of the previous day.

Day 4

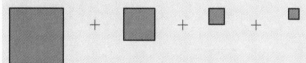

The length of the new square is half that of the previous day.

This pattern of growth continues.
Investigate the area covered by square sea shells on different days.
Investigate further.

For a square of side a

Day 1   Area $= a^2$

Day 2   Area $= a^2 + \dfrac{a^2}{4}$

Day 3   Area $= a^2 + \dfrac{a^2}{4} + \dfrac{a^2}{16}$

Day 4   Area $= a^2 + \dfrac{a^2}{4} + \dfrac{a^2}{16} + \dfrac{a^2}{64}$

Day …   Area $= a^2 + \dfrac{a^2}{4} + \dfrac{a^2}{16} + \dfrac{a^2}{64} + \ldots$

$\qquad\qquad\quad = a^2\left(1 + \dfrac{1}{4} + \dfrac{1}{16} + \dfrac{1}{64} + \ldots\right)$

$\qquad\qquad\quad = \ldots$

$\qquad\qquad\quad = \dfrac{4}{3}a^2$

This work involves summing an infinite series, which is beyond the mathematical content of the GCSE examination and so will get you higher marks.

Extending Task 4: 'Square sea shells'
For an equilateral triangle of side a:

Day 1 $\qquad$ Area $= \dfrac{1}{2} a^2 \sin 60°$

$\qquad\qquad\qquad\quad = \dfrac{1}{2} a^2 \dfrac{\sqrt{3}}{2}$

$\qquad\qquad\qquad\quad = \dfrac{\sqrt{3}}{4} a^2$

For a regular n-sided polygon of length a:

Day 1 $\qquad$ Area $= \dfrac{na^2}{4 \tan \left(\dfrac{180}{n} \right)^°}$

> **Moderator comment:**
> Remember that for the higher levels, you will need to explore the task extensively using higher level mathematics, use mathematical symbols concisely and efficiently to argue your case and provide a rigorous justification, argument or proof.

In this unit we have tried to give you some hints on approaching investigative coursework to gain your best possible mark.

This mathematics is often useful in investigative tasks:

- creating tables
- drawing and interpreting graphs
- recognising square numbers
- recognising triangular numbers
- finding the nth term of a linear sequence.

Summary

These are the grade criteria your coursework will be marked by:

Testing general statements (grade D)

To achieve this level you must:

- break down the task by solving it in an orderly manner
- interpret and explain your diagrams and calculations
- confirm your general statement by further testing.

Posing questions and justifying (grade C)

To achieve this level you must:

- extend the task by introducing your own relevant questions
- begin to use mathematical symbols to explain your work
- make a justification for your general statement
- provide a sophisticated justification.

Making further progress (grade B)

To achieve this level you must:

- extend the task by following through alternative approaches
- use mathematical symbols consistently to explain your work
- develop your justification further.

Justifying a number of mathematical features (grade A)

To achieve this level you must:

- develop the task by including a number of mathematical features
- use mathematical symbols accurately to argue your case
- provide a sophisticated justification.

Exploring extensively (grade A*)

To achieve this level you must:

- explore the task extensively using higher level mathematics
- use mathematical symbols concisely and efficiently to argue your case and provide a rigorous justification, argument or proof.

Module 2: Statistical coursework

In your statistical coursework you will need to:

- provide a well considered hypothesis and provide a plan of action to carry out the task
- decide what data is needed and collect results for the task using an appropriate sample size and sampling method
- consider the most appropriate way to represent your results and write down any observations you make
- consider the most appropriate statistical calculations to use and interpret your findings in terms of the original hypothesis

- develop the task by posing your own questions – you may need to collect further data to move the task on
- extend the task by using techniques and calculations from the content of the higher tier
- make your conclusions clear – always link them to the original hypothesis recognising limitations and suggesting improvements.

Assessing the task

It may be helpful for you to know how the work is marked. This way, you can make sure you are familiar with the sort of things that examiners are looking for.

Statistical work is marked under three headings:

1. **Specifying the problem and planning**
2. **Collecting, processing and representing the data**
3. **Interpreting and discussing the results**

Each strand assesses a different aspect of your coursework as follows:

1. Specifying the problem and planning
This strand is about choosing a problem and deciding what needs to be done, then doing it. The strand requires you to provide clear aims, consider the collection of data, identify practical problems and explain how you might overcome them.

For the higher marks you need to decide upon a suitable sampling method, explain what steps were taken to avoid possible bias and provide a well structured report.

2. Collecting, processing and representing the data
This strand is about collecting data and using appropriate statistical techniques and calculations to process and represent the data. Diagrams should be appropriate and calculations mostly correct.

For the higher marks you will need to accurately use higher level statistical techniques and calculations from the higher tier GCSE syllabus content.

3. Interpreting and discussing the results
This strand is about commenting, summarising and interpreting your data. Your discussion should link back to the original problem and provide an evaluation of the work undertaken.

For the higher marks you will need to provide sophisticated and rigorous interpretations of your data and provide an analysis of how significant your findings are.

This unit uses a series of statistical tasks to demonstrate how each of these strands can be achieved.

Planning your work

Statistical coursework requires careful planning if you are to gain good marks. Before undertaking any statistical investigation, it is important that you plan your work and decide exactly what you are going to investigate – do not be too ambitious!

Before you start you should:

- decide what your investigation is about and why you have chosen it
- decide how you are going to collect the information
- explain how you intend to ensure that your data is representative
- detail any presumptions which you are making.

Getting started – setting up your hypothesis

A good starting point for any statistical task is to consider the best way to collect the data and then to write a clear hypothesis you can test.

Moderator comment:
Your statistical task must always start with a 'hypothesis' where you state exactly what you are investigating and what you expect to find.

Task 1

WHAT THE PAPERS SAY

Choose a passage from two different newspapers and investigate the similarities and differences between them.

Write down a hypothesis to test.

Design and carry out a statistical experiment to test the hypothesis.

Investigate further.

Note:
You are reminded that any coursework submitted for your GCSE examination must by your own. If you copy from someone else then you may be disqualified from the examination.

First consider different ways to compare the newspapers:

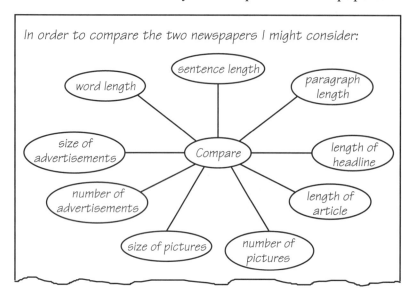

Now formulate your hypothesis.
Here are some possible hypotheses:

> For my statistics coursework I am going to investigate the hypothesis that tabloid papers use shorter words than broadsheet newspapers.

> My hypothesis is that word lengths in the tabloid newspaper will be shorter than word lengths in the broadsheet newspaper.

> My hypothesis is that sentence lengths in the tabloid newspaper will be shorter than sentence lengths in the broadsheet newspaper.

> My hypothesis is that the number of advertisements in the tabloid newspaper will be greater than the number of advertisements in the broadsheet newspaper.

Remember:
It does not matter whether your hypotheses are true or false. You will still gain marks if your hypothesis turns out to be false.

Now you try …
See if you can add some hypotheses of your own.
Explain how you would proceed with the task.

Choosing the right sample

Once you have decided your aims and set up a hypothesis then it is important to consider how you will test your hypothesis.

Sampling techniques include:

- **Random sampling** is where each member of the population has an equally likely chance of being selected. An easy way to do this would be to give each person a number and then choose the numbers randomly, e.g. out of a hat.

- **Systematic sampling** is the same as random sampling except that there is some system involved such as numbering each person and then choosing every 20th number.

- **Stratified sampling** is where each person is placed into some particular group or category (stratum) and the sample size is proportional to the size of the group or category in the population as a whole.

- **Convenience sampling** or opportunity sampling is one which involves simply choosing the first person to come along ... although this method is not particularly random!

> **Moderator comment:**
> It is important to choose an appropriate sample, give reasons for your choice and explain what steps you will take to avoid bias.

> **Moderator comment:**
> You should **always** say why you chose your sampling method.

Here are some ways you could test the hypotheses for Task 1:

Sampling method	Reason
To investigate my hypothesis I am going to choose a similar article from each type of newspaper and count the lengths of the first 100 words.	I decided to choose similar articles because the words will be describing similar information and so it will be easier to make comparisons.
To investigate my hypothesis I will choose every tenth page from each type of newspaper and calculate the percentage area covered by pictures.	I decided to choose every tenth page from each type of newspaper because the types of articles vary throughout the newspaper (for example news headlines at the front and sports at the back of the paper).

Collecting primary data

If you are collecting primary data then remember:

- **Observation** involves collecting information by observation and might involve participant observation (where the observer takes part in the activity), or systematic observation (where the observation happens without anyone knowing).
- **Interviewing** involves a conversation between two or more people. Interviewing can be formal (where the questions follow a strict format) or informal (where they follow a general format but can be changed around to suit the questioning).
- **Questionnaires** are the most popular way of undertaking surveys. Good questionnaires are:
  - simple, short, clear and precise,
  - attractively laid out and quick to complete.

  The questions are:
  - not biased or offensive
  - written in a language that is easy to understand
  - relevant to the hypothesis being investigated
  - accompanied by clear instructions on how to answer the questions.

> **Moderator comment:**
> It is important to carefully consider the collection of reliable data. Appropriate methods of collecting primary data might include observation, interviewing, questionnaires or experiments.

> **Moderator comment:**
> It is always a good idea to undertake a small scale 'dry run' to check for problems. This 'dry run' is called a pilot survey and can be used to improve your questionnaire or survey before it is undertaken.

Avoiding bias

You must be very careful to avoid any possibility of bias in your work. For example, in making comparisons it is important to ensure that you are comparing like with like.

Jean undertook Task 1: 'What the papers say'.
She collected this data from two newspapers by measuring (observation).

	Tabloid	Broadsheet
Number of advertisements	35	30
Number of pages	50	30
Area of each page	1000 cm^2	2000 cm^2

Her hypothesis is:

The tabloid newspaper has more advertisements than the broadsheet newspaper.

A quick glance at the table may make her claim look true: the broadsheet has 30 adverts but the tabloid has 35, which is more.

However, the sizes of the newspapers are different so Jean is not really comparing like with like.

To ensure Jean compares like with like she should take account of:

- the area of the pages
- the number of pages, etc.

Number of adverts per page:
Tabloid $= 35 \div 50 = 0.7$
Broadsheet $= 30 \div 30 = 1$

This shows that:

> The broadsheet newspaper has more advertisements per page than the tabloid newspaper.

The total area of the pages is:
Tabloid $= 50 \times 1000 \text{ cm}^2 = 50\,000 \text{ cm}^2$
Broadsheet $= 30 \times 2000 \text{ cm}^2 = 60\,000 \text{ cm}^2$

So the percentage coverage is:
Tabloid $= 35 \div 50\,000 \times 100\% = 0.07\%$
Broadsheet $= 30 \div 60\,000 \times 100\% = 0.05\%$

This shows that:

> The tabloid newspaper has more advertisements per area of coverage than the broadsheet newspaper.

Note:
This still doesn't take account of the size of the adverts!

Methods and calculations

Once you have collected your data, you need to use appropriate statistical methods and calculations to process and represent your data.

You can represent the data using statistical calculations such as the mean, median, mode, range and standard deviation.

Task 2

> **GUESSING GAME**
>
> Dinesh asked a sample of people to estimate the length of a line and the weight of a packet.
>
> Write down a hypothesis about estimating lengths and weights and carry out your own experiment to test your hypothesis.
>
> Investigate further.

Note:
You are reminded that any coursework submitted for your GCSE examination must be your own. If you copy from someone else then you may be disqualified from the examination.

Maurice and Angela decide to explore the hypothesis that:

> *Students are better at estimating the length of a line than the weight of a packet.*

To test the hypothesis they collect data from 50 children, detailing their estimations of the length of a line and the weight of a packet.

Here are their findings:

A table to show the estimations for the length of a line and the weight of a packet

	Length (cm)	Weight (g)
Mean	15.9	105.2
Median	15.5	100
Mode	14	100
Range	8.6	28
SD	1.2	2.1

Moderator comment:
It is important to consider whether information on all of these statistical calculations is essential.

Note: the actual length of the line is 15 cm and the weight of the packet is 100 g.

Now you try ...

Using the table:

- What do you notice about the average of the length and weight?
- What do you notice about the spread of the length and weight?
- Does the information support the hypothesis?

Graphical representation

The data for Task 2 was sorted into different categories and represented as a table.

It may be useful to show your results using graphs and diagrams as sometimes it is easier to see trends.

Graphical representations might include stem and leaf diagrams, box and whisker diagrams and cumulative frequency graphs.
Once you have drawn a graph you should say what you notice from the graph:

Other statistical representations might include pie charts, bar charts, scatter graphs and histograms.

Moderator comment:
You should only use appropriate diagrams and graphs.

> From my representation, I can see that the estimations for the line are generally more accurate than the estimations for the weight because:
> - the mean is closer to the actual value for the lengths and
> - the standard deviation is smaller for the lengths.
>
> However, on closer inspection:
> - the percentage error on the mean is smaller for the weights
> - the median and mode are better averages to use for the weights.

Remember to consider the possibility of bias in your data:

The percentage error for the lengths is

$$\frac{0.9}{15} \times 100 = 6\%$$

The percentage error for the weights is

$$\frac{5.2}{100} \times 100 = 5.2\%$$

Using secondary data

You may use secondary data in your coursework.

Secondary data is data that is already collected for you.

Moderator comment:
If you use secondary data there must be enough 'to allow sampling to take place' – about 50 pieces of data.

Task 3

> **GENDER DIFFERENCES IN EXAMINATIONS**
>
> Jade is conducting a survey in the GCSE examination results for Year 11 students at her school.
>
> She has collected data on last year's results, and wants to compare the performance of boys and girls at the school.

Jade explores the hypothesis that:

Year 11 girls do better in their GCSE examinations than boys.

To test the hypothesis she decided to concentrate on the core subjects and her findings are shown in the table:

A table to show the performance of Year 11 girls and boys in their GCSE examinations

		%A*–C	%A*–G	APS
English	Girls	62	93	4.9
	Boys	46	89	4.3
	All	54	91	4.6
Mathematics	Girls	47	91	4.2
	Boys	45	89	4.2
	All	46	90	4.2
Science	Girls	48	91	4.4
	Boys	45	88	4.3
	All	47	90	4.4

Note: APS means 'average points score'.

> **Now you try ...**
> - What do you notice about the percentages of A*–C grades?
> - What do you notice about the percentages of A*–G grades?
> - What do you notice about the average point scores (APS)?
> - Does the information support the hypothesis?

Jade could use comparative bar charts to represent the data as it will allow her to make comparisons more easily.

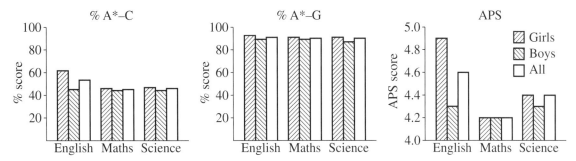

Summarising and interpreting data

Jade summarises her findings like this:

> From my graph, I can see that Year 11 girls do better in their GCSE examinations than boys in terms of A*–C grades, A*–G grades and average point scores.
>
> The performance of Year 11 girls is significantly better than boys in English although less so in mathematics.

Moderator comment:
You should refer to your original hypothesis when you summarise your results.

Moderator comment:
In your conclusion you should also suggest limitations to your investigation and explain how these might be overcome.
You may wish to discuss:
- sample size
- sampling methods
- biased data
- other difficulties.

Hint:
You need to appreciate that the data is more secure if the sample size is 500 rather than 50.
You might wish to carry out a significance test on the data.

Extending the task

To gain better marks in your coursework you should extend the task in light of your findings.

In your extension you should:

- give a clear hypothesis
- collect further data if necessary
- present your findings using charts and diagrams as appropriate
- summarise your findings referring to your hypothesis.

Extending Task 3: 'Gender differences in examinations'

Jade extends the task by looking at the performance of individual students in combinations of different subjects.

> I am now going to extend my task by looking at the performance of individual students in English and mathematics. My hypothesis is that there will be no correlation between the two subjects.

She represents the data on a scatter graph:

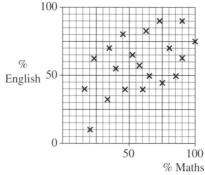

Note:
You should only draw a line of best fit on the diagram to show the correlation if you make some proper use of it (for example, to calculate a student's likely English mark given their mathematics mark).

She summarises her findings:

> From my scatter graph, I can see that there is some correlation between the performance of individual students in English and mathematics.

Note:
The strength of the correlation could be measured using higher level statistical techniques such as Spearman's Rank Correlation.

She extends the task further:

> I shall now look at the performance of individual students in mathematics and science. My hypothesis is that there will be a correlation between the two subjects.

She presents the data on a scatter graph.

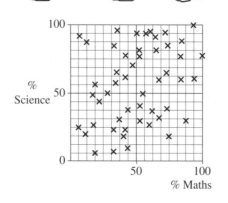

> *From my scatter graph, I can see that there is no correlation between the performance of individual students in mathematics and science.*

Extending Task 2: 'Guessing game'

Maurice extends 'Guessing game' like this:

> *I am now going to extend my task by looking to see whether people who are good at estimating lengths are also good at estimating weights. My hypothesis is that there will be a strong correlation between people's ability at estimating lengths and estimating weights.*

He draws a scatter graph:

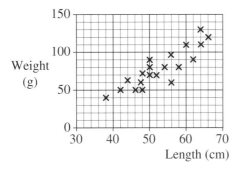

He summarises his findings:

> *From my scatter graph, I can see that there is a strong positive correlation between people's ability at estimating lengths and estimating weights.*

The strength of the correlation can be measured using higher level statistical techniques such as Spearman's Rank Correlation or Product Moment Correlation.

I am now going to extend my investigation by using Spearman's Rank Correlation to calculate the rank correlation coefficient.

Person	Length	Weight	Rank Length	Rank Weight	Difference D	D²
Jane	15 cm	102 g	22	18	4	16
Suresh	16 cm	108 g	11	12	−1	1
					Total	2250

$$r = 1 - \frac{6(\Sigma D^2)}{n(n^2 - 1)}$$

$$r = 1 - \frac{6(2250)}{50(50^2 - 1)}$$

$$r = 1 - \frac{13500}{50(2499)}$$

$$r = 1 - \frac{13500}{124950}$$

$$r = 1 - 0.108043\ldots$$

$$r = 0.891956\ldots$$

$$r = 0.89 \text{ (2dp)}$$

The value for the rank correlation coefficient is quite close to 1 so there is a strong positive correlation between my results.

This tells me that there is a strong correlation between people's ability at estimating lengths and estimating weights, and confirms my original hypothesis.

Moderator comment:
The development of the task to include higher level statistical analysis must be 'appropriate and accurate'.

50 people were in the survey so $n = 50$.

Using a computer

It is quite acceptable that calculations and representations are generated by computer, as long as any such work is accompanied by some analysis and interpretation.

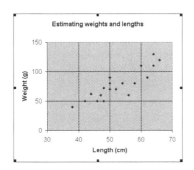

Remember:
Make sure that your computer-generated scatter graph has labelled axes and a title to make it quite clear what it is showing.

> From my computer-generated scatter graph, I can see that there is a strong correlation between people's ability at estimating lengths and estimating weights.

In this unit we have tried to give you some hints on approaching statistical coursework to gain your best possible mark.

These statistics are often useful in investigative tasks:

- calculating averages (mean, median and mode)
- finding the range
- pie charts, bar charts, stem and leaf diagrams
- constructing a cumulative frequency graph
- finding the interquartile range
- histograms
- calculating the standard deviation
- drawing a scatter graph and line of best fit
- sampling techniques
- discussing bias.

Summary

These are the grade criteria your coursework will be marked by:

Intermediate statistical task (grade C)

To achieve this level you must:

- set out clear aims and include a plan designed to meet those aims
- ensure that the sample size is of an appropriate size (about 50)
- give reasons for your choice of sample

- collect data and make use of statistical techniques and calculations

    For example: pie charts, bar charts, stem and leaf diagrams, mean, median, mode (of grouped data), scattergraphs and cumulative frequency

- summarise and correctly interpret your diagrams and calculations
- consider your strategies and how successful they were.

Higher statistical task (grade A)

To achieve this level you must:

- set out clear aims for a more demanding problem
- include a plan which is specifically designed to meet those aims
- ensure that sample size is considered and limitations discussed
- collect relevant data and use statistical techniques and calculations

    For example: pie charts, bar charts, stem and leaf diagrams, mean, median, mode (of grouped data), scatter graphs, cumulative frequency, histograms and sampling techniques

- summarise and correctly interpret your diagrams and calculations
- use your results to respond to your original question
- consider your strategies, limitations and suggest possible improvements.

ANSWERS MODULE 1

D1 Data handling

Check in D1
1. (a) 30 (b) 645 (c) 16% 2. (a) 48, 72 (b) 2 : 3

Exercise 1A
1. (a) 5, 6, 4, 9 (b) 7, 9, 7, 19 (c) 8, 6·5, 9, 11 (d) 3·5, 3·5, 4, 8
2. 45 or 2 3. 4
4. (a) False (b) Possible (c) False (d) Possible
5. (a) mean = £33 920, median = £22 500, mode = £22 500
   (b) the mean is skewed by one large figure
6. (a) mean = 157·1 kg, median = 91 kg
   (b) mean; No: Over three-quarters of the cattle are below the mean weight
7. (a) mean = 74·5 cm, median = 91 cm (b) Yes
8. 78 kg 9. 35·2 cm 10. (a) 2 (b) 9
11. (a) 20·4 m (b) 12·8 m (c) 1·66 m 12. 55 kg
13. 4, 4, 5, 7, 10 14. 2, 4, 4, 4, 6
15. 12 16. $3\frac{2}{3}$ 17. (a) N (b) mean = $N^2 + 2$; median = N^2 (c) 2

Exercise 1B
1. (a) 24·3p (b) 25p (c) 25·6p
2. averages: 9, 8·8, 9·2, 9·2, 10, 11·2, 12·8, 13·8, 14·8, 15, 14·6, 14·8, 14·8, 13·6, 13, 13
3. (a) because she is open three days per week
   (b) averages: 161, 160·67, 152·33, 154·33, 153·33, 156·33, 151, 150, 147, 146
   (d) attendances are reducing
4. (a) five-point
   (b) averages: 1800, 1800, 1900, 2000, 2100, 2200, 2300, 2200, 2300, 2400, 2500
   (c) takings are increasing
5. (b) 86·25, 85·18, 85·1, 89·89, 88·38, 86·57, 86·33, 94·2, 92
   (d) cost of phone calls is increasing
   (e) The September bill is always lowest, perhaps because it covers the school holidays.

Exercise 1C
1. 3·38 2. (a) mean = 6·62; median = 8; mode = 3 (b) mode
3. (a) mean = 3·025; median = 3; mode = 3 (b) mean = 17·75; median = 17; mode = 17
4. (b) About 11 5. (a) 68·25 (b) 55–69
6. 3·8 7. (a) 180/181 8. (a) 9 (b) 9 (c) 15
9. (a) 5 (b) 10 (c) 10

Exercise 1D
2. (a) 20% (b) 20–24 and 30–34
4. (a) 62
   (b) Sport B has more heavy people. Sport A has much smaller range of weights compared to sport B.
5. (a) Plants with fertiliser are significantly taller (b) no significant effect
6. Picture height represents weight, but picture area makes the difference look much larger.

7. Profit axis does not start at zero.
8. (a) £3000 (b) £4000 (c) £6000 (d) £11 000
9. eggs 270°; milk 12°; butter 23·4°; cheese 54°; salt/pepper 0·6°
10. 18°, 54°, 54°, 234°
11. (a) 22·5% (b) $x = 45°, y = 114°$
12. $x = 8$

Exercise 1E

3. (a) 50 (b) 15 (c) 50 kg
5. (a) 13 (b) 78
   (c) The men generally had a lower pulse rate and their pulse rates were more varied than those of the women.

Exercise 1F

2. (a) strong positive (b) no correlation (c) weak negative
3. (a) no correlation (b) strong negative (c) no correlation (d) weak positive
4. (b) 9
5. 44
6. 6·8 km
7. (b) Bogota; High altitude (c) About 73°C
8. For discussion

Exercise 1I

1. 16, 16, 16, 6, 6
2. 10, 25, 10, 5, 50

Exercise 1J

1. Pupils who do not eat school meals will not be questioned.
2. Ex-directory customers should be included.
3. His friends may not be typical.
4. People going on ferry may not be representative of the population.
5. Might not be typical of the whole week.
6. OK.
7. People are unlikely to include those at work/school etc.
8. All these people likely to have a video. Not representative of whole population.

Exercise 1K

1. (a) 4, 9·5 (b) 5·5
2. (a) 1·5, 5·5 (b) 4
3. (a) 10, 21 (b) 11
4. (a) 1·5, 6 (b) 4·5
5. (a) 28, 6, 12 (b) 4·8, 1·5, 2·7 (c) 33, 12, 24

Exercise 1L

1. (a) 43 (b) 28, 57 (c) 29 (d) 28
2. (a) 100 (b) 250 (c) 2250 h (d) 750 h
3. (b) (i) 45 (ii) 17
4. (b) (i) about 160·5 cm (ii) about 13 cm
5. (a) about 124 (b) IQR 52 Boris is more consistent
6. (b) (i) 7·6 h (ii) (about) 3·3 h (c) 90%
7. (b) France 18·5, Britain 24·5
   (c) Results from Britain bunched together more closely with a higher median.
8. (b) 10·2 cm (c) Many failed to germinate
   (d) Median is better as it discounts those which do not germinate
9. (c) 44 million (d) A: 21, B: 59 (e) A has much younger population

Exercise 1M
Frequency densities are given.
1. 0·25, 1, 1·4, 0·3
2. 1, 0·6, 1·2, 1·7, 1·3, 0·5
3. 0·5, 1·4, 2, 1, 0·2
4. 1·8, 9·2, 7, 1·3, 0·8
5. 0·4, 1·2, 1·5, 2, 0·9, 0·2
6. 1·1, 1·8, 0·35, 0·25, 0
7. 1·6, 3·4, 2·8, 1·1

Exercise 1N
1. (a) (i) 10   (ii) 20   (b) 65
2. 55
3. 135
4. 28
5. 44
6. f.d.: 0·2, 0·3, 1·2, 1, 0·4
8. (a) 12   (b) 23   (c) 29   (d) 25

Check out D1
1. (a) 23   (b) 21   (c) 22
2. (a) four-point   (b) 89·5, 90·25, 91·5, 93·25, 94·25
   (e) about £99
3. 2
5. (b) about 33
7. (a) No, people leaving a newsagent are more likely to have bought a newspaper.
8. 18 from year 7, 20 from year 8, 20 from year 9, 18 from year 10, 25 from year 11
9. (b) (i) about 28   (ii) about 11   (c) about 13   (e) f.d.: 0, 0·8, 3·2, 5·2, 3·4, 0·2

Revision exercise D1
1. (a) 33·6 minutes   (b) cf: 12, 32, 47, 50   (d) about 40
2. (a) 480   (b) there are three terms
   (d) about 580   (e) using answer to (d) gives 700
3. (b) 1–3   4. (a) 6   (b) median = 100, IQR = 40
5. (a) 9   (b) 66 inches   6. (b) 4 wine glasses, 6 ornaments, 1 bottle

D2 Probability

Check in D2
1. (a) $\frac{43}{55}$ (b) $\frac{3}{10}$ (c) $\frac{9}{10}$
2. (a) 0·48   (b) 0·19   (c) 0·12   (d) 0·005   (e) 0·06

Exercise 2A
1. B   2. C   3. A   4. B or C   5. C or D
6. A   7. B   8. C   9. D
13. Mike with a large number of spins he should get zero with a probability of about $\frac{1}{10}$

Exercise 2B
1. (a) $\frac{1}{13}$ (b) $\frac{1}{52}$ (c) $\frac{1}{4}$
2. (a) $\frac{1}{9}$ (b) $\frac{1}{3}$ (c) $\frac{4}{9}$ (d) $\frac{2}{9}$
3. (a) $\frac{5}{11}$ (b) $\frac{2}{11}$ (c) $\frac{4}{11}$
4. (a) $\frac{4}{17}$ (b) $\frac{3}{17}$ (c) $\frac{11}{17}$
5. (a) $\frac{2}{9}$ (b) $\frac{2}{9}$ (c) $\frac{1}{9}$ (d) 0   (e) $\frac{5}{9}$
6. (a) (i) $\frac{5}{13}$ (ii) $\frac{6}{13}$ (b) (i) $\frac{5}{12}$ (ii) $\frac{1}{12}$
7. (a) $\frac{x}{12}$ (b) 3
8. $\frac{1}{7}$
9. (a) (i) $\frac{1}{4}$ (ii) $\frac{1}{4}$ (iii) $\frac{1}{4}$ (b) $\frac{1}{4}$ (c) $\frac{6}{27} = \frac{2}{9}$

Exercise 2C

1. (a) 150 (b) 50 2. 25 3. 50 4. 40 5. (a) $\frac{3}{8}$ (b) 25
6. (a) $\frac{1}{2}$ (b) $\frac{1}{2}$ 7. (a) 15 (b) 105 8. (a) 0·2146 (b) 1073

Exercise 2D

1. (a) 8 ways (b) $\frac{1}{8}$ 2. 16 ways 3. (b) 4 (c) $\frac{1}{9}$
4. (a) $\frac{1}{4}$ (b) $\frac{1}{4}$ 5. WX, WY, WZ, XY, XZ, YZ, $\frac{1}{6}$ 6. (a) 30 outcomes (b) $\frac{1}{2}$
7. (a) $\frac{1}{12}$ (b) $\frac{5}{36}$ (c) $\frac{2}{3}$ (d) $\frac{1}{12}$ (e) $\frac{1}{36}$ 8. Yes
9. No. Y has the best chance. 10. lose 20p 11. (a) $\frac{1}{144}$ (b) $\frac{1}{18}$

Exercise 2F

1. (a) Yes (b) No (c) Yes (d) No (e) No 2. AF, BC, DE
3. (a) $\frac{5}{11}$ (b) $\frac{7}{22}$ (c) $\frac{15}{22}$ (d) $\frac{17}{22}$ 4. $\frac{1}{18}$
5. (a) 0·3 (b) 0·9 6. $\frac{264}{360} = \frac{11}{15}$
7. (a) (i) 0·24 (ii) 0·89 (b) 575 8. (a) (i) exclusive (ii) not exclusive (b) $\frac{11}{15}$
9. (a) (i) exclusive (ii) exclusive (iii) not exclusive (b) $\frac{3}{4}$
10. (a) 0·8 (b) 0·7 (c) not exclusive 11. not mutually exclusive events

Exercise 2G

1. (a) $\frac{1}{13}, \frac{1}{6}$ (b) $\frac{1}{78}$ 2. (a) $\frac{1}{2}$ (b) $\frac{1}{2}$ (c) $\frac{1}{4}$ 3. $\frac{1}{10}$
4. (a) $\frac{1}{78}$ (b) $\frac{1}{104}$ (c) $\frac{1}{24}$ 5. (a) $\frac{1}{16}$ (b) $\frac{25}{144}$
6. (a) $\frac{1}{121}$ (b) $\frac{9}{121}$ 7. $\frac{8}{1125}$
8. (a) $\frac{1}{9}$ (b) $\frac{4}{27}$ 9. $\frac{1}{24}$ 10. $\frac{1}{128}$
11. (a) $\left(\frac{1}{6}\right)^{20}$ (b) $\left(\frac{5}{6}\right)^n$ (c) $1 - \left(\frac{5}{6}\right)^n$

Exercise 2H

1. (a) $\frac{49}{100}$ (b) $\frac{9}{100}$ 2. (a) $\frac{9}{64}$ (b) $\frac{15}{64}$ 3. (a) $\frac{7}{15}$ (b) $\frac{1}{15}$
4. (a) $\frac{1}{12}$ (b) $\frac{1}{6}$ (c) $\frac{1}{3}$ (d) $\frac{2}{9}$ 5. (a) $\frac{1}{216}$ (b) $\frac{125}{216}$ (c) $\frac{25}{72}$ (d) $\frac{91}{216}$
6. (a) $\frac{1}{64}$ (b) $\frac{5}{32}$ (c) $\frac{27}{64}$ 7. (a) $\frac{1}{6}$ (b) $\frac{1}{30}$ (c) $\frac{1}{30}$ (d) $\frac{29}{30}$
8. (a) 6 (b) $\frac{1}{3}$ 9. (a) $\frac{4}{9}$ (b) $\frac{1}{24}$ 10. (a) $\frac{3}{20}$ (b) $\frac{9}{20}$
11. (a) $\frac{3}{20} \times \frac{2}{19} \times \frac{1}{18} \left(=\frac{1}{1140}\right)$ (b) $\frac{5}{20} \times \frac{4}{19} \times \frac{3}{18} \left(=\frac{1}{114}\right)$ (c) $\frac{5}{20} \times \frac{4}{19} \times \frac{3}{18} \times \frac{2}{17} \left(=\frac{1}{969}\right)$

Exercise 2I

1. (a) $\frac{10 \times 9}{1000 \times 999}$ (b) $\frac{990 \times 989}{1000 \times 999}$ (c) $\frac{2 \times 10 \times 990}{1000 \times 999}$ 2. (a) $\frac{3}{20}$ (b) $\frac{7}{20}$ (c) $\frac{1}{2}$
3. (a) 5 (b) $\frac{1}{64}$ 4. (a) $\frac{1}{220}$ (b) $\frac{1}{22}$ (c) $\frac{3}{11}$ (d) 5
5. (a) $\frac{3}{5}$ (b) $\frac{1}{3}$ (c) $\frac{2}{15}$ (d) $\frac{2}{21}$ (e) $\frac{1}{7}$ 6. (a) 0·00781 (b) 0·511
7. (a) $\frac{21}{506}$ (b) $\frac{455}{2024}$ (c) $\frac{945}{2024}$
8. (a) $\dfrac{x}{x+y}$ (b) $\dfrac{x(x-1)}{(x+y)(x+y-1)}$ (c) $\dfrac{2xy}{(x+y)(x+y-1)}$ (d) $\dfrac{y(y-1)}{(x+y)(x+y-1)}$
9. (a) $\dfrac{x}{z}$ (b) $\dfrac{x(x-1)}{z(z-1)}$ (c) $\dfrac{2x(z-x)}{z(z-1)}$ 10. (a) $\frac{1}{125}$ (b) $\frac{1}{125}$ (c) $\frac{1}{10\,000}$ (d) $\frac{3}{500}$
11. (a) $\frac{1}{49}$ (b) $\frac{1}{7}$ 12. (a) $\frac{1}{52}$ (b) 1

Exercise 2J

1. $\frac{3}{10}$
2. $\frac{9}{140}$
3. (a) $\frac{1}{5}$ (b) $\frac{18}{25}$ (c) $\frac{1}{20}$ (d) $\frac{2}{25}$ (e) $\frac{77}{100}$
4. (a) 0·07 (b) 0·64 (c) 0·29
5. $\frac{4}{35}$
6. (a) $\frac{19}{25}$ (b) 38 out of 50
7. 27

Check out D2

1. $\frac{1}{13}$
2. $\frac{1}{9}$
3. The events cannot occur at the same time.
4. $\frac{1}{6}$
5. $\frac{7}{12}$

Revision exercise D2

1. (a) $\frac{16}{49}$ (b) $\frac{40}{49}$
2. $\frac{28}{75}$
3. (a) $\frac{7}{10}$ (b) 58
4. (a) $\frac{1}{2}$ (b) $\frac{1}{6}$ (c) $\frac{1}{6}$
5. (a) 0·18 (b) 0·12 (c) 0·54
6. (a) $\frac{1}{15}$ (b) $\frac{2}{5}$ (c) $\frac{2}{15}$
7. (a) 90 (b) 0·21
8. (a) $\frac{3}{5}$ (b) $\frac{3}{4}$ (c) $\frac{1}{2}$
9. (b) $\frac{21}{50}$
10. (a) 0·72 (b) 0·26

Module 1 Practice calculator test

1. 5·12 minutes
2. (b) about £44
3. (b) $\frac{16}{25}$
4. (b) 5 detached, 11 semi-detached, 6 terraced, 3 other

Module 1 Practice non-calculator test

1. (b) 120 g (c) (i) 0 (ii) 20
2. (a) 100 (b) cf: 15, 35, 60, 75, 85, 95, 100 (c) 12 36 (d) 93
3. $\frac{15}{28}$
4. (a) B
   (b) (i) Mean or median
       (ii) There are two modes. The mean would be distorted by the large value.

Answers Module 3

N1 Number 1

Check in N1
1. (a) (i) 241 (ii) 2410 (b) (i) 0·031 (ii) 0·0031
2. (a) 91p (b) 11p

Exercise 1A
1. (a) 1, 2, 3, 6 (b) 1, 3, 5, 15 (c) 1, 2, 3, 6, 9, 18 (d) 1, 3, 7, 21 (e) 1, 2, 4, 5, 8, 10, 20, 40
2. 2, 3, 5, 7, 11, 13, 17, 19
3. [e.g.] $2 + 3 = 5, 2 + 5 = 7 \ldots$
4. 3, 11, 19, 23, 29, 31, 37, 47, 59, 61, 67, 73
5. (a) $2^3 \times 3 \times 5^2$ (b) $3^2 \times 7 \times 11$ (c) $2^5 \times 7 \times 11$
   (d) $2 \times 3^3 \times 5 \times 13$ (e) $2^5 \times 5^3$ (f) $2^2 \times 5 \times 7^2 \times 23$

6.
	Prime	Multiple of 3	Factor of 16
Number greater than 5	7	9	8
Odd number	5	3	1
Even number	2	6	4

7. (a) 7 (b) 50 (c) 1 (d) 5 8. (c) 18
9. (a) 30 (b) 60 (c) 30
10. 1, 4, 9, 16, 25, 36, 49, 64, 81, 100, 121, 144, 169, 196, 225
11. 1, 8, 27, 64, 125 12. 101, 151, 293 are prime
14. (a) $1008 = 2^4 \times 3^2 \times 7$, $840 = 2^3 \times 3 \times 5 \times 7$ (b) 168 (c) 7
15. (a) $19\,800 = 2^3 \times 3^2 \times 5^2 \times 11$, $12\,870 = 2 \times 3^2 \times 5 \times 11 \times 13$ (b) 990 (c) 22
16. 8
17. (a) (i) 1, 8, 27, 64 (ii) cube numbers (iii) 100^3
    (b) (i) 9, 36, 100 (ii) square numbers (iii) $55^2 (= 3025)$
18. 2, 3, 5, 7, 89, 641 19. Always a multiple of 4. 20. Not always a multiple of n.

Exercise 1B
1. (a) 7, 22, 44, 22, $5\frac{1}{2}$ (b) 10, 25, 50, 28, 7 (c) 16, 31, 62, 40, 10 (d) $\frac{1}{2}$, $15\frac{1}{2}$, 31, 9, $2\frac{1}{4}$
2. (a) 4, 16, 48, 38, 19 (b) 5, 25, 75, 65, $32\frac{1}{2}$ (c) 6, 36, 108, 98, 49 (d) 8, 64, 192, 182, 91
3. ×4, square root, −10, ×⁻2 4. reciprocal, +1, square, ÷3 5. +3, cube, ÷⁻2, +100

Exercise 1C
1. 7·91   2. 22·22   3. 7·372   4. 0·066   5. 466·2
6. 1·22   7. 1·67   8. 1·61   9. 16·63   10. 24·1
11. 26·7 12. 3·86   13. 0·001 14. 1·56   15. 0·0288
16. 2·176 17. 0·02   18. 0·0001 19. 355   20. 20
21. £12·80 22. 34 000 23. 18   24. 0·569 25. 2·335
26. 1620 27. 8·8   28. 1200   29. 0·001 75 30. 13·2
31. 200   32. 0·804 33. 0·8   34. 0·077 35. 0·0009
36. 0·01 37. (b) (i) 20 (ii) 184 (iii) 0·63 (iv) 540 (v) 3
38. (a) 7·4 (b) 0·35 (c) 0·07 (d) 0·1
39. (a) T (b) T (c) T (d) F (e) F (f) T (g) T (h) F (i) T
40. (a) > (b) > (c) < (d) < (e) > (f) >

Exercise 1D
1. 805
2. 459
3. 650
4. 1333
5. 2745
6. 1248
7. 4522
8. 30 368
9. 28 224
10. 8568
11. 46 800
12. 66 281
13. 57 602
14. 89 516
15. 97 525

Exercise 1E
1. 32
2. 25
3. 18
4. 13
5. 35
6. 22 r 2
7. 23 r 24
8. 18 r 10
9. 27 r 18
10. 13 r 31
11. 35 r 6
12. 25 r 28
13. 64 r 37
14. 151 r 17
15. 2961 r 15

Exercise 1F
1. £47·04
2. 46
3. 7592
4. 21, 17p change
5. 8
6. £80·64
7. £14 million
8. £85
9. £21 600
10. 54
11. (a) 6·72 (b) 2·795 (c) 3·976
12. (a) 4·9 (b) 3·5 (c) 2·0
13. (b) 608

Exercise 1G
1. (a) $\frac{3}{8}$ (b) $\frac{9}{10}$ (c) $\frac{11}{15}$
2. (a) $\frac{5}{6}$ (b) $\frac{5}{12}$ (c) $\frac{7}{12}$ (d) $\frac{9}{20}$ (e) $1\frac{11}{20}$ (f) $1\frac{1}{2}$ (g) $1\frac{7}{12}$ (h) $3\frac{7}{12}$
3. (a) $\frac{1}{6}$ (b) $\frac{5}{12}$ (c) $\frac{7}{15}$ (d) $\frac{1}{10}$ (e) $1\frac{1}{12}$ (f) $\frac{1}{6}$

4.

+	$\frac{1}{4}$	$\frac{3}{8}$	$\frac{1}{3}$
$\frac{1}{2}$	$\frac{3}{4}$	$\frac{7}{8}$	$\frac{5}{6}$
$\frac{1}{8}$	$\frac{3}{8}$	$\frac{1}{2}$	$\frac{11}{24}$
$\frac{1}{5}$	$\frac{9}{20}$	$\frac{23}{40}$	$\frac{8}{15}$

5. (a) $\frac{8}{15}$ (b) $\frac{5}{42}$ (c) $\frac{15}{26}$ (d) $1\frac{1}{6}$ (e) $\frac{5}{8}$ (f) $9\frac{1}{6}$ (g) $\frac{4}{15}$ (h) $\frac{3}{14}$

6. (a) $\frac{5}{12}$ (b) $4\frac{1}{2}$ (c) $1\frac{2}{3}$ (d) $\frac{5}{16}$ (e) $1\frac{7}{8}$ (f) $2\frac{5}{8}$ (g) $1\frac{9}{26}$ (h) 8
7. 123 cm
8. $\frac{1}{5}$
9. (a) $\frac{9}{16}$ (b) $\frac{7}{20}$ (c) $1\frac{1}{5}$ (d) $1\frac{1}{6}$ (e) $\frac{1}{10}$ (f) $\frac{7}{15}$ (g) $\frac{20}{23}$ (h) $1\frac{3}{5}$
10. 5
11. 3
12. $1\frac{16}{17}$
13. (a) $a = 1$, $b = 2$ (b) $a = 1$, $b = 5$ (c) (1, 13) or (3, 10) or (5, 7) or (7, 4) or (9, 1)

14.

×	$\frac{2}{5}$	$\frac{3}{4}$	$\frac{2}{3}$
$\frac{1}{3}$	$\frac{2}{15}$	$\frac{1}{4}$	$\frac{2}{9}$
$\frac{1}{2}$	$\frac{1}{5}$	$\frac{3}{8}$	$\frac{1}{3}$
$\frac{1}{4}$	$\frac{1}{10}$	$\frac{3}{16}$	$\frac{1}{6}$

15. 144 ml
16. (a) $\frac{15}{16}$, $\frac{31}{32}$ (b) $\dfrac{2^{n+1} - 1}{2^{n+1}}$

Exercise 1H
1. 70·56
2. 118·958
3. 451·62
4. 33678·8
5. 0·6174
6. 1068
7. 19·53
8. 18 914·4
9. 38·72
10. 0·009 79

11. 2·4   12. 11   13. 41   14. 8·9   15. 4·7
16. 56   17. 0·0201   18. 30·1   19. 1·3   20. 0·31
21. 210·21   22. 294   23. 282·131   24. 35   25. 242

Exercise 1I

1. £8000   2. £6   3. 20 min   4. B   5. B
6. B   7. C   8. B   9. A
10. (a) 89·89   (b) 4·2   (c) 358·4   (d) 58·8   (e) 0·3   (f) 2·62
11. (a) 4·5   (b) 462   (c) 946·4   (d) 77·8   (e) 0·2   (f) 21
12. £5000   13. About 20   14. Yes   15. 50

Exercise 1J

1. (a) 0·7   (b) 8·81   (c) 0·726   (d) 1·18   (e) 0·9   (f) 8·22
   (g) 0·075   (h) 11·726   (i) 20·2   (j) 6·67   (k) 0·3   (l) 0·0725
2. (a) 2·7   (b) 189   (c) 3   (d) 0·36   (e) 1·7   (f) 0·0416
   (g) 0·04   (h) 800   (i) 9300   (j) 8·1   (k) 6·10   (l) 8

Exercise 1K

1. 1:3   2. 1:6   3. 1:50   4. 1:1·6   5. 1:0·75   6. 1:0·375
7. 2·4:1   8. 2·5:1   9. 0·8:1   10. £15, £25   11. £36, £84
12. 15 kg, 75 kg, 90 kg   13. 46 min, 69 min, 69 min   14. £39   15. 5:3
16. (a) 1:3   (b) 2:7   (c) 3:5   17. £200
18. 3:7   19. $\frac{1}{7}x$ 20. 6   21. £120   22. 300 g

Exercise 1L

1. 12·3 km   2. 4·71 km   3. 50 cm   4. 64 cm   5. 5·25 cm
6. 40 m by 30 m; 12 cm²; 1200 m²   7. 1 m², 6 m²   8. 0·32 km²

Exercise 1M

1. £1·10   2. 6 h   3. $2\frac{1}{2}$ l   4. 60 km   5. 119 g
6. 260 min   7. $2\frac{1}{4}$ weeks   8. 6 men   9. (a) 12   (b) 2100
10. 4   11. 5·6 days   12. 9600   13. $\dfrac{nx}{x+1000}$

Exercise 1N

1. ⁻4   2. ⁻12   3. ⁻11   4. ⁻3   5. ⁻5   6. 4   7. −5   8. ⁻8
9. 19   10. ⁻17   11. ⁻4   12. ⁻5   13. ⁻11   14. 6   15. ⁻4   16. 6
17. 0   18. ⁻18   19. ⁻3   20. ⁻11   21. ⁻12   22. 4   23. 4   24. 0
25. ⁻8   26. ⁻3   27. 3   28. ⁻12   29. 18   30. ⁻5   31. ⁻66   32. 98

Exercise 1O

1. ⁻8   2. ⁻7   3. 1   4. 1   5. 9   6. 11   7. ⁻8
8. 42   9. 4   10. 15   11. ⁻7   12. ⁻9   13. ⁻1   14. ⁻7
15. 0   16. 11   17. ⁻14   18. 0   19. 17   20. 3   21. ⁻1
22. ⁻3   23. 12   24. ⁻9   25. 3   26. 0   27. 8   28. 2

29. (a)

+	⁻5	1	6	⁻2
3	⁻2	4	9	1
⁻2	⁻7	⁻1	4	⁻4
6	1	7	12	4
⁻10	⁻15	⁻9	⁻4	⁻12

(b)

+	⁻3	2	⁻4	7
5	2	7	1	12
⁻2	⁻5	0	⁻6	5
10	7	12	6	17
⁻6	⁻9	⁻4	⁻10	1

Exercise 1P
1. ⁻6   2. ⁻4   3. ⁻15   4. 9   5. ⁻8   6. ⁻15   7. ⁻24   8. 6
9. 12   10. ⁻18   11. ⁻21   12. 25   13. ⁻60   14. 21   15. 48   16. ⁻16
17. ⁻42   18. 20   19. ⁻42   20. ⁻66   21. ⁻4   22. ⁻3   23. 3   24. ⁻5
25. 4   26. ⁻4   27. ⁻4   28. ⁻1   29. ⁻2   30. 4   31. ⁻16   32. ⁻2
33. ⁻4   34. 5   35. ⁻10   36. 11   37. 16   38. ⁻2   39. ⁻4   40. ⁻5
41. 64   42. ⁻27   43. ⁻600   44. 40   45. 2   46. 36   47. ⁻2   48. ⁻8

49. (a)

×	4	⁻3	0	⁻2
⁻5	⁻20	15	0	10
2	8	⁻6	0	⁻4
10	40	⁻30	0	⁻20
⁻1	⁻4	3	0	2

(b)

×	⁻2	5	⁻1	⁻6
3	⁻6	15	⁻3	⁻18
⁻3	6	⁻15	3	18
7	⁻14	35	⁻7	⁻42
2	⁻4	10	⁻2	⁻12

Test 1
1. ⁻16   2. 64   3. ⁻15   4. ⁻2   5. 15   6. 18   7. 3
8. ⁻6   9. 11   10. ⁻48   11. ⁻7   12. 9   13. 6   14. ⁻18
15. ⁻10   16. 8   17. ⁻6   18. ⁻30   19. 4   20. ⁻1

Test 2
1. ⁻16   2. 6   3. ⁻13   4. 42   5. ⁻4   6. ⁻4   7. ⁻12
8. ⁻20   9. 6   10. 0   11. 36   12. ⁻10   13. ⁻7   14. 10
15. 6   16. ⁻18   17. ⁻9   18. 15   19. 1   20. 0

Test 3
1. 100   2. ⁻20   3. ⁻8   4. ⁻7   5. ⁻4   6. 10   7. 9
8. ⁻10   9. 7   10. 35   11. ⁻20   12. ⁻24   13. ⁻10   14. ⁻7
15. ⁻19   16. ⁻1   17. ⁻5   18. ⁻13   19. 0   20. 8

Test 4
1. ⁻5   2. 2   3. ⁻20   4. 4   5. $-\frac{1}{2}$ 6. ⁻5
7. 12   8. 0   9. ⁻8   10. 1   11. ⁻1000   12. ⁻1

Exercise 1Q
1. 6   2. 17   3. $6^2 + 7^2 + 42^2 = 43^2$, $x^2 + (x+1)^2 + [x(x+1)]^2 = (x^2 + x + 1)^2$
4. (a) 87p   (b) 240p   (c) 72p   (d) 15p   (e) 9p

5. 200 g **6.** (a) £150 (b) £0 (c) £8 **7.** $\frac{1}{66}$ **8.** 5
9. (a) $54 \times 9 = 486$ (b) $57 \times 8 = 456$ (c) $52 \times 2 = 104$
10. (a) $\frac{1}{3}$ (b) $a = 2, b = 8$

Exercise 1R

1.

×	$\frac{2}{3}$	$\frac{3}{4}$	$\frac{1}{5}$
$\frac{1}{2}$	$\frac{1}{3}$	$\frac{3}{8}$	$\frac{1}{10}$
$\frac{1}{4}$	$\frac{1}{6}$	$\frac{3}{16}$	$\frac{1}{20}$
$\frac{2}{5}$	$\frac{4}{15}$	$\frac{3}{10}$	$\frac{2}{25}$

2. (a) $66667^2 = 4\,444\,488\,889$
  (b) $44\,444\,448\,888\,889$
  (c) $66\,666\,667$
3. (a) 100, 1 (b) 100, 2
4. $40 = 2^3 \times 5$, $63 = 3^2 \times 7$, $25 = 5^2$, 71 is prime
5. 36, 49, 64 **6.** (a) 1 (b) 15
7. 50 **8.** 106

Exercise 1S

1. £7055 **2.** 10
3. (a) 1, 3, 5, 7, 9, 11 (b) Step 1. Write down the number 2.
5. (a) 66 666 (b) 82 (c) 455 551
6.

8	1		9
	3	7	1
2	5	5	
7		6	8

7. (a) $\frac{3}{5}$ (b) $\frac{75}{16}$ **8.** 500

Exercise 1T

1. (a) 4 (b) 5 **2.** 21 **3.** 0·525 kg **5.** (a) $\frac{1}{994}$, $\frac{1}{949}$, $\frac{1}{499}$ (b) $\frac{91}{94}$
6. (a) £6576 (b) (i) 8 (ii) 99

Operator squares

1. 0·49, 3.5×10, 40 **2.** 5·2, $4.56 \times 5 \rightarrow 22.8$, 0·64
3. 0·7, $16 - (^-19)$, 112, $^-37.8$ **4.** $^-12$, $7 \rightarrow 11$, $^-3$
5. $-\frac{1}{5}$, $-\frac{1}{20}$, $-3\frac{1}{5}$, $-1\frac{1}{4}$ **6.** 0·86, $\div$, $\frac{1}{3} - 0.625$, 0·336

Check out N1

1. 216
2. (a) $\frac{3}{10}$ (b) $6\frac{1}{20}$ (c) $2\frac{5}{8}$
3. (a) (i) 37·35 (ii) 37·349 (b) (i) 2490 (ii) 2500
4. 30, 24 **5.** (a) $^-8$ (b) $^-24$ (c) 96

Revision exercise N1

1. (a) 400 (b) (i) 0·0049 (ii) $3\frac{5}{12}$
2. (a) $2^2 \times 3^2$ (b) 12
3. (a) 15 000 (b) $\frac{7}{15}$
4. (a) $2^3 \times 3^2$ (b) 144
5. (a) (i) 21·9164 (ii) 21 916·4 (b) $\frac{7}{20}$

N2 Number 2

Check in N2
1. (a) 6 (b) 59
2. (a) 2×3^2 (b) $2^2 \times 3^3$ (c) 2×5^2
3. (a) $^-4$ (b) $\frac{5}{11}$
4. (a) 748 cm (b) 7.431
5. (a) 75 in (b) 72 oz
6. (a) 77 lb (b) 80 in
7. 5 h 15 min
8. (a) continuous (b) discrete (c) discrete

Exercise 2A
1. (a) 0·25 (b) 0·4 (c) 0·375 (d) $0·41\dot{6}$ (e) $0·1\dot{6}$ (f) $0·\dot{2}8571\dot{4}$
2. (a) $\frac{1}{5}$ (b) $\frac{9}{20}$ (c) $\frac{9}{25}$ (d) $\frac{1}{8}$ (e) $1\frac{1}{20}$ (f) $\frac{7}{1000}$
3. (a) 25% (b) 10% (c) 72% (d) 7·5% (e) 2% (f) $33\frac{1}{3}$%
4. (a) 45%; $\frac{1}{2}$; 0·6 (b) 4%; $\frac{6}{16}$; 0·38 (c) 11%; 0·111; $\frac{1}{9}$ (d) 0·3; 32%; $\frac{1}{3}$
5. 0·58 6. 1·42 7. 0·65 8. 0·32 9. 0·07
10. 0·69 11. 40% 12. 25% 13. No 14. 96·8%
15. (a) 58·9% (b) 55% 16. Prices will still rise but not as quickly.

Exercise 2B
1. (a) £15 (b) 900 kg (c) $2·80 (d) 125 people
2. £32 3. 13·2p 4. 52·8 kg
5. (a) 0·53 (b) 0·03 (c) 0·085 (d) 1·22
6. (a) £1·02 (b) £21·58 (c) £2·22 (d) £0·53
7. £248·57 8. £26 182 9. £8425·60 10. (b) £6825
11. £201·40, £166 500, £2700, £199·50, £55·25 12. £6·30
13. (a) 256 g (b) 72p (c) 14 mm (d) 30 min (e) 14·4 sec
14. (b) (i) $1·08 \times P \times 0·9$ (ii) 15%, 6% (iii) $1·25 \times Q \times 0·95$

Exercise 2C
1. (a) 25%, profit (b) 25%, profit (c) 10%, loss (d) 20%, profit
   (e) 30%, profit (f) 7·5%, profit (g) 12%, loss (h) 54%, loss
2. 28% 3. $44\frac{4}{9}$% 4. 46·9% 5. 12% 6. $5\frac{1}{3}$% 7. 44%
8. 27·05% 9. 20% 10. 14·3% 11. 21% 12. (a) 25 000 (b) 76·3%

Exercise 2D
1. £6200 2. £25 500 3. (a) £50 (b) £200 (c) £54 (d) £150 (e) £2400
4. 210 000 5. 85 kg 6. 8 cm × 10 cm 7. 350 g 8. £56
9. 400 kg 10. 200 11. 29 000 12. 500 cm 13. £500 14. £480 million

Exercise 2E
1. (a) £2200 (b) £2420 (c) £2662 2. (a) £5600 (b) £7024·64
3. £13 107·96 4. (a) £36 465·19 (b) £40 202·87
5. (a) £95 400 (b) £107 191 (c) £161 176 6. (a) £14 033 (b) £734 (c) £107 946
7. £9211·88 8. No. It should be £193·48. 9. 11 years
10. (b) $x = 9$ (c) 9 years 11. £30 000 at 8% compound interest

Exercise 2F

1. 3·041
2. 1460
3. 0·03083
4. 47·98
5. 130·6
6. 0·4771
7. 0·3658
8. 37·54
9. 8·000
10. 0·6537
11. 0·03716
12. 34·31
13. 0·7195
14. 3·598
15. 0·2445
16. 2·043
17. 0·3798
18. 0·7683
19. $^{-}$0·5407
20. 0·07040
21. 2·526
22. 0·09478
23. 0·2110
24. 3·123
25. 2·230
26. 128·8
27. 4·268
28. 3·893
29. 0·6290
30. 0·4069
31. 9·298
32. 0·1010
33. 0·3692
34. 1·125
35. 1·677
36. 0·9767
37. 0·8035
38. 0·3528
39. 2·423
40. 1·639
41. 0·0004659
42. 0·3934
43. $^{-}$0·7526
44. 2·454
45. 40000
46. 0·07049
47. 405400
48. 471·3
49. 20810
50. $2·218 \times 10^6$
51. $1·237 \times 10^{-24}$
52. 3·003
53. 0·03581
54. 47·40
55. $^{-}$1748
56. 0·01138
57. 1757
58. 0·02635
59. 0·1651
60. 5447

Exercise 2I

1. 3^4
2. $4^2 \times 5^3$
3. 3×7^3
4. $2^3 \times 7$
5. 10^{-3}
6. $2^{-2} \times 3^{-3}$
7. $15^{\frac{1}{2}}$
8. $3^{\frac{1}{3}}$
9. a^8
9. $10^{\frac{1}{3}}$
11. 11^{-1}
12. $5^{\frac{3}{2}}$
13. T
14. F
15. T
16. T
17. F
18. F
19. F
20. T
21. T
22. T
23. F
24. F
25. F
26. T
27. T
28. T
29. T
30. T
31. T
32. F
33. 5^6
34. 6^5
35. 2^{13}
36. 3^4
37. 6^5
38. 8
39. 6^4
40. 2^5
41. 2^{19}
42. 10^{100}
43. 5^{-5}
44. 3^{-10}
45. 3^7
46. 2^7
47. 7^2
48. 5^{-1}

Exercise 2J

1. 3^6
2. 5^{12}
3. 7^{10}
4. x^6
5. 2^{-2}
6. 1
7. 7^2
8. y^2
9. 2^{21}
10. 10^{99}
11. x^7
12. y^{13}
13. z^4
14. z^{100}
15. m
16. e^{-5}
17. y^2
18. w^6
19. y
20. x^{10}
21. 1
22. w^{-5}
23. w^{-5}
24. x^7
25. a^8
26. k^3
27. 1
28. x^{29}
29. y^2
30. x^6
31. z^4
32. t^{-4}
33. $4x^6$
34. $16y^{10}$
35. $6x^4$
36. $10y^5$
37. $15a^4$
38. $8a^3$
39. 3
40. $4y^2$
41. $\frac{5}{2}y$
42. $32a^4$
43. $108x^5$
44. $4z^{-3}$
45. $2x^{-4}$
46. $\frac{5}{2}y^5$
47. 1
48. $21w^{-3}$
49. $2n^4$
50. $2x$

Exercise 2K

1. 27
2. 1
3. $\frac{1}{9}$
4. 25
5. 2
6. 4
7. 9
8. 2
9. 27
10. 3
11. $\frac{1}{3}$
12. $\frac{1}{2}$
13. 1
14. $\frac{1}{5}$
15. 10
16. 8
17. 32
18. 4
20. (a) $\frac{1}{2}$ (b) $-\frac{1}{2}$ (c) $\frac{3}{2}$
21. (a) $\frac{1}{9}$ (b) $\frac{1}{8}$ (c) 100^6
22. 10
23. 1000
24. $\frac{1}{1000}$
25. $\frac{1}{9}$
26. 1
27. $1\frac{1}{2}$
28. $\frac{1}{25}$
29. $\frac{1}{10}$
30. $\frac{1}{4}$
31. $\frac{1}{4}$
32. 100000
33. 1
34. $\frac{1}{32}$
35. 0·1
36. 0·2
37. 1·5
38. 1
39. 9
40. $1\frac{1}{2}$
41. $\frac{3}{10}$
42. 64
43. $\frac{1}{100}$
44. $1\frac{2}{3}$
45. $\frac{1}{100}$
46. (a) $\frac{1}{n}$

Exercise 2L

1. 4×10^3
2. 5×10^2
3. 7×10^4
4. 6×10
5. $2 \cdot 4 \times 10^3$
6. $3 \cdot 8 \times 10^2$
7. $2 \cdot 56 \times 10^3$
8. 7×10^{-3}
9. 4×10^{-4}
10. $3 \cdot 5 \times 10^{-3}$
11. $4 \cdot 21 \times 10^{-1}$
12. $5 \cdot 5 \times 10^{-5}$
13. 1×10^{-2}
14. $5 \cdot 64 \times 10^5$
15. $1 \cdot 9 \times 10^7$
16. $1 \cdot 1 \times 10^9$
17. $1 \cdot 67 \times 10^{-24}$
18. $2 \cdot 5 \times 10^{-10}$
19. $6 \cdot 023 \times 10^{23}$
20. £$3 \cdot 6 \times 10^6$
21. c, a, b
22. 16
23. (a) 8×10^{11} (b) 5×10^{-7} (c) $5 \cdot 5 \times 10^{-4}$ (d) 4×10^9 (e) 3×10^8 (f) 4×10^{12}
24. (a) 23 000 (b) 0·03 (c) 560 (d) 800 000 (e) 0·0022
    (f) 900 000 000 (g) 0·6 (h) 7000 (i) 3 140 000
25. (a) $6 \cdot 5 \times 10^5$ (b) $2 \cdot 7 \times 10^3$ (c) $3 \cdot 54 \times 10^6$ (d) 7×10^{-3}
26. (a) 6×10^3 (b) $1 \cdot 1 \times 10^{-1}$ (c) $4 \cdot 5 \times 10^8$ (d) $8 \cdot 5 \times 10^{-6}$ (e) 6×10^9 (f) 5×10^{-1}

Exercise 2M

1. 6×10^7
2. 2×10^7
3. $1 \cdot 5 \times 10^{10}$
4. 7×10^9
5. 9×10^8
6. $1 \cdot 6 \times 10^9$
7. $4 \cdot 5 \times 10$
8. $1 \cdot 2 \times 10^{14}$
9. $1 \cdot 86 \times 10^{11}$
10. 2×10^{-10}
11. $7 \cdot 5 \times 10^{-12}$
12. $2 \cdot 5 \times 10^9$
13. $2 \cdot 7 \times 10^{16}$
14. $3 \cdot 5 \times 10^{-6}$
15. $8 \cdot 1 \times 10^{-21}$
16. $6 \cdot 3 \times 10^{13}$
17. $1 \cdot 5 \times 10^{16}$
18. 8×10^3
19. $2 \cdot 8 \times 10^{16}$
20. 8×10^{-3}
21. 2×10
22. 5×10^{-20}
23. 5×10^{-4}
24. 3×10^{-11}
25. $5 \cdot 57 \times 10^9$
26. Asia (0·068 people/m^2)
27. $2 \cdot 4528 \times 10^7$ km
28. 50 min
29. $2 \cdot 4 \times 10^9$ kg
30. (a) $9 \cdot 46 \times 10^{12}$ km (b) 144 million km
31. (a) 6×10^{101} (b) 20·5 sec (c) $6 \cdot 3 \times 10^{91}$ years

Exercise 2N

1. Rational: $(\sqrt{17})^2$; $3 \cdot 14$; $\frac{\sqrt{12}}{\sqrt{3}}$; $3^{-1} + 3^{-2}$; $\frac{22}{7}$; $\sqrt{2 \cdot 25}$
3. (a) A 4, B $\sqrt{41}$ (b) A rational, B irrational (c) Both rational (d) B
4. (a) 6π cm, irrational (b) 6 cm, rational (c) 36 cm^2, rational
   (d) 9π cm^2, irrational (e) $36 - 9\pi$ cm^2, irrational
5. (a) $\frac{2}{9}$ (b) $\frac{29}{99}$ (c) $\frac{541}{999}$
8. (a) $\frac{35}{99}$ (b) $\frac{5}{11}$ (c) $\frac{41}{495}$ (d) $\frac{145}{333}$ (e) $\frac{287}{990}$ (f) $\frac{7}{22}$

Exercise 2O

1. AF, EG, BD, CH
2. (a) T (b) T (c) F (d) T (e) F (f) T
3. (a) $3 + 2\sqrt{2}$ (b) $7 - 4\sqrt{3}$ (c) 18 (d) $14 - 6\sqrt{5}$ (e) $6 + 4\sqrt{2}$ (f) 8
4. (a) T (b) T (c) F (d) T (e) F (f) T
5. (a) 4 (b) $4\sqrt{2}$ (c) $10\sqrt{3}$ (d) $9\sqrt{2}$ (e) $5\sqrt{5}$ (f) $5\sqrt{3}$
   (g) $5\sqrt{5}$ (h) $\sqrt{3}$ (i) 2 (j) $1\frac{1}{2}$ (k) $2\frac{1}{2}$ (l) $1\frac{1}{3}$
6. (a) $2\sqrt{3}$ (b) (i) $4\sqrt{2}$ (ii) $4\sqrt{3}$ (iii) $3\sqrt{2}$ (iv) $\frac{2}{3}\sqrt{3}$

Exercise 2P

1. 7
2. 13
3. 13
4. 22
5. 1
6. $^-1$
7. 18
8. $^-4$
9. $^-3$
10. 37
11. 0
12. $^-4$
13. $^-7$
14. $^-2$
15. $^-3$
16. $^-8$
17. $^-30$
18. 16
19. $^-10$
20. 0
21. 7
22. $^-6$
23. $^-2$
24. $^-7$
25. $^-5$
26. 3
27. 4
28. $^-8$
29. $^-2$
30. 2
31. 0
32. 4
33. $^-4$
34. $^-3$
35. $^-9$
36. 4

Exercise 2Q

1. 9
2. 27
3. 4
4. 16
5. 36
6. 18
7. 1
8. 6
9. 2
10. 8
11. ⁻7
12. 15
13. ⁻23
14. 3
15. 32
16. 36
17. 144
18. ⁻8
19. ⁻7
20. 13
21. 5
22. ⁻16
23. 84
24. 17
25. 6
26. 0
27. ⁻25
28. ⁻5

Exercise 2R

1. ⁻20
2. 16
3. ⁻42
4. ⁻4
5. ⁻90
6. ⁻160
7. ⁻2
8. ⁻81
9. 4
10. 22
11. 14
12. 5
13. 1
14. $\sqrt{5}$
15. 4
16. ⁻$6\frac{1}{2}$
17. 54
18. 25
19. 4
20. 312
21. 45
22. 22
23. 14
24. ⁻36
25. ⁻7
26. 1
27. 901
28. ⁻30
29. ⁻5
30. $7\frac{1}{2}$
31. ⁻7
32. ⁻$\frac{3}{13}$
33. $1\frac{1}{3}$
34. ⁻$\frac{5}{36}$

Exercise 2S

1. 21
2. 1·62
3. 396
4. 650
5. 63·8
6. 9×10^{12}
7. $10\frac{1}{2}$
8. 800
9. (a) 1245 km/h (b) 5 °C (c) 1008 km/h
10. (a) $ac + ab - a^2$ (b) 30
11. (a) $r - p + q$ (b) 4·1

Exercise 2T

1. (a) $2\frac{1}{2}$ h (b) $3\frac{1}{8}$ h (c) 75 s (d) 4 h
2. 156 km/h
3. (a) 75 km/h (b) 4×10^6 m/s (c) 3 km/h
4. (a) 10 000 m (b) 56 400 m (c) 4500 m (d) 50 400 m
5. (a) 4·45 h (b) 23·6 km/h
6. (a) 8 m/s (b) 7·6 m/s (c) 102·63 s
7. 1230 km/h
8. 3 h
9. 100 s
10. $1\frac{1}{2}$ minutes
11. 600 m
12. 5 cm/s
13. 60 s

Exercise 2U

1. 10 g/cm³
2. 240 g
3. 35 m³
4. 0·6 kg
5. 250 people/km²
6. (a) 6000 people/km² (b) 5.58×10^{10}
7. 1000 kg
8. 3000 kg/m³
9. 0·8 g/cm³
10. (a) 20 m/s (b) 108 km/h (c) 1·2 cm/s (d) 90 m/s

Exercise 2V

1. 195·5 cm
2. 36·5 kg
3. 3·25 kg
4. 95·55 m
5. 28·65 s
6. (a) 1·5 °C, 2·5 °C (b) 2·25 g, 2·35 g (c) 63·5 m, 64·5 m (d) 13·55 s, 13·65 s
7. B
8. C
9. (a) Not necessarily (b) 1 cm
10. (a) 16·5, 17·5 (b) 255·5, 256·5 (c) 2·35, 2·45 (d) 0·335, 0·345 (e) 2·035, 2·045 (f) 11·95, 12·05 (g) 81·35, 81·45 (h) 0·25, 0·35 (i) 0·65, 0·75 (j) 51 500, 52 500
11. No, max. card length 11·55 cm min. envelope length 11·5 cm

Exercise 2W

1. (a) 7·5, 8·5, 10·5 cm (b) 26·5 cm
2. 46·75 cm²
3. (a) 7 (b) 5 (c) 10 (d) 4 (e) 2 (f) 5 (g) 2 (h) 24
4. (a) 13 (b) 11 (c) 3 (d) 12·5
5. (i) 10·5 (ii) 4·3
6. (i) 11 (ii) 1 (iii) 0·6
7. 56 cm²
8. 55·706 604
9. £10·98
10. $33\frac{1}{3}$%
11. 41·6%

Exercise 2X
1. 7·15%   2. 4·23%   3. 12·9%   4. 0·177%   5. 6·15%
6. 0·0402%   7. 1·55%   8. 8·00%   9. 0·28%   10. Second

Exercise 2Y
1. (a) Fr 218   (b) $114·80   (c) Ptas 52 000   (d) DM 4·65   (e) lire 6670   (f) €1·386
2. (a) £45·87   (b) £1524·39   (c) £2·42   (d) £584·42   (e) £172·41   (f) £3·65
3. £3·84   4. £15 000, USA £15 172, France £17 800   5. 3·25 francs to the pound
6. 20   7. (a) 256   (b) 65 536   8. $4·68 \times 10^5$
9. (a) $\frac{3}{19}$ (b) $\frac{1}{6}$ 10. £10 485·76   11. (a) 1·25 m/min   (b) 2·08 cm/s
12. 77·5%   13. 225 mm   14. (a) 35   (b) 5·14
15. 43·5 mm   16. (b) 12   17. 210 kg
18. (a) 10:45   (b) 51 min   19. 1105   20. (a) £699

Exercise 2Z
1. (a) 0·0018 km/h   (b) $1·8 \times 10^{-3}$ 2. 13   3. $£6·3 \times 10^{10}$
4. (a) £9·79   (b) £32   (c) £44·20   (d) £45
5. (a) 323 g   (b) 23   (c) 67p   (d) 29
6. 6 m   7. approx 190 ml   8. 400   9. 18:52   10. £118
11. 37   12. 7^{77} 13. $\frac{43}{99}$

Check out N2
1. (a) (i) £19·44   (ii) 0·67%   (b) £1·80   2. (a) £4868·95   (b) 12%   3. 7.52
4. (a) 2×3^3 (b) 10^3 or 0·001   (c) $2^{\frac{5}{2}}$ 5. (a) $2·187 \times 10^{-4}$ (b) 5×10^3 (c) $4·195 \times 10^4$
6. (a) $\frac{5}{11}$ (b) $\frac{163}{198}$ 7. (a) $6\sqrt{3}$ (b) $11\sqrt{3}$ (c) 3
8. 42·4 mph   9. (a) 11·35 degrees   (b) 10·25 degrees

Revision exercise N2
1. 1900   2. 8 h 24 min   3. (a) 5   (b) $\frac{1}{7}$ (c) 3^{-2} (d) 4
4. £9408   5. (a) £561·25   (b) 12·6%   6. (a) $\frac{1}{9}$ (b) 2^5
7. $\frac{7}{100}$ 8. 44%   9. 44%   10. 49 mph   11. 38·309 645 25 m³   12. 29p

N3 Algebriac techniques for Module 3

Check in N3
1. $2(m + n - 1)$ 2. $4mn - 2m + 2n - 1$ 3. (a) 54   (b) $\frac{1}{400}$
4. (a) 5   (b) 2·04   5. $y = 1, ^-3, ^-5, ^-5, ^-3, 1$

Exercise 3B
1. (a) If today is Monday, then tomorrow is Tuesday.
   (b) If it is raining, then there are clouds in the sky.
   (c) If Abraham Lincoln was born in 1809, then Abraham Lincoln is dead.
2. (a) T   (b) T   (c) T   (d) F   (e) T   (f) F   (g) T   (h) T   (i) F   (j) T

Exercise 3D
1. (a) $S = ke$ (b) $v = kt$ (c) $x = kz^2$ (d) $y = k\sqrt{x}$ (e) $T = k\sqrt{L}$
2. (a) 9   (b) $2\frac{2}{3}$ 3. (a) 35   (b) 11   4. (a) 75   (b) 4

5. 1, 4; 3, 12; 4, 16; $5\frac{1}{2}$, 22   6. 1, 4; 2, 32; 4, 256; $1\frac{1}{2}$, $13\frac{1}{2}$
7. 333 N/cm²   8. 180 m; 2 s   9. 675 J; $\sqrt{\frac{4}{3}}$ cm
10. $p \propto w^3$ 11. $15^4 : 1$ (50 625 : 1)

Exercise 3E

1. (a) $x = \frac{k}{y}$ (b) $s = \frac{k}{t^2}$ (c) $t = \frac{k}{\sqrt{q}}$ (d) $m = \frac{k}{w}$ (e) $z = \frac{k}{t^2}$
2. (a) 6   (b) $\frac{1}{2}$ 3. (a) 6   (b) $1\frac{1}{2}$ 4. (a) 1   (b) 4
5. (a) 36   (b) ±4   6. (a) 6   (b) 16
7. 2, 8; 4, 4; 1, 16; $\frac{1}{4}$, 64   8. 2, 25; 5, 4; 20, $\frac{1}{4}$; 10, 1
9. (a) 6   (b) 50   10. 2·5 m³; 200 N/m²   11. 3 h; 48 men   12. 6 cm
13. 80   14. (a) 20 min   (b) 2 min

Exercise 3F

1. (a) ⁻0·4, 2·4   (b) ⁻0·8, 3·8   (c) ⁻1, 3   (d) ⁻0·4, 2·4
2. (a) ⁻2·6, 1·6   (b) ⁻0·75, 2·75   (c) ⁻2, 2   (d) 0·7
3. ⁻0·3, 3·3   4. 0·3, 3·7   5. (c) (i) 1, 4   (ii) 0·3, 3·7
6. (a) $y = 3$ (b) $y = ⁻2$ (c) $y = x + 4$ (d) $y = x$ (e) $y = 6$
7. (a) $y = 6$ (b) $y = 0$ (c) $y = 4$ (d) $y = 2x$ (e) $y = 2x + 4$
8. (b) (i) 0·3, ⁻3·3   (ii) 1·5, ⁻4·5   (iii) ⁻3, 1
9. (a) ⁻1·65, 3·65   (b) ⁻1·3, 2·3   (c) ⁻1·45, 3·45   10. (a) 1·7, 5·3   (b) 0·2, 4·8
11. (a) 3·35   (b) 2·4, 7·6   (c) 4·25
12. (a) ±3·74   (c) ±2·83   13. (a) 1   (b) 2   (c) 2   (d) 1   (e) 2   (f) 3

Check out N3

1. $9 + 6\sqrt{2}$
3. (a) $y = 16, 10, 6, 4, 4, 6, 10, 16$ (b) No solutions   (c) 0·6, 3·4
4. (a) $y = \frac{28}{x^2}$ (b) 1·12   (c) ±1·6

Revision exercise N3

1. (b) $110 - 12\sqrt{6}$
2. (a) $v = 0.0003h^3$ (b) 18·8 cm
4. (a) $F = \frac{640\,000}{d^2}$ (b) 4 N   (c) 200 m
5. (a) $F = 0.0027W^3$ (b) 21 600   (c) 121 m
6. (a) $E = \frac{24}{S}$ (b) 2 m   (c) 6
7. (a) 6, ⁻4   (c) ⁻1, 4   (d) ⁻0·24, 4·24
8. (a) ⁻3·8, 1·8   (b) ⁻2·9, 3·9
9. (a) $A = 0.75W^2$ (b) 12 cm²

Module 3 Practice calculator test

1. 1280   2. 89%   3. £17·34   4. £3685·19
5. (b) 0·2, 4·8   (c) ⁻0·6, 4·6   6. 38%
7. (a) $\frac{6}{11}$ (b) $\frac{83}{110}$ 8. 19·9 mph

Module 3 Practice non-calculator test

1. £55   2. 30p   3. (a) $\frac{29}{35}$ (b) $2\frac{11}{15}$ (c) 5×10^{11}
5. (a) $A = 1\frac{2}{3}W^2$ (b) 60 ft²   (c) 1·2 ft   6. (a) (i) 4   (ii) 27   (b) 2^{-15}
7. (a) $3\sqrt{2}$ (b) $3\sqrt{3}$

Answers Module 5

AS1 Algebra 1

Check in AS1
1. (a) 2^{15} (b) 3^{10} 2. (a) $^-2$ (b) 13 (c) $^-21$ (d) 40 3. (a) 21 (b) $^-20$

Exercise 1A
1. (a) $2x - 6$ (b) $(x+4)^2$ (c) $\dfrac{2(x-5)}{3}$ (d) $(w-x)^2$ (e) $\dfrac{(n+p)^3}{a}$ (f) $[3(t-1)]^2$

2. nw kg 3. $\dfrac{x}{n}$ pence 4. $\dfrac{y}{n}$ kg 5. $£\dfrac{p}{5}$ 6. $y+n+6$

7. $£w(n+r)$ 8. $\dfrac{xn}{t}$ metres 9. $\dfrac{100n}{x}$ 10. $\dfrac{100n}{x+1}$ pence

Exercise 1B
1. $5x+8$ 2. $9x+5$ 3. $7x+4$ 4. $7x+4$ 5. $7x+7$
6. $8x+12$ 7. $12x-6$ 8. $2x+5$ 9. $2x-5$ 10. $2x-5$
11. $13a+3b-1$ 12. $10m+3n+8$ 13. $3p-2q-8$ 14. $2s-7t+14$ 15. $2a+1$
16. $x+y+7z$ 17. $5x-4y+4z$ 18. $5k-4m$ 19. $4a+5b-9$ 20. $a-4x-5e$
21. (a) $3x^2+1$ (b) $2a+5ab$ (c) x^3+4x^2-7x (d) $4a^2-7a$
22. B and D   23. x^2+3x+3 24. $3x^2+6x+5$ 25. $2x^2+6x-7$ 26. $3x^2+x+12$
27. $2x^2+x+3$ 28. $2x^2-x$ 29. x^2-4x-2 30. $3x^2-2x-2$ 31. $4y^2+5x^2+x$
32. $x+12$ 33. $5y^2-6y+2$ 34. $3ab-3b$ 35. $2cd-2d^2$ 36. $4ab-2a^2+2a$
37. $2x^3+5x^2$ 38. $11+x^2+x^3$ 39. $3xy$ 40. p^2-q^2

Exercise 1C
1. (a) $6x$ (b) $15a$ (c) ^-10t (d) $1000a$ (e) $2x^2$ (f) $4x^2$ (g) $2x^3$ (h) $20t^2$
2. (a) $4x+2$ (b) $10x+15$ (c) a^2-3a (d) $2n^2+2n$ (e) $^-4x-6$ (f) $2x^2+2xy$
   (g) $^-3a+6$ (h) $pq-p^2$ (i) $ab+a^2$ (j) $2b^2-2b$ (k) $2n^2+4n$ (l) $6x^2+9x$
3. AG, BD, CE, FH
4. (a) $6ab$ (b) $3x^4$ (c) $2xy$ (d) $10pq$ (e) $15xy$ (f) $18x^3$
   (g) $24a^4$ (h) $2a^2b$ (i) $3xy^2$ (j) $5c^2d$ (k) a^2b^2 (l) $2x^2y^2$
   (m) $3d^3$ (n) $10x^2y$ (o) $^-6a^2$ (p) $2a^2b$
5. $6x^2+12x$
6. (a) $7x+10$ (b) $8x+2$ (c) $5a-3$ (d) $11a+17$ (e) $x+4$ (f) $x+6$
7. (a) $2x-2$ (b) $x-1$ (c) $x+2$ (d) $x-3$ (e) $x+3$ (f) $x+3$
8. (a) $2x^2+4x+6$ (b) $2x^2+2x+5$ (c) $5x^2+2x$ (d) a^2+2a

Exercise 1D
1. x^2+4x+3 2. x^2+5x+6 3. $y^2+9y+20$ 4. x^2+x-12
5. $x^2+3x-10$ 6. x^2-5x+6 7. $a^2-2a-35$ 8. $z^2+7z-18$
9. x^2-9 10. k^2-121 11. $2x^2-5x-3$ 12. $3x^2-2x-8$
13. $2y^2-y-3$ 14. $49y^2-1$ 15. $x^2+8x+16$ 16. x^2+4x+4
17. x^2-4x+4 18. $4x^2+4x+1$ 19. $2x^2+6x+5$ 20. $2x^2+2x+13$
21. $5x^2+8x+5$ 22. $2y^2-14y+25$ 23. $10x-5$ 24. $^-8x+8$
25. (a) $4x(x+5)$ (b) $(2x+5)^2$ (c) 25 square units
26. $x^3+6x^2+11x+6$ 27. (a) $x^3+8x^2+19x+12$ (b) x^3+3x^2+3x+1 (c) x^3+3x^2-4
28. (d) Yes. $n=10, 11, 21, 22$ etc.

Exercise 1E
1. $x(x+5)$
2. $x(x-6)$
3. $x(7-x)$
4. $y(y+8)$
5. $y(2y+3)$
6. $2y(3y-2)$
7. $3x(x-7)$
8. $2a(8-a)$
9. $3c(2c-7)$
10. $3x(5-3x)$
11. $7y(8-3y)$
12. $x(a+b+2c)$
13. $x(x+y+3z)$
14. $y(x^2+y^2+z^2)$
15. $ab(3a+2b)$
16. $xy(x+y)$
17. $2a(3a+2b+c)$
18. $m(a+2b+m)$
19. $2k(x+3y+2z)$
20. $a(x^2+y+2b)$
21. $x(7x+1)$
22. $4y(y-1)$
23. $p(p-2)$
24. $2a(3a+1)$
25. $4(1-2x^2)$
26. $5x(1-2x^2)$
27. $\pi(4r+h)$
28. $\pi r(r+2)$
29. $\pi r(3r+h)$
30. $x(3y+2)$
31. $xk(x+k)$
32. $ab(a^2+2b)$
33. $bc(a-3b)$
34. $ae(2a-5e)$
35. $ab(a^2+b^2)$
36. $x^2y(x+y)$
37. $2xy(3y-2x)$
38. $3ab(b^2-a^2)$
39. $a^2b(2a+5b)$
40. $ax^2(y-2z)$
41. $2ab(x+b+a)$
42. $yx(a+x^2-2yx)$
43. (a) $2x+8y$ (b) $10a-2x$

Exercise 1G
1. 8
2. 9
3. 7
4. 10
5. $\frac{1}{3}$
6. 10
7. $1\frac{1}{2}$
8. $^-1$
9. $^-1\frac{1}{2}$
10. $\frac{1}{3}$
11. $\frac{99}{100}$
12. 0
13. 1000
14. $^-\frac{1}{1000}$
15. 1
16. $^-7$
17. $^-5$
18. $1\frac{1}{6}$
19. $^-3$
20. $^-1\frac{1}{2}$
21. 2
22. 1
23. $3\frac{1}{2}$
24. 2
25. $^-1$
26. $10\frac{2}{3}$
27. 1·1
28. $^-1$
29. 2
30. 35
31. 130
32. 14
33. $\frac{2}{3}$
34. $1\frac{1}{8}$
35. $\frac{3}{10}$
36. $^-1\frac{1}{4}$
37. 10
38. 27
39. 20
40. $2\frac{1}{2}$
41. $1\frac{1}{3}$

Exercise 1H
1. $^-1\frac{1}{2}$
2. 2
3. $^-\frac{2}{5}$
4. $^-\frac{1}{3}$
5. $1\frac{2}{3}$
6. 6
7. $^-\frac{2}{5}$
8. $^-3\frac{1}{5}$
9. $\frac{1}{2}$
10. $^-4$
11. 18
12. 5
13. 4
14. 3
15. $2\frac{3}{4}$
16. $^-\frac{7}{22}$
17. $\frac{1}{4}$
18. 1
19. 4
20. $^-11$
21. $\frac{1}{3}$
22. $\frac{1}{5}$
23. $1\frac{2}{3}$
24. $^-3$
25. $\frac{5}{11}$
26. $^-2$
27. $^-7$
28. $^-7\frac{2}{3}$
29. 2
30. 3
31. 2
32. 3
33. $\frac{3}{5}$
34. $1\frac{1}{8}$
35. $^-1$
36. 1
37. 1
38. $1\frac{5}{7}$

Exercise 1I
1. $3\frac{1}{3}$ cm
2. 12 cm
3. 91, 92, 93
4. 21, 22, 23, 24
5. 57, 59, 61
6. 506, 508, 510
7. $12\frac{1}{2}$
8. 20
9. $18\frac{1}{2}$, $27\frac{1}{2}$
10. 20°, 60°, 100°
11. 5, 15, 8
12. 5 cm
13. $59\frac{2}{3}$ kg, $64\frac{2}{3}$ kg, $72\frac{2}{3}$ kg
14. 7 cm
15. (b) $14+8x$ (c) 0·75 m
16. $5\frac{2}{5}$
17. $53\frac{1}{3}$ or $56\frac{2}{3}$
18. 26, 58
19. 8 km
20. 400 m
21. 21
22. £3600
23. 15
24. 2 km
25. 7, 8, 9
26. 26 cm

Exercise 1J
1. $\frac{1}{4}$
2. $^-3$
3. 4
4. $^-7\frac{2}{3}$
5. $^-43$
6. 11
7. $^-\frac{1}{2}$
8. 0
9. 1
10. $^-1\frac{2}{3}$
11. 10, 8, 6
12. 13, 12, 5
13. 4 cm
14. 5 m

Exercise 1K
1. 12 cm
2. 15×5
3. (a) 26×13 (b) 16×8 (c) 32×16 (d) 9×4.5 (e) 6.5×3.25
4. (a) 19×18 (b) 6.5×5.5 (c) 8.7×7.7

Answers   501

Exercise 1L
1. (a) 9·5 (b) 7·6   2. (a) 5·1 (b) 3·8 (c) 6·7 (d) 3·4
3. (a) 3·2 (b) 2·7 (c) 3·6   4. 2·15   5. 8·1   6. 9·56
7. (a) $x + 10$ (c) $34 \times 44 \times 6$ 8. (a) $3\,\text{cm}^2$ (b) 3·8
9. (a) $x - 1$ (b) 1·62   10. 6·3   11. 38·5
12. (a) $2x\sqrt{1 - x^2}$ (b) 0·71   13. (b) 1·5

Exercise 1M
1. $w = b + 4$ 2. $w = 2b + 6$ 3. $w = 2b - 12$ 4. $m = 2t + 1$ 5. $m = 3t + 2$
6. $s = t + 2$ 7. (a) $p = 5n - 2$ (b) $k = 7n + 3$ (c) $w = 2n + 11$ 8. $m = 8c + 4$
9. (a) $y = 3n + 1$ (b) $h = 4n - 3$ (c) $k = 3n + 5$
10. (a) $t = 2n + 4$ (b) $e = 3n + 11$ (c) $e = \dfrac{3t + 10}{2}$
11. Number of white squares $= n^2 - n$

Exercise 1N
1. (a) $10n$ (b) $5n$ (c) $n + 2$ (d) $2n + 1$ (e) $30n$ (f) $6n - 1$ (g) $3n - 2$
2. (a) $10 \to 23$, $n \to 2n + 3$ (b) $20 \to 61$, $n \to 3n + 1$
3. (b) 51; $5n + 1$ 4. (b) A $3n - 2$; B $4n + 2$; C $7n - 2$
5. (a) 80   (b) 76   (c) $8n$ (d) $8n - 4$ 6. (a) $n \to n(n + 1)$ (b) $n \to n^2 + 1$
7. $n^2 + 4$ 8. 10^2, n^2 9. 10×11, $n(n + 1)$ 10. $\dfrac{10}{11}$, $\dfrac{n}{n+1}$
11. $\dfrac{5}{10^2}$, $\dfrac{5}{n^2}$ 12. 10×12, $n(n + 2)$
13. (a) 3, 5, 7   (b) 2, 7, 12   (c) 98, 96, 94
    (d) $\frac{1}{3}, \frac{2}{4}, \frac{3}{5}$ (e) 10, 100, 1000   (f) 2, 6, 12
14. (a) $t = n(n + 2)$ (b) $t = n^2 + 2n$
15. (a) $\dfrac{1}{n^2}$ (b) 2^n (c) 3×2^n (d) $10^n - 1$ (e) $2n(n + 1)$ (f) $\dfrac{n^2}{3^n}$
16. number of diagonals $= \dfrac{n(n - 3)}{2}$ 17. (a) 5   (b) $a = {}^-2, b = 5$ (c) $c = 3, d = 5$

Exercise 1O
For questions **1** to **9** three points on each graph are given.
1. $({}^-3, {}^-5), (0, 1), (1, 3)$ 2. $({}^-2, {}^-10), (0, {}^-4), (3, 5)$ 3. $({}^-2, 10), (0, 8), (4, 4)$
4. $({}^-2, 14), (0, 10), (4, 2)$ 5. $({}^-3, 1), (0, 2·5), (3, 4)$ 6. $({}^-3, {}^-15), (0, {}^-6), (3, 3)$
7. $({}^-3, 2·5), (0, 4), (3, 5·5)$ 8. $({}^-2, {}^-7), (0, {}^-3), (4, 5)$ 9. $({}^-2, 18), (0, 12), (4, 0)$

10.
x	0	50	100	150	200	250	300
C	35	45	55	65	75	85	95

(a) 180   (b) $C = 0·2x + 35$

11.
h	0	1	2	3
C	18	33	48	63

(a) $2·5h$ (b) $C = 15h + 18$

12.
m	0	5000	10 000
C	200	450	700

(a) (i) £560   (ii) 2400 miles

13.
s	0	80	160
C	50	530	1010

(a) £188   (b) 158 km/h

Exercise 1P
1. (a) (4, 10) (b) (6, 7) (c) (1, 3) (d) (1, 3)   2. (a) 5   (b) 13   (c) $\sqrt{74}$
3. (a) (3, 6)   (b) $\sqrt{180}$ 4. $(4a, a)$

Exercise 1Q
1. AB: $\frac{1}{5}$;   BC: $\frac{5}{2}$;   AC: $^-\frac{4}{3}$ 2. PQ: $\frac{4}{5}$;   PR: $^-\frac{1}{6}$;   QR: $^-5$
3. (a) 3   (b) $\frac{3}{2}$ (c) 4   (d) 5   4. $3\frac{1}{2}$
5. (a) $\dfrac{n+4}{2m-3}$ (b) $n = {}^-4$ (c) $m = 1\frac{1}{2}$
6. (a) 1, $^-1$; $\frac{1}{2}$, $^-2$; 3, $^-\frac{1}{3}$ (b) $^-1$ (c) $^-1$ (d) the same
7. (from top) 5, $^-\frac{1}{2}$, $^-\frac{2}{3}$, $\frac{1}{5}$, $^-\frac{4}{3}$ 8. $t = {}^-1$

Exercise 1R
1. 1, 3   2. 1, $^-2$ 3. 2, 1   4. 2, $^-5$ 5. 3, 4   6. $\frac{1}{2}$, 6   7. 3, $^-2$
8. 2, 0   9. $\frac{1}{4}$, $^-4$ 10. $^-1$, 3   11. $^-2$, 6   12. $^-1$, 2   13. $^-2$, 3   14. $^-3$, $^-4$
15. $\frac{1}{2}$, 3   16. $^-\frac{1}{3}$, 3   17. 4, $^-5$ 18. $\frac{3}{2}$, $^-4$ 19. 10, 0   20. 0, 4
21. A: $y = 3x - 4$;   B: $y = x + 2$ 22. C: $y = \frac{2}{3}x - 2$ or $3y = 2x - 6$;   D: $y = {}^-2x + 4$
23. (a) C and D, F and G   (b) A and E, B and H   (c) I
24. (a) A(0, $^-8$), B(4, 0)   (b) 2   (c) $y = 2x - 8$

Exercise 1S
1. $y = 3x + 7$ 2. $y = 2x - 9$ 3. $y = {}^-x + 5$ 4. $y = 2x - 1$ 5. $y = 3x + 5$
6. $y = {}^-x + 7$ 7. $y = \frac{1}{2}x - 3$ 8. $y = 2x - 3$ 9. $y = 3x - 11$ 10. $y = {}^-x + 5$

Exercise 1T
1. $a = 2.5$, $c = 1.5$; $Z = 2.5X + 1.5$ 2. $m = 2$, $c = 3$ 3. $n = {}^-2.5$, $k = 15$
4. $m = 0.2$, $c = 3$ 5. $a = 12$, $b = 3.5$; $V = \dfrac{12}{T} + 3.5$ 6. $k = 0.092$

Exercise 1U
1. (a) 45 min   (b) 09:15   (c) 60 km/h   (d) 100 km/h   (e) 57·1 km/h
2. (a) 09:00   (b) 64 km/h   (c) 40 km/h   (d) 70 km   (e) 80 km/h
3. (b) 11:05   4. (b) 12:42   5. (b) 12:35   6. $1\frac{1}{8}$
7. (a) C   (b) A   (c) D   (d) B   8. D   11. AZ, BX, CY
12. (a) 6 gallons   (b) 40 mpg; 30 mpg   (c) $33\frac{1}{3}$ mpg; $5\frac{1}{2}$ gallons

Exercise 1V
1. (a) (3, 7)   (b) (1, 3)   (c) (11, $^-1$)
2. (2, 4)   3. (2, 3)   4. (3, 1)   5. (1, 5)   6. (5, 3)
7. (a) (4, 0)   (b) (1, 6)   (c) ($^-2$, $^-3$)   (d) (8, $^-1$)   (e) ($^-0.6$, 1·2)

Exercise 1W
1. $x = 2$, $y = 1$ 2. $x = 4$, $y = 2$ 3. $x = 3$, $y = 1$ 4. $x = {}^-\frac{1}{3}$, $y = {}^-2\frac{1}{3}$
5. $x = 3$, $y = 2$ 6. $x = 5$, $y = -2$ 7. $x = 2$, $y = 1$ 8. $x = 5$, $y = 3$
9. $x = 3$, $y = {}^-1$ 10. $m = \frac{1}{2}$, $n = 4$ 11. $w = 2$, $x = 3$ 12. $x = 6$, $y = 3$
13. $x = \frac{1}{2}$, $z = {}^-3$ 14. $m = 1\frac{15}{17}$, $n = \frac{11}{17}$ 15. $c = 1\frac{16}{23}$, $d = {}^-2\frac{12}{23}$

Exercise 1X

1. $x = 2, y = 4$
2. $x = 1, y = 4$
3. $x = 2, y = 5$
4. $x = 3, y = 7$
5. $x = 5, y = 2$
6. $a = 3, b = 1$
7. $x = {}^-2, y = 3$
8. $x = 4, y = 1$
9. $x = \frac{5}{7}, y = 4\frac{3}{7}$
10. $x = 1, y = 2$
11. $x = 2, y = 3$
12. $x = 4, y = {}^-1$
13. $x = 1, y = 2$
14. $a = 4, b = 3$
15. $x = 4, y = 3$
16. $x = 5, y = {}^-2$
17. $x = 3, y = {}^-1$
18. $x = 5, y = 0 \cdot 2$

Exercise 1Y

1. $5\frac{1}{2}, 9\frac{1}{2}$
2. 6, 3 or $2\frac{2}{5}, 5\frac{2}{5}$
3. 4, 10
4. $10 \cdot 5, 7 \cdot 5$
5. $a = 2, c = 7$
6. $m = 4, c = {}^-3$
7. $a = 30, b = 5$
8. TV £200, video £450
9. 60 m, 80 m
10. 150 m, 350 m
11. 2p × 15, 5p × 25
12. 20
13. $\frac{5}{7}$
14. boy 10, mouse 3
15. $a = 1, b = 2, c = 5$

Exercise 1Z

1. (a) $({}^-1, 3)$ and $(4, 8)$ (b) $(3, 13)$ and $(5, 27)$ 2. (a) $(3, 7)$ and $(5, 17)$ (b) $(\frac{1}{2}, \frac{1}{2})$ and $(4, 32)$
3. $(2, 3)$ and $({}^-3, {}^-2)$ 4. $(4, 2)$ and $({}^-2, {}^-4)$ 5. $(4, 8)$ and $(1, {}^-4)$
6. curve and line do not intersect   7. $(2, 8)$

Check out AS1

1. (a) $6x^3 - 14x$ (b) $30x - 20$
2. (a) $x(x - 8)$ (b) $(4x + 5y)(4x - 5y)$
3. (a) $x = {}^-9$ (b) $x = \frac{53}{3}$
4. $x = {}^-3 \cdot 2$ or $x = 2 \cdot 2$ 5. $7n - 3$ 7. ${}^-1, (0, 5)$
8. 11:56 pm   9. $x = 2, y = 3$

Revision exercise AS1

1. (a) $x = 1$ (b) $x = 4, y = {}^-3$
2. (a) (i) e^5 (ii) f^4 (iii) $9h^2k^6$ (b) (i) $3p^2 + 6p$ (ii) $x^2 - 8xy + 7y^2$
3. (a) (i) 1   (ii) $6x^7$ (iii) $32p^5q^{10}$
   (b) $y^2 - 14y + 49$
   (c) $p = 3, q = 5$
4. $x = 4 \cdot 3$
5. (a) $18p - 6p^2$ (b) $2p^2 + p - 3$ (c) $3a^2(2b - 1)$
6. $x = 2 \cdot 3$
7. $x = 3, y = 1 \cdot 6$
8. (a) (i) p^4 (ii) q^{10} (b) (i) $c = \frac{1}{2}P - a - b$ (ii) $4 \cdot 5$
9. (a) $y = 4$ (b) $\frac{1}{2}$ (c) $y = \frac{1}{2}x + 1$
10. (a) $n^2 + 1$ (b) No

AS2 Shape, space and measures 1

Check in AS2

1. $312 \cdot 5$
2. (a) 60   (b) 40
3. 16, 20
4. (a) $314 \cdot 2$ (b) 45
5. (a) $r = \sqrt{(A\pi)}$ (b) $h = \dfrac{A - 2\pi r^2}{2\pi r}$
6. (a) $17\frac{1}{2}$ (b) $y = \frac{243}{25}$

Exercise 2A

1. 70°
2. 70°
3. 48°
4. 40°, 140°
5. 60°
6. 122°, 116°
7. 135°
8. 28°
9. 20°
10. 70°, 30°
11. 36°, 36°
12. 60°, 40°
13. 72°
14. 98°
15. 80°
16. 95°, 50°
17. 87°, 74°
18. 65°, 103°
19. 70°, 60°
20. 65°
21. 46°
24. 136°
25. 80°
26. 66°
27. $27\frac{1}{2}°$

Exercise 2B

1. C (5, 1), D (0, 0)
3. (a) 72° (b) 108° (c) 80°
4. (a) 40° (b) 30° (c) 110°
5. (a) 116° (b) 32° (c) 58°
6. 53°
7. (a) 26° (b) 26° (c) 77°
8. 150°
9. 110°
10. (a) 54° (b) 72° (c) 36°
11. (a) 60° (b) 15° (c) 75°

Exercise 2C

1. (a) $a = 80°, b = 70°, c = 65°, d = 86°, e = 59°$ 2. (a) $a = 36°$ (b) 144°
3. (a) (i) 40° (ii) 20° (iii) 8° (iv) 6° (b) (i) 140° (ii) 160° (iii) 172° (iv) 174°
4. $p = 101°, q = 79°, x = 70°, m = 70°, n = 130°$ 5. 24 6. 9 7. 20

Exercise 2D

1. (a) 1080° (b) 1440°
2. (a) 3240° (b) 162°
3. (a) 270° (b) 100° (c) $a = 119°, b = 25°$
4. (a) 150° (b) 30°
5. 22
6. 20

Exercise 2E

1. Yes, S.S.S.
2. Yes, S.A.S.
3. No
4. Yes, A.A.S.
5. No
6. Yes, A.A.S
7. ABD, DCA, DEB, EAC

Exercise 2H

1. 10 cm
2. 4·1 cm
3. 10·6 cm
4. 5·7 cm
5. 4·2 cm
6. 9·9 cm
7. 4·6 cm
8. 5·2 cm
9. 9·8 cm
10. 9·8 cm
11. 3·5 m
12. 40·3 km
13. 12·7 cm
14. 5·7 cm
15. (a) 5 cm (b) 40 cm²
16. 9·5 cm
17. 32·6 cm
18. (a) $\sqrt{200}$ cm (b) 14·3 cm²

Exercise 2I

1. $\sqrt{29}$
2. Yes
3. $\sqrt{44}$
4. $\sqrt{31}$
5. $\sqrt{76}$
6. $\sqrt{32}$
7. $\sqrt{44}$
8. $\sqrt{5}$
9. (a) 5 cm (b) 7·8 cm
10. 6·3 m
11. (a) 7·5 (b) 12·5 (c) 14·9
12. 24 cm
13. 10 feet
14. (a) 8·60 (b) 16·4
15. $x = 4$ m, 20·6 m
16. (a) 13, 25, 9 (d) 11, 60, 61
17. 29·5 cm

Exercise 2J

1. 42 cm²
2. 22 cm²
3. 103 cm²
4. 60·5 cm²
5. 143 cm²
6. 9 cm²
7. 13 m
8. 15 cm
9. 2500
10. (cm²) (a) 25 (b) 21 (c) 11·8 (d) 19·7 (e) 21 (f) 16·4
11. 24 cm²
12. 40 cm²
13. 32 cm²
14. 46 cm²
15. 47 cm²
16. $81\frac{3}{4}$ cm²

Exercise 2K

1. 80 m²
2. (a) 8·5 m (b) 10 m (c) 4 m
3. 45 cm²
4. 2·4 cm
5. (a) $\frac{1}{3}$ (b) $\frac{4}{9}$ (c) 25 cm²
7. 1100 m
8. 6 square units

Answers 505

9. 14 square units **10.** 10 cm **11.** (a) 60° (b) 23·4 cm² **12.** 124 cm²
13. 57·1 cm² **14.** 10·7 cm **15.** 4·1 m **16.** 7·2 cm
17. (a) $\dfrac{360°}{n}$ (b) (i) $\tfrac{1}{2}\sin\left(\dfrac{360°}{n}\right)$ (ii) $\dfrac{n}{2}\sin\left(\dfrac{360°}{n}\right)$ (c) Area tends towards π
18. 18·7 cm

Exercise 2L

1. 31·4 cm, 78·5 cm² **2.** 18·8 cm, 28·3 cm² **3.** 129 m, 982 m² **4.** 56·6 cm, 190 cm²
5. 53·7 m, 198 m² **6.** 28·1 m, 54·7 m² **7.** 20·6 cm, 24·6 cm² **8.** 25·1 cm, 43·4 cm²
9. 20·3 cm, 24·6 cm² **10.** 37·3 cm, 92·9 cm² **11.** 25·1 cm, 13·7 cm² **12.** 25·1 cm, 25·1 cm²
13. 18·8 cm, 12·6 cm²

Exercise 2M

1. 2·19 cm **2.** 31·8 cm **3.** 2·65 km **4.** 9·33 cm **5.** 17·8 cm
6. 14·2 mm **7.** 497 000 km² **8.** 21·5 cm² **9.** 30; (a) 1508 cm² (b) 508 cm²
10. 5305 **11.** 29 **12.** (a) 40·8 m² (b) 6
13. 5·39 cm **14.** 118 m² **15.** (a) 33·0 cm (b) 70·9 cm²
16. 796 m² **17.** (a) side = 4 (b) $r = 2$ (c) side = $4\sqrt{3}$ **18.** 57·5° **19.** Yes

Exercise 2N

1. 2 : 1 **2.** 112 cm² **3.** $\tfrac{7}{16}$ **4.** 0·586 m **5.** 20%
6. 7·172 cm **7.** $112\tfrac{1}{2}°$ **8.** 60° **9.** 55·4 cm² **10.** 8 cm
11. 4 : 1 **12.** (a) 2·41 (b) 1·85 (c) $1 + \sqrt{2}$

Exercise 2O

1. (a) 2·1 cm (b) 7·9 cm (c) 8·2 cm **2.** 15·4 cm
3. (a) 8·7 cm² (b) 40·6 cm² (c) 18·8 cm²
4. (a) 7·1 cm² (b) 19·5 cm² **5.** 17·8 cm **6.** 74·2 cm³
7. 5·9 cm² **8.** (a) 4·0 cm (b) 75°
9. (a) 12 cm (b) 30° **10.** (a) 30° (b) 10·5 cm
11. (a) 85·9° (b) 57·3° (c) 6·3 cm **12.** 30·6 cm² **13.** 57·3°
14. 36° **15.** (a) 6·1 cm (b) 27·6 m (c) 28·6 cm² **16.** 1850 m
17. (a) 18 cm (b) 38·2° **18.** (b) 66·8° **19.** $x = 38·1°$

Exercise 2P

1. (a) 14·5 cm (b) 72·6 cm² (c) 24·5 cm² (d) 48·1 cm²
2. (a) 5·08 cm² (b) 82·8 m² (c) 5·14 cm²
3. (a) 60°, 9·06 cm² (b) 106·3°, 11·2 cm² **4.** 3 cm
5. (a) 13·5 cm² (b) 405 cm³ **6.** (a) 130 cm² (b) 184 cm² **7.** 19·6 cm²
8. $0·313\, r^2$ **9.** (a) 8·37 cm (b) 54·5 cm (c) 10·4 cm **10.** 81·2 cm²

Exercise 2Q

1. (a) 30 cm³ (b) 168 cm³ (c) 110 cm³ (d) 94·5 cm³ (e) 754 cm³ (f) 283 cm³
2. (a) 503 cm³ (b) 760 m³ **3.** 10 litres **4.** 7·1 kg
5. 7·5 g/cm³ **6.** 358 m³ **7.** 3·98 cm **8.** 6·37 cm
9. 1·89 cm **10.** 9·77 cm **11.** 7·38 cm **12.** 106 cm/s
13. 1570 cm³, 12·57 kg **14.** 1·19 cm **15.** 53 times **16.** 191 cm

Exercise 2R

1. $40\,\text{cm}^3$
2. $33.5\,\text{cm}^3$
3. $66.0\,\text{cm}^3$
4. $89.8\,\text{cm}^3$
5. $339\,\text{cm}^3$
6. $144\,\text{cm}^3$
7. $4.71\,\text{kg}$
8. $262\,\text{cm}^3$
9. $359\,\text{cm}^3$
10. $235\,\text{cm}^3$
11. $5\,\text{m}$
12. $0.0036\,\text{mm}$
13. 10 balls
14. $415\,\text{cm}^3$
15. $106\,\text{s}$
16. $488\,\text{cm}^3$
17. $37\tfrac{1}{2}$ million

Exercise 2S

1. (a) $3.91\,\text{cm}$ (b) $2.43\,\text{cm}$ (c) $7.16\,\text{cm}$
2. $6.45\,\text{cm}$
3. $23.9\,\text{cm}$
4. (a) 125 (b) 2.7×10^7
5. (a) $0.36\,\text{cm}$ (b) $0.427\,\text{cm}$
6. (a) $6.69\,\text{cm}$ (b) $39.1\,\text{cm}$
7. $4.19\,\text{cm}^3$
8. $53.6\,\text{cm}^3$
9. $74.5\,\text{cm}^3$
10. $123\,\text{cm}^3$
11. (a) $16\pi\,\text{cm}$ (b) $8\,\text{cm}$ (c) $6\,\text{cm}$
12. $2720\,\text{cm}^3$

Exercise 2T

1. (All cm^2) (a) 36π (b) 40π (c) 60π (d) 1.4π
2. (a) 40π (b) 28π (c) 48π (d) 24π
3. £3870
4. £331
5. $303\,\text{cm}^2$
6. $675\,\text{cm}^2$
7. $1.64\,\text{cm}$
8. $2.12\,\text{cm}$
9. $3.46\,\text{cm}$
10. $94.0\,\text{cm}^3$
11. $44.6\,\text{cm}^2$
12. $377\,\text{cm}^2$
13. $l = 20\,\text{cm},\ r = 10\,\text{cm}$
14. (a) $3.72\,\text{cm}$ (b) $41.9\,\text{cm}^2$
15. $123\,\text{cm}^2$

Exercise 2U

1. C only
2. $m = 12\,\text{cm}$
3. $x = 9\,\text{cm}$
4. $a = 2.5\,\text{cm},\ e = 3\,\text{cm}$
5. $x = 6\tfrac{3}{4}\,\text{cm}$
6. $3.2\,\text{cm}$
7. $t = 5.25\,\text{cm},\ y = 5.6\,\text{cm}$
8. $7.7\,\text{cm}$
9. No
10. (a) Yes (b) No (c) No (d) Yes (e) Yes (f) No (g) No (h) Yes
11. (b) $11.2\,\text{cm}$ (c) $4.2\,\text{cm}$
12. $6\,\text{cm}$
13. $6\,\text{cm}$
14. $4\tfrac{1}{2}\,\text{cm}$

Exercise 2V

1. $16\,\text{m}$
2. $3.75\,\text{m}$
3. $10.8\,\text{m}$
4. $2\,\text{cm},\ 6\,\text{cm}$
5. $m = 4.5\,\text{cm},\ n = 6\,\text{cm}$
6. $v = 5\tfrac{1}{3},\ w = 6\tfrac{2}{3}$
7. $3\tfrac{2}{3}\,\text{cm},\ 1\tfrac{1}{11}\,\text{cm}$
8. $0.618;\ 1.618 : 1$
9. 5

Exercise 2W

1. $16\,\text{cm}^2$
2. $27\,\text{cm}^2$
3. $11\tfrac{1}{4}\,\text{cm}^2$
4. $14\tfrac{1}{2}\,\text{cm}^2$
5. $128\,\text{cm}^2$
6. $12\,\text{cm}^2$
7. $8\,\text{cm}$
8. $18\,\text{cm}$
9. $4\tfrac{1}{2}\,\text{cm}$
10. $7\tfrac{1}{2}\,\text{cm}$
11. $9,\ 12,\ 15$
12. (a) $500\,000\,\text{cm}^2$ (b) $50\,\text{m}^2$
13. $270\,\text{min}$
14. 150
15. 360
16. $5r$
17. (a) $16\tfrac{2}{3}\,\text{cm}^2$ (b) $10\tfrac{2}{3}\,\text{cm}^2$
18. (a) $25\,\text{cm}^2$ (b) $21\,\text{cm}^2$
19. $24\,\text{cm}^2$
20. Less (for the same mass)
21. $6.29\,\text{cm}$
22. $\sqrt{2}$
23. $\tfrac{4}{9}$

Exercise 2X

1. $480\,\text{cm}^3$
2. $540\,\text{cm}^3$
3. $160\,\text{cm}^3$
4. $4500\,\text{cm}^3$
5. $81\,\text{cm}^3$
6. $11\,\text{cm}^3$
7. $16\,\text{cm}^3$
8. $85\tfrac{1}{3}\,\text{cm}^3$
9. $4\,\text{cm}$
10. $21\,\text{cm}$
11. $6.6\,\text{cm}$
12. $4\tfrac{1}{2}\,\text{cm}$
13. $168\tfrac{3}{4}\,\text{cm}^3$
14. $106.3\,\text{cm}^3$
15. $12\,\text{cm}$
16. (a) $2:3$ (b) $8:27$
17. $8:125$
18. $60\,\text{cm}$
19. £12.80
20. $21\,\text{m},\ 62.5\,\text{m}^3,\ 700\,\text{cm}^2,\ 12,\ 92.5\,\text{m}^2$
21. $54\,\text{kg}$
22. $240\,\text{cm}^2$

Check out AS2
1. F angles 3. 120° 5. (b) 30° 6. 3·9 cm 7. 47 cm^2
8. (a) 39 cm (b) 121 cm^2 9. (a) 9·3 cm (b) 23·2 cm^2 (c) 11·2 cm^2
10. (a) (i) 452 cm^2 (ii) 167 cm^2 (iii) 188 cm^2 (b) (ii) 905 cm^3 (ii) 154 cm^3 (iii) 188 cm^3
11. 4·68 cm^2

Revision exercise AS2
1. (a) 30 (b) Yes 2. (a) $x = 4, y = 9$ (b) 95 mm 3. (a) 135° 4. 6 km
6. 16·2 litres 7. (a) trapezium (b) rhombus (c) rectangle 9. 9800 cm^3 10. 100°

AS3 Algebra 2
Check in AS3
1. (a) $6x + 10x$ (b) $20x^2 - 8x^3$ 2. (a) $9x^2 + y^2 + 6xy$ (b) $49x^2 - 28x + 4$
4. gradient = 2, intercept = 5 5. (a) x^{10} (b) x^{-4} (c) x^{12} (d) 1 (e) $\frac{1}{2}$ (f) $\frac{1}{8}$

Exercise 3A
1. $\frac{B}{A}$ 2. $\frac{T}{N}$ 3. $\frac{K}{M}$ 4. $\frac{4}{y}$ 5. $\frac{T+N}{9}$
6. $\frac{B-R}{A}$ 7. $\frac{R+T}{C}$ 8. $\frac{N-R^2}{L}$ 9. $\frac{R-S^2}{N}$ 10. 2
11. $S - B$ 12. $N - D$ 13. $T - N^2$ 14. $N + M - L$ 15. $A + R$
16. $E + A$ 17. $F + B$ 18. $F^2 + B^2$ 19. $L + B$ 20. $N + T$
21. 2 22. $4\frac{1}{2}$ 23. $\frac{N-C}{A}$ 24. $\frac{L-D}{B}$ 25. $\frac{F-E}{D}$
26. $\frac{H+F}{N}$ 27. $\frac{Q-m}{V}$ 28. $\frac{n+a+m}{t}$ 29. $\frac{c-b}{V^2}$ 30. $\frac{r+6}{n}$
31. $2\frac{2}{3}$ 32. $\frac{C-AB}{A}$ 33. $\frac{F-DE}{D}$ 34. $\frac{a-hn}{h}$ 35. $\frac{q+bd}{b}$
36. $\frac{n-rt}{r}$

Exercise 3B
1. 12 2. 10 3. BD 4. TB 5. RN
6. bm 7. 26 8. $BT + A$ 9. $AN + D$ 10. $B^2N - Q$
11. $ge + r$ 12. $4\frac{1}{2}$ 13. $\frac{DC-B}{A}$ 14. $\frac{pq-m}{n}$ 15. $\frac{vS+t}{r}$
16. $\frac{qt+m}{z}$ 17. $\frac{bc-m}{A}$ 18. $\frac{AE-D}{B}$ 19. $\frac{nh+f}{e}$ 20. $\frac{qr-b}{g}$
21. 4 22. $^-2$ 23. 2 24. $A - B$ 25. $C - E$
26. $D - H$ 27. $n - m$ 28. $q - t$ 29. $s - b$ 30. $r - v$
31. $m - t$ 32. 2 33. $\frac{T-B}{X}$ 34. $\frac{M-Q}{N}$ 35. $\frac{V-T}{M}$
36. $\frac{N-L}{R}$ 37. $\frac{v^2-r}{r}$ 38. $\frac{w-t^2}{n}$ 39. $\frac{n-2}{q}$ 40. $\frac{1}{4}$
41. $-\frac{1}{7}$ 42. $\frac{B-DE}{A}$ 43. $\frac{D-NB}{E}$ 44. $\frac{h-bx}{f}$ 45. $\frac{v^2-Cd}{h}$
46. $\frac{NT-MB}{M}$ 47. $\frac{mB+ef}{fN}$ 48. $\frac{TM-EF}{T}$ 49. $\frac{yx-zt}{y}$ 50. $\frac{k^2m-x^2}{k^2}$

Exercise 3C

1. $\frac{1}{2}$
2. $1\frac{2}{3}$
3. $\frac{B}{C}$
4. $\frac{T}{X}$
5. $\frac{v}{t}$
6. $\frac{n}{\sin 20°}$
7. $\frac{7}{\cos 30°}$
8. $\frac{B}{x}$
9. $\frac{vs}{m}$
10. $\frac{mb}{t}$
11. $\frac{B-DC}{C}$
12. $\frac{Q+TC}{T}$
13. $\frac{V+TD}{D}$
14. $\frac{L}{MB}$
15. $\frac{N}{BC}$
16. $\frac{m}{cd}$
17. $\frac{tc-b}{t}$
18. $\frac{xy-z}{x}$
19. 1
20. $\frac{5}{6}$
21. $\frac{A}{C-B}$
22. $\frac{V}{H-G}$
23. $\frac{r}{n+t}$
24. $\frac{b}{q-d}$
25. $\frac{m}{t+n}$
26. $\frac{b}{d-h}$
27. $\frac{d}{C-e}$
28. $\frac{m}{r-e^2}$
29. $\frac{n}{b-t^2}$
30. $\frac{d}{mn-b}$
31. $\frac{N-2MP}{2M}$
32. $\frac{B-6Ac}{6A}$
33. $\frac{m^2}{n-p}$
34. $\frac{q}{w-t}$

Exercise 3D

1. 4
2. 11
3. $D^2 - C$
4. $\frac{c^2-b}{a}$
5. $\frac{b^2+t}{g}$
6. $d - t^2$
7. $n - c^2$
8. $c - g^2$
9. $\frac{D-B}{A}$
10. $\pm\sqrt{g}$
11. $\pm\sqrt{B}$
12. $\pm\sqrt{(M+A)}$
13. $\pm\sqrt{(C-m)}$
14. $\pm\sqrt{\frac{n}{m}}$
15. $\frac{at}{z}$
16. $\pm\sqrt{(a-n)}$
17. $\pm\sqrt{(B^2+A)}$
18. $\pm\sqrt{(t^2-m)}$
19. $\frac{M^2-A^2B}{A^2}$
20. $\frac{N}{B^2}$
21. $\pm\sqrt{(a^2-t^2)}$
22. $\frac{4}{\pi^2} - t$
23. $\pm\sqrt{\left(\frac{C^2+b}{a}\right)}$
24. $\pm\sqrt{(x^2-b)}$

Exercise 3E

1. $3\frac{2}{3}$
2. 3
3. $\frac{D-B}{2N}$
4. $\frac{E+D}{3M}$
5. $\frac{2b}{a-b}$
6. $\frac{e+c}{m+n}$
7. $\frac{3}{x+k}$
8. $\frac{C-D}{R-T}$
9. $\frac{z+x}{a-b}$
10. $\frac{nb-ma}{m-n}$
11. $\frac{d+xb}{x-1}$
12. $\frac{a-ab}{b+1}$
13. $\frac{d-c}{d+c}$
14. $\frac{M(b-a)}{b+a}$
15. $\frac{n^2-mn}{m+n}$
16. $\frac{m^2+5}{2-m}$
17. $\frac{2+n^2}{n-1}$
18. $\frac{e-b^2}{b-a}$
19. $\frac{3x}{a+x}$
20. $\frac{e-c}{a-d}$
21. $\frac{d}{a-b-c}$
22. $\frac{ab}{m+n-a}$
23. $\frac{s-t}{b-a}$
24. $2x$
25. $\frac{v}{3}$
26. $\frac{a(b+c)}{b-2a}$
27. $\frac{5x}{3}$
28. $-\frac{4z}{5}$
29. $\frac{mn}{p^2-m}$
30. $\frac{mn+n}{4+m}$

Exercise 3F

1. $-\left(\frac{by+c}{a}\right)$
2. $\pm\sqrt{\left(\frac{e^2+ab}{a}\right)}$
3. $\frac{n^2}{m^2} + m$
4. $\frac{a-b}{1+b}$
5. $3y$
6. $\frac{a}{e^2+c}$
7. $-\left(\frac{a+lm}{m}\right)$
8. $\frac{t^2g}{4\pi^2}$
9. $\frac{4\pi^2 d}{t^2}$
10. $\pm\sqrt{\frac{a}{3}}$
11. $\pm\sqrt{\left(\frac{t^2e-ba}{b}\right)}$
12. $\frac{1}{a^2-1}$
13. $\frac{a+b}{x}$

14. $\pm\sqrt{(x^4 - b^2)}$ **15.** $\dfrac{c-a}{b}$ **16.** $\dfrac{a^2-b}{a+1}$ **17.** $\pm\sqrt{\left(\dfrac{G^2}{16\pi^2} - T^2\right)}$ **18.** $-\left(\dfrac{ax+c}{b}\right)$
19. $\dfrac{1+x^2}{1-x^2}$ **20.** $\pm\sqrt{\left(\dfrac{a^2m}{b^2} + n\right)}$ **21.** $\dfrac{P-M}{E}$ **22.** $\dfrac{RP-Q}{R}$ **23.** $\dfrac{z-t^2}{x}$ **24.** $(g-e)^2 - f$

Exercise 3G

1. (a) $3 < 7$ (b) $0 > ^{-}2$ (c) $3 \cdot 1 > 3 \cdot 01$ (d) $^{-}3 > ^{-}5$ (e) $100\,\text{mm} < 1\,\text{m}$ (f) $1\,\text{kg} > 1\,\text{lb}$
2. (a) $x > 2$ (b) $x \leq 5$ (c) $x < 100$ (d) $^{-}2 \leq x \leq 2$ (e) $x > ^{-}6$ (f) $3 < x \leq 8$
3. (a) ●→ at 7 (b) ←● at 2.5 (c) ○—○ from 1 to 7
  (d) ●—● from 0 to 4 (e) ○—● from $^{-}1$ to 5
4. (a) $A \geq 16$ (b) $3 < A \leq 70$ (c) $150 \leq T \leq 175$ (d) $h \geq 1 \cdot 75\,\text{m}$
5. (a) True (b) True (c) True
7. $3 \leq x < 5$

Exercise 3H

1. $x > 13$ **2.** $x < ^{-}1$ **3.** $x < 12$ **4.** $x \leq 2\tfrac{1}{2}$
5. $x > 3$ **6.** $x \geq 8$ **7.** $x < \tfrac{1}{4}$ **8.** $x \geq ^{-}3$
9. $x < ^{-}8$ **10.** $x < 4$ **11.** $x > ^{-}9$ **12.** $x < 8$
13. $x > 3$ **14.** $x \geq 1$ **15.** $x < 1$ **16.** $x > 2\tfrac{1}{3}$
17. $x < ^{-}3$ **18.** $x > 7\tfrac{1}{2}$ **19.** $x > 5$ **20.** $5 \leq x \leq 9$
21. $^{-}1 < x < 4$ **22.** $\tfrac{11}{2} \leq x \leq 6$ **23.** $\tfrac{4}{3} < x < 8$ **24.** $^{-}8 < x < 2$
25. $\tfrac{1}{4} < x < \tfrac{6}{5}$ **26.** 24

Exercise 3I

1. $x > 8$ **2.** 1, 2, 3, 4, 5, 6 **3.** 7, 11, 13, 17, 19 **4.** 4, 9, 16, 25, 36, 49
5. $^{-}4, ^{-}3, ^{-}2, ^{-}1$ **6.** 2, 3, 4, … 12 **7.** 2, 3, 5, 7, 11 **8.** 2, 4, 6, … 18
9. 1, 2, 3, 4 **10.** 5 **11.** (a) 16, (b) $^{-}16$, (c) 20, (d) $^{-}5$
12. $>$ **13.** $\tfrac{1}{2}$ (or others) **14.** 19 **15.** 17
16. $x > 3\tfrac{2}{3}$ **17.** (b) (i) $^{-}10 < x < 10$ (ii) $^{-}9 < x < 9$ (iii) $x > 6$ or $x < ^{-}6$
18. $^{-}5 < x < 5$ **19.** $^{-}4 \leq x \leq 4$ **20.** $x > 1, x < ^{-}1$ **21.** $x \geq 6, x \leq ^{-}6$
22. all values except zero **23.** $^{-}2 < x < 2$
24. (a) $2 \leq pq \leq 40$ (b) $\tfrac{1}{2} \leq \dfrac{p}{q} \leq 10$ (c) $^{-}2 \leq p - q \leq 9$ (d) $3 \leq p + q \leq 14$
25. 7 **26.** 5 **27.** 6 **28.** 3, 4, 5 **29.** $30° < x < 90°$ **30.** $0 < x < 75 \cdot 5°$

Exercise 3J

1. $x \geq 3$ **2.** $y \leq 2\tfrac{1}{2}$ **3.** $1 \leq x \leq 6$ **4.** $x < 7, y < 5$
5. $y \geq x$ **6.** $x + y \leq 10$ **7.** $2x - y \leq 3$ **8.** $y \leq x, x \leq 8, y \geq ^{-}2$
27. A: $x + y \leq 5, y \geq x + 1$ B: $x + y \leq 5, y \leq x + 1$
  C: $x + y \geq 5, y \leq x + 1$ D: $x + y \geq 5, y \geq x + 1$

Exercise 3K

1. (a) quadratic, negative x^2 (b) cubic, positive x^3 (c) reciprocal
  (d) cubic, negative x^3 (e) quadratic, positive x^2 (f) exponential
3. (i) c (ii) b (iii) a (iv) e (v) f (vi) d
4. (a) 1 (b) 2 (c) 1 (d) 3

Exercise 3M

For questions 1–9 some points on the curves are given.

1. (1, 12), (2, 6), (3, 4), (4, 3), (6, 2), (10, 1·2)
2. (1, 9), (2, $4\frac{1}{2}$), (3, 3), (6, $1\frac{1}{2}$), (9, 1)
3. (0, 12), (1, 6), (2, 4), (3, 3), (5, 2), (8, $1\frac{1}{3}$)
4. (¯4, ¯1), (¯2, ¯$1\frac{1}{3}$), (0, ¯2), (2, ¯4), (3, ¯8), ($3\frac{1}{2}$, ¯16)
5. (¯$3\frac{1}{2}$, ¯7), (¯3, ¯3), (¯2, ¯1), (0, 0), (4, $\frac{1}{2}$)   6. (¯3, $\frac{1}{27}$), (¯2, $\frac{1}{9}$), (¯1, $\frac{1}{3}$), (0, 1), (1, 3), (2, 8), (3, 27)
7. (¯4, 16), (¯3, 8), (¯2, 4), (¯1, 1), (0, 1), (1, $\frac{1}{2}$), (2, $\frac{1}{4}$), (3, $\frac{1}{8}$)
8. (a) $7\frac{1}{4}$ (b) 3·8 and ¯0·8   9. (a) 0·75   (b) 1·2   10. (a) 3·1   (b) 3·3

Exercise 3N

1. 15 m × 30 m   2. (a) 2·5 s   (b) 31·3 m   (c) 2 < t < 3
6. (a) 2000   (b) 270   (c) 1·6 ⩽ x ⩽ 2·4
7. 96 min   8. (b) After 76/77 years i.e. 2056/57   9. (a) 0, 2·9   (b) ¯0·65, 1·35, 5·3

Exercise 3O

1. 3   2. 4   3. ¯1   4. ¯2   5. 3   6. 3
7. 1   8. $\frac{1}{5}$ 9. 0   10. ¯4   11. 2   12. ¯5
13. 1   14. $\frac{1}{18}$ 15. (a) (0, 1)   (b) $x = 0$ 16. (a) ¯3   (b) 1   (c) 5
17. (a) $2\frac{1}{2}$ (b) 1000   (c) $\frac{7}{4}$ 18. (a) 4   (b) ¯1   (c) 0
19. 2, 4   20. (a) 3·60   (b) 5·44   21. (b) 1, 7, 9   22. 21
23. (a) 512   (b) 6 h   (c) 2^{21} 24. (a) 976   (b) 20
25. (a) £2519·42   (b) 9 years

Check out AS3

1. (a) $s = \dfrac{v^2 - u^2}{2a}$ (b) $v = \dfrac{uf}{f - u}$ 2. (a) $x \geqslant 7.5$ (b) $x \geqslant 5, x \leqslant \ ^-5$
3. (a) (iii)   (b) (i)   (c) (ii)   4. (¯1, 12), (0, 3), (1, ¯2), (2, ¯3), (3, 0), (4, 7), (5, 18)
5. (a) $x = 4$ (b) $x = \ ^-4$

Revision exercise AS3

1. (a) ¯1, 0, 1, 2   (b) $y > 2, y < -2$ 2. (a) $p = \dfrac{t + 50}{7}$ (b) $x = \dfrac{y}{k + 3}$
3. (a) $x = \ ^-1.7, \ ^-0.3, 3.1$ (b) $x = 3.7$ 4. $x = \dfrac{5y}{1 - y}$ 5. (a) $x^{\frac{7}{2}}$ (b) $n = \frac{1}{8}$ 6. $x < \ ^-\frac{1}{2}$

AS4 Shape, space and measures 2

Check in AS4

1. $p = \frac{4}{37}$ 2. 2463 cm³

Exercise 4B

1. (a), (b) and (d)
5. (a) prism, triangle cross-section; 6 vertices, 5 faces
   (b) square-based pyramid; 5 vertices, 5 faces
7. (a) 6   (b) 4   (c) $a = 10, b = 6, c = 10, d = 10$ (d) 64 cm³
8. $a = \sqrt{2}, b = \sqrt{2}, c = \sqrt{3}, x = \sqrt{2}, y = \sqrt{3}$

Exercise 4C

1. (a) 1 (b) none
2. (a) 1 (b) none
3. (a) 4 (b) 4
4. (a) 2 (b) 2
5. (a) 0 (b) 6
6. (a) 0 (b) 2
7. (a) 0 (b) 2
8. (a) 4 (b) 4
9. (a) 0 (b) 4
10. (a) 4 (b) 4
11. (a) 6 (b) 6
12. (a) infinite (b) infinite

Exercise 4D

1. 3
2. (a) 1 (b) 1 (c) 2
3. 9
5. 4

Exercise 4E

1. 4·54
2. 3·50
3. 3·71
4. 6·62
5. 8·01
6. 31·9
7. 45·4
8. 4·34
9. 17·1
10. 13·2
11. 38·1
12. 3·15
13. 516
14. 79·1
15. 5·84
16. 2·56
17. 18·3
18. 8·65
19. 11·9
20. 10·6
21. 5, 5·55
22. 13·1, 27·8
23. 41·3
24. 8·82
25. 20·4
26. 26·2

Exercise 4F

1. 36·9°
2. 44·4°
3. 48·2°
4. 60°
5. 36·9°
6. 50·2°
7. 29·0°
8. 56·4°
9. 38·9°
10. 43·9°
11. 41·8°
12. 39·3°
13. 34·8°
14. 60·0°
15. 42·0°
16. 36·9°
17. 51·3°
18. 19·6°
19. 17·9°
20. 32·5°
21. 59·6°
22. 17·8°

Exercise 4G

1. 19·5°
2. 4·12 m
3. 7°
4. 91·2°
5. 1·29 m → 2·11 m
6. 36·4°
7. 10·3 cm
8. 2·6 m
9. (a) 26·0 km (b) 23·4 km
10. (a) 88·6 km (b) 179·3 km
11. (a) 484 km (b) 858 km (c) 985 km, 060·6°
12. 71°
13. 67·1 m
14. 83·2 km
15. 60°
16. No
17. 72°, 8·2 cm

Exercise 4H

1. (a) 3·9 m (b) 8·6 m
2. 56·3°
3. 35·5°
4. 71·6°
5. 91·8°
6. 180 m
7. 11·1 m, 11·1 s, 222 m
8. 2·9 m
9. $(\sin A)^2 + (\cos A)^2 = 1$
10. (a) $\dfrac{360°}{n}$ (b) $\dfrac{180°}{n}$ (c) $AB = 2\sin\left(\dfrac{180}{n}\right)$ (d) $2n\sin\left(\dfrac{180}{n}\right)$ (e) perimeter → 2π
11. 27·1 cm

Exercise 4I

1. (a) 13 cm (b) 13·6 cm (c) 17·1°
2. (a) 4·04 m (b) 38·9° (c) 11·2 m (d) 19·9°
3. (a) 8·49 cm (b) 8·49 cm (c) 10·4 cm (d) 35·3°
4. (a) 14·1 cm (b) 18·7 cm (c) 69·3°
5. (a) 11·3 cm, 12·4 cm (b) 23·8°
6. (a) 10 cm, 9·43 cm (b) 26·6° (c) 32·5°
7. (a) 4·47 m (b) 7·48 m (c) 15·5°
8. (a) 58·0°
9. (a) 57·5° (b) 61·0°
10. (a) $h \tan 65°$ (b) $h \tan 57°$ (c) 22·7 m
11. 22·6 m

Exercise 4J

1. (a) L (b) V (c) A (d) L (e) V (f) A
2. (a) 2 (b) 2 (c) 3 (d) 1 (e) 2 (f) 3
3. (a) 2 (b) 3 (c) 2 (d) 1 (e) 3 (f) 1 (g) 0 (h) 2
4. (a) $wh + \dfrac{\pi}{6}wh$ (b) $2h + w + \tfrac{5}{4}w$
5. (a) A (b) L (c) V (d) A (e) F (f) A (g) V (h) A
6. (a) L (b) I (c) V (d) A (e) V (f) A
   (g) I (h) V (i) L (j) L (k) I (l) I
7. 2
9. $\dfrac{\pi}{2}a$ is a length, not an area.

Exercise 4K

3. (c) (8, 8) (8, ⁻6) (⁻8, 6)
4. (f) (1, 1) (⁻3, ⁻1) (⁻3, ⁻3)
5. (c) (⁻2, 1) (⁻2, ⁻1) (1, ⁻2)
6. (e) (⁻5, 2) (⁻5, 6) (⁻3, 5)

Exercise 4L

1. (a) $\begin{pmatrix}7\\3\end{pmatrix}$ (b) $\begin{pmatrix}0\\-9\end{pmatrix}$ (c) $\begin{pmatrix}9\\10\end{pmatrix}$ (d) $\begin{pmatrix}-10\\3\end{pmatrix}$ (e) $\begin{pmatrix}-1\\13\end{pmatrix}$
   (f) $\begin{pmatrix}10\\0\end{pmatrix}$ (g) $\begin{pmatrix}-9\\-4\end{pmatrix}$ (h) $\begin{pmatrix}-10\\0\end{pmatrix}$

12. (a) Rotation 90° clockwise, centre (0, ⁻2)
    (b) Reflection in $y = x$
    (c) Translation $\begin{pmatrix}3\\7\end{pmatrix}$
    (d) Enlargement, scale factor 2, centre (⁻5, 5)
    (e) Translation $\begin{pmatrix}-7\\-3\end{pmatrix}$
    (f) Reflection in $y = x$

13. (a) Rotation 90° clockwise, centre (4, ⁻2)
    (b) Translation $\begin{pmatrix}8\\2\end{pmatrix}$
    (c) Reflection in $y = x$
    (d) Enlargement, scale factor $\tfrac{1}{2}$, centre (7, ⁻7)
    (e) Rotation 90° anticlockwise, centre (⁻8, 0)
    (f) Enlargement, scale factor 2, centre (⁻1, ⁻9)
    (g) Rotation 90° anticlockwise, centre (7, 3)

Exercise 4M

1. (c) Reflection in $x = 4$
2. (c) Rotation 90° clockwise, centre (0, 0)
3. (a) (⁻4, 4) (b) (2, ⁻2)
   (c) (0, 0) (d) (0, 4)
4. (a) (⁻2, 5) (b) (⁻4, 0)
   (c) (2, ⁻2) (d) (1, ⁻1)
5. (a) Reflection in y-axis
   (b) Rotation 180°, centre (⁻2, 2)
   (c) Rotation 90° clockwise, centre (2, 2)

Answers 513

6. (a) Rotation 90° clockwise, centre (0, 0) (b) Translation $\begin{pmatrix} -2 \\ 5 \end{pmatrix}$
   (c) Rotation 90° anticlockwise, centre (2, ⁻4) (d) Rotation 90° anticlockwise, centre $(-\frac{1}{2}, 3\frac{1}{2})$
7. (a) Rotation 90° anticlockwise, centre (2, 2) (b) Enlargement, scale factor $\frac{1}{2}$, centre (8, 6)
   (c) Rotation 90° clockwise, centre $(-\frac{1}{2}, -3\frac{1}{2})$
8. A^{-1}: reflection in $x = 2$ B^{-1}: B
   C^{-1}: translation $\begin{pmatrix} 6 \\ -2 \end{pmatrix}$ D^{-1}: D
   E^{-1}: E F^{-1}: translation $\begin{pmatrix} -4 \\ -3 \end{pmatrix}$
   G^{-1}: 90° rotation anticlockwise, centre (0, 0) H^{-1}: enlargement, scale factor 2, centre (0, 0)

Exercise 4N

1. d 2. 2c 3. 3c 4. 3d 5. 5d 6. 3c 7. ⁻2d
8. ⁻2c 9. ⁻3c 10. ⁻c 11. c + d 12. c + 2d 13. 2c + d 14. 3c + d
15. 2c + 2d 16. $\overrightarrow{QI}$ 17. $\overrightarrow{QU}$ 18. $\overrightarrow{QH}$ 19. $\overrightarrow{QB}$ 20. $\overrightarrow{QF}$ 21. $\overrightarrow{QJ}$
22. (a) 2a + b (b) 2a + 2b (c) ⁻a − b (d) 4a + 2b (e) 2a − 2b (f) 2a + b
23. (a) $\overrightarrow{CO}$ (b) $\overrightarrow{TN}$ (c) $\overrightarrow{FT}$ (d) $\overrightarrow{KC}$
24. (a) ⁻a (b) a + b (c) 2a − b (d) ⁻a + b
25. (a) a + b (b) a − 2b (c) ⁻a + b (d) ⁻a − b
26. (a) ⁻a − b (b) 3a − b (c) 2a − b (d) ⁻2a + b
27. (a) a − 2b (b) a − b (c) 2a (d) ⁻2a + 3b

Exercise 4O

1. (a) a (b) ⁻a + b (c) 2b (d) ⁻2a (e) ⁻2a + 2b (f) ⁻a + b
   (g) a + b (h) b (i) ⁻b + 2a (j) ⁻2b + a
2. (a) a (b) ⁻a + b (c) 3b (d) ⁻2a (e) ⁻2a + 3b (f) ⁻a + $\frac{3}{2}$b
   (g) a + $\frac{3}{2}$b (h) $\frac{3}{2}$b (i) ⁻b + 2a (j) ⁻3b + a
3. (a) 2a (b) ⁻a + b (c) 2b (d) ⁻3a (e) ⁻3a + 2b (f) ⁻$\frac{3}{2}$a + b
   (g) $\frac{3}{2}$a + b (h) $\frac{1}{2}$a + b (i) ⁻b + 3a (j) ⁻2b + a
4. (a) $\frac{1}{2}$a (b) ⁻a + b (c) 4b (d) ⁻$\frac{3}{2}$a (e) ⁻$\frac{3}{2}$a + 4b (f) ⁻a + $\frac{8}{3}$b
   (g) $\frac{1}{2}$a + $\frac{8}{3}$b (h) ⁻$\frac{1}{2}$a + $\frac{8}{3}$b (i) $\frac{3}{2}$a − b (j) a − 4b
5. $\frac{1}{2}$s − $\frac{1}{2}$t 6. $\frac{1}{3}$a + $\frac{2}{3}$b 7. 2m + 2n
8. (a) b − a (b) b − a (c) 2b − 2a (d) b − 2a (e) b − 2a (f) 2b − 3a
9. (a) y − z (b) $\frac{1}{2}$y − $\frac{1}{2}$z (c) $\frac{1}{2}$y + $\frac{1}{2}$z (d) ⁻x + $\frac{1}{2}$y + $\frac{1}{2}$z
   (e) ⁻$\frac{2}{3}$x + $\frac{1}{3}$y + $\frac{1}{3}$z (f) $\frac{1}{3}$x + $\frac{1}{3}$y + $\frac{1}{3}$z
10. (a) (i) 2b − 2a (ii) 2c − 2b (iii) b (iv) c − a (v) c − a
    (b) parallel and equal in length (c) parallelogram
11. (a) (i) a − b (ii) $\frac{1}{3}$a − $\frac{1}{3}$b (iii) $\frac{2}{3}$b − $\frac{1}{6}$a $[= \frac{1}{6}(4b − a)]$ (iv) 2b − $\frac{1}{2}$a $[= \frac{1}{2}(4b − a)]$
    (b) CD is parallel to DE and both lines pass through point D

Exercise 4P

4. sin 70° = sin 110°, cos 60° = cos 300°, sin 50° = sin 130°
5. tan 45° = tan 225°, sin 180° = cos 270°, cos 30° = cos 330°, tan 60° = tan 240°
6. 162° 7. 153° 8. (a) 140° (b) 110° (c) 50°
9. 290° 10. 315° 11. (a) 350° (b) 304° (c) 60°
12. 220° 13. 160° 14. 82° 15. 315° 16. 240° 17. 250°

Exercise 4Q

1. 58·0°, 122·0°
2. 20·5°, 159·5°
3. 53·1°, 306·9°
4. 45°, 225°
5. (a) 19·8°, 160·2° (b) 72°, 108° (c) 30°, 150° (d) 60°, 120°
6. (a) 46·1°, 133·9° (b) 72·5°, 287·5° (c) 78·7°, 258·7° (d) 220·5°, 319·5°
7. 30°, 150°, 210°, 330°
8. 45°, 135°, 225°, 315°
10. (a) $0 < x < 180°$ (b) $90° < x < 270°$
12. (a) 90°, 450°, 810° etc. (b) 90°, 270°, 450°, 630° etc.
13. (a) 0 (b) 1
16. (a) 45°, 225° (b) 30°, 60°, 210°, 240° (c) 15°, 75°, 135°, 195°, 255°, 315°
17. (a) 40°, 140° (b) 30°, 150°
18. (a) (i) 16°, 111° (ii) 153° (b) 2·24 (c) 63°
19. (a) 48°, 205° (b) 37°, 217°

Exercise 4R

1. 6·38 m
2. 12·5 m
3. 5·17 cm
4. 40·4 cm
5. 7·81 m
6. 6·68 m
7. 8·61 cm
8. 9·97 cm
9. 8·52 cm
10. 15·2 cm
11. 35·8°
12. 42·9°
13. 32·3°
14. 37·8°
15. 35·5°, 48·5°
16. 68·8°
17. 50·6°
18. 39·1°
19. 72·5°, 107·5°

Exercise 4S

1. 6·24
2. 6·05
3. 5·47
4. 9·27
5. 10·1
6. 8·99
7. 5·87
8. 4·24
9. 11·9
10. 154
11. 25·2°
12. 71·4°
13. 115·0°
14. 111·1°
15. 24·0°
16. 92·5°
17. 99·9°
18. 38·2°
19. 137·8°
20. 34·0°

Exercise 4T

1. (a) 50·2 km (b) 054·7°
2. 35·6 km
3. 25·2 m
4. (a) 10·3 cm (b) 11·6 cm (c) 4·86 cm (d) 37·2° (e) 94·1° (f) 42·6°
5. 92·9°
6. 14·5 cm
7. (a) 9·8 km (b) 085·7°
8. (a) 29·6 km (b) 050·5°
9. 141 km
10. 378 km, 048·4°
11. 101·5°
12. 9 m
14. 9·64 m
15. 131·8°
16. (a) 5·66 cm (b) 4·47 cm (c) 3·66 cm
17. 70·2°

Exercise 4U

1. $a = 27°, b = 30°$
2. $c = 20°, d = 45°$
3. $c = 58°, d = 41°, e = 30°$
4. $f = 40°, g = 55°, h = 55°$
5. $a = 32°, b = 80°, c = 43°$
6. $c = 34°, y = 34°$
7. 43°
8. 92°
9. 42°
10. $c = 46°, d = 44°$
11. $e = 49°, f = 41°$
12. $g = 76°, h = 52°$
13. 48°
14. 32°
15. 22°
16. $a = 36°, x = 36°$

Exercise 4V

1. $a = 94°, b = 75°$
2. $c = 101°, d = 84°$
3. $x = 92°, y = 116°$
4. $c = 60°, d = 45°$
5. 37°
6. 118°
7. $e = 36°, f = 72°$
8. 35°
9. 18°
10. 90°
11. 30°
12. $22\frac{1}{2}°$
13. $n = 58°, t = 64°, w = 45°$
14. $a = 32°, b = 40°, c = 40°$
15. $a = 18°, c = 72°$
16. 55°
17. $e = 41°, f = 41°, g = 41°$
18. 8°
19. $x = 30°, y = 115°$
20. $x = 80°, z = 10°$

Exercise 4W

1. $a = 18°$
2. $53°$
3. $77°$
4. $x = 40°, y = 65°, z = 25°$
5. $c = 30°, e = 15°$
6. $f = 50°, g = 40°$
7. $h = 70°, k = 40°, i = 40°$
8. $m = 108°, n = 36°$
9. $x = 50°, y = 68°$
10. $a = 74°, b = 32°$
11. $e = 36°$
12. $k = 63°, m = 54°$
13. $k = 50°, m = 50°, n = 80°, p = 80°$
14. (a) p (b) 2p (c) $90 - 2p$
15. $x = 70°, y = 20°, z = 55°$

Exercise 4X

1. $137°$
2. $120°$
3. $c = d = 30°, e = 27°$
4. $49°$
5. $18°$
6. $110°$
7. $i = 52°, j = 128°$
8. $k = 45°, l = 135°$
9. $m = 35°, n = 50°$
10. $69°$
11. $78°$
12. $r = 200°, s = 100°$
13. $45°$
14. $b = 30°, c = 60°$
15. $e = 19°$
16. $g = 58°, h = 90°$
17. $j = 96°, k = 68°$
18. $m = 28°$

Check out AS4

2. (a) 3 (b) 3
3. $73.7°$
4. $2\pi rh$
6. $3b - 2a$
7. (a) $^-0.940$ (b) $60°, 300°$
8. (a) 3.7 cm (b) $43°, 2.8$ cm
9. $44.4°$
10. $a = 37°, b = 53°, c = 37°, d = 53°$

Revision exercise AS4

1. Rotation $90°$ clockwise about $(^-2, ^-3)$
2. (a) $2(v + 2w + x + y + z)$ (b) $\frac{1}{2}z(x + y)w$
3. $39°$
4. (a) (i) $\mathbf{b} - \mathbf{a}$ (ii) $\mathbf{a} - 3\mathbf{b}$ (iii) $4\mathbf{a} + 2\mathbf{b}$ (b) $3\mathbf{a} + \frac{3}{2}\mathbf{b}$
5. 20 cm
6. 2.2 m
7. (a) $116°$ (b) $^-64°, ^-116°$
8. 177.3 m

AS5 Algebra 3

Check in AS5

1. (a) $y = 0.25, 0.5, 1, 2, 4, 8$ (b) $y = 6, 2, 0, 0, 2, 6$
2. $x = 4, y = 3$

Exercise 5A

1. (a) 2 (b) $^-4$
2. (a) 3 (accept $2.4 \to 3.6$) (b) $^-5$ (accept $^-3.5 \to ^-6.5$) (c) 1.5
3. 3.3 (accept $2.8 \to 3.8$)
4. (a) $^-2\frac{1}{2}$ (b) $^-\frac{2}{5}$ (c) gradient $\to 0$

Exercise 5B

1. 22 square units
2. (a) about 360/370 (b) about 130/140
3. 7.47
4. 13.4
5. (a) 58 (b) greater
6. (a) $x = 1, x = 4$ (b) about $7\frac{1}{2}$ square units

Exercise 5C

1. (a) $1\frac{1}{2}$ m/s^2 (b) 675 m
2. (a) 600 m (b) 225 m (c) $^-2$ m/s^2
3. (a) 600 m (b) $387\frac{1}{2}$ m (c) 0 m/s^2
4. (a) 20 m/s (b) 750 m
5. (a) 8 s (b) 496 m (c) 12.4 m/s
6. A5, B3, C1, D2, E4, F6
7. (a) 0.75 m/s^2 (b) 680 m (both approximate)
8. (a) 0.35 m/s^2 (b) 260 m (both approximate)
9. (a) 50 m/s (b) 20 s
10. (a) 20 m/s (b) 20 s

Exercise 5D

1. $(x+2)(x+5)$
2. $(x+3)(x+4)$
3. $(x+3)(x+5)$
4. $(x+3)(x+7)$
5. $(x+2)(x+6)$
6. $(y+5)(y+7)$
7. $(y+3)(y+8)$
8. $(y+5)(y+5)$
9. $(y+3)(y+12)$
10. $(a+2)(a-5)$
11. $(a+3)(a-4)$
12. $(z+3)(z-2)$
13. $(x+5)(x-7)$
14. $(x+3)(x-8)$
15. $(x-2)(x-4)$
16. $(y-2)(y-3)$
17. $(x-3)(x-5)$
18. $(a+2)(a-3)$
19. $(a+5)(a+9)$
20. $(b+3)(b-7)$
21. $(x-4)(x-4)$
22. $(y+1)(y+1)$
23. $(y-7)(y+4)$
24. $(x-5)(x+4)$
25. $(x-20)(x+12)$
26. $(x-15)(x-11)$
27. $(y+12)(y-9)$
28. $(x-7)(x+7)$
29. $(x-3)(x+3)$
30. $(x-4)(x+4)$
31. (a) $2(x-3)(x+5)$ (b) $3(x+2)(x+5)$ (c) $3(x+3)(x+5)$
    (d) $2(n-5)(n+2)$ (e) $5(a-2)(a+3)$ (f) $4(x-4)(x+4)$

Exercise 5E

1. $(2x+3)(x+1)$
2. $(2x+1)(x+3)$
3. $(3x+1)(x+2)$
4. $(2x+3)(x+4)$
5. $(3x+2)(x+2)$
6. $(2x+5)(x+1)$
7. $(3x+1)(x-2)$
8. $(2x+5)(x-3)$
9. $(2x+7)(x-3)$
10. $(3x+4)(x-7)$
11. $(2x+1)(3x+2)$
12. $(3x-2)(x-3)$
13. $(y-2)(3y-5)$
14. $(2y+3)(3y-1)$
15. $(5x+2)(2x+1)$
16. $(6x-1)(x-3)$
17. $(4x+1)(2x-3)$
18. $(3x+2)(4x+5)$

Exercise 5F

1. $(y-a)(y+a)$
2. $(m-n)(m+n)$
3. $(x-t)(x+t)$
4. $(y-1)(y+1)$
5. $(x-3)(x+3)$
6. $(a-5)(a+5)$
7. $(x-\frac{1}{2})(x+\frac{1}{2})$
8. $(x-\frac{1}{3})(x+\frac{1}{3})$
9. $(2x-y)(2x+y)$
10. $(a-2b)(a+2b)$
11. $(5x-2y)(5x+2y)$
12. $(3x-4y)(3x+4y)$
13. $\left(2x-\frac{z}{10}\right)\left(2x+\frac{z}{10}\right)$
14. $x(x-1)(x+1)$
15. $a(a-b)(a+b)$
16. $x(2x-1)(2x+1)$
17. $2x(2x-y)(2x+y)$
18. $y(y-3)(y+3)$
19. $1\,200\,000$
20. $12\,000\,000$
21. 103×97

Exercise 5G

1. $^-3, ^-4$
2. $^-2, ^-5$
3. $3, ^-5$
4. $2, ^-3$
5. $2, 6$
6. $^-3, ^-7$
7. $2, 3$
8. $5, ^-1$
9. $^-7, 2$
10. $-\frac{1}{2}, 2$
11. $\frac{2}{3}, ^-4$
12. $1\frac{1}{2}, ^-5$
13. $\frac{2}{3}, 1\frac{1}{2}$
14. $\frac{1}{4}, 7$
15. $\frac{3}{5}, -\frac{1}{2}$
16. $7, 8$
17. $\frac{5}{6}, \frac{1}{2}$
18. $7, ^-9$
19. $^-1, ^-1$
20. $3, 3$
21. $^-5, ^-5$
22. $7, 7$
23. $-\frac{1}{3}, \frac{1}{2}$
24. $^-1\frac{1}{4}, 2$
25. $13, ^-5$
26. $^-3, \frac{1}{6}$
27. $\frac{1}{10}, ^-2$
28. $1, 1$
29. $\frac{2}{9}, -\frac{1}{4}$
30. $-\frac{1}{4}, \frac{3}{5}$
31. $\pm 1, \pm 2$
32. $\pm 2, \pm 3$
33. $\pm \frac{1}{2}, \pm 2$
34. $1, 2$

Exercise 5H

1. $0, 3$
2. $0, ^-7$
3. $0, 1$
4. $0, \frac{1}{3}$
5. $4, ^-4$
6. $7, ^-7$
7. $\frac{1}{2}, -\frac{1}{2}$
8. $\frac{2}{3}, -\frac{2}{3}$
9. $0, ^-1\frac{1}{2}$
10. $0, 1\frac{1}{2}$
11. $0, 5\frac{1}{2}$
12. $\frac{1}{4}, -\frac{1}{4}$
13. $\frac{1}{2}, -\frac{1}{2}$
14. $0, \frac{5}{8}$
15. $0, \frac{1}{12}$
16. $0, 6$
17. $0, 11$
18. $0, 1\frac{1}{2}$
19. $0, 1$
20. $0, 4$

Exercise 5I

1. $-\frac{1}{2}, -5$
2. $-\frac{2}{3}, ^-3$
3. $-\frac{1}{2}, -\frac{2}{3}$
4. $\frac{1}{3}, 3$
5. $\frac{2}{5}, 1$
6. $\frac{1}{3}, 1\frac{1}{2}$
7. $^-0.63, ^-2.37$
8. $^-0.27, ^-3.73$
9. $0.72, 0.28$
10. $6.70, 0.30$
11. $0.19, ^-2.69$
12. $0.85, ^-1.18$
13. $0.61, ^-3.28$
14. $^-1\frac{2}{3}, 4$
15. $^-1\frac{1}{2}, 5$

16. 3·56, ⁻0·56 **17.** 0·16, ⁻3·16 **18.** $-\frac{1}{2}, 2\frac{1}{3}$ **19.** $-\frac{1}{3}, -8$ **20.** $1\frac{2}{3}, -1$
21. 2·28, 0·22 **22.** ⁻0·35, ⁻5·65 **23.** $-\frac{2}{3}, \frac{1}{2}$ **24.** ⁻0·58, 2·58 **25.** (a) 0·2, 4·8
26. If $b^2 - 4ac > 0$

Exercise 5J

1. ⁻3, 2 **2.** ⁻3, ⁻7 **3.** $-\frac{1}{2}, 2$ **4.** 1, 4 **5.** $-1\frac{2}{3}, \frac{1}{2}$ **6.** ⁻0·39, ⁻4·28
7. ⁻0·16, 6·16 **8.** 3 **9.** 2, ⁻$1\frac{1}{3}$ **10.** ⁻3, ⁻1 **11.** 0·66, ⁻22·66
12. ⁻7, 2 **13.** $\frac{1}{4}, 7$ **14.** $-\frac{1}{2}, \frac{3}{5}$ **15.** 0, $3\frac{1}{2}$ **16.** $-\frac{1}{4}, \frac{1}{4}$ **17.** ⁻2·77, 1·27
18. $-\frac{2}{3}, 1$ **19.** $-\frac{1}{2}, 2$ **20.** 0, 3 **21.** Yes
22. (a) 3, $-\frac{1}{2}$ (b) 6, ⁻4 **23.** (a) ⁻1 (b) 0·6258 (c) 0·5961 (d) 0·2210

Exercise 5K

1. $(x+4)^2 - 16$ **2.** $(x-6)^2 - 36$ **3.** $(x+\frac{1}{2})^2 - \frac{1}{4}$ **4.** $(x+2)^2 - 3$ **5.** $(x-3)^2$
6. $(x+1)^2 - 16$ **7.** $(x+8)^2 - 59$ **8.** $(x-5)^2 - 25$ **9.** $(x+\frac{3}{2})^2 - \frac{9}{4}$
10. (a) 2, 6 (b) ⁻3, ⁻7 (c) 5, ⁻1
11. (a) 0·65, ⁻4·65 (b) 3·56, ⁻0·56 (c) 0·08, ⁻12·08
15. (a) 3 (b) ⁻2 (c) $\frac{1}{3}$

Exercise 5L

1. 8, 11 **2.** 11, 13 **3.** 12 cm **4.** $x = 11$ **5.** 8, 9, 10
6. (d) 3, 4, 5 **7.** 157 km **8.** 4 **9.** 10 cm × 24 cm **10.** 4·3 s
11. 2·5 cm **12.** 1 **13.** 9 cm or 13 cm **14.** $n(n+2)$; 15th term
15. 6 cm **16.** $\frac{40}{x}$ h, $\frac{40}{x-2}$ h, 10 km/h **17.** 4 km/h **18.** 20 mph
19. ⁻1 or 4 **20.** (c) $x = 595$ mm, $y = 841$ mm (d) 297·5 mm

Exercise 5M

1. (a) $\frac{5}{7}$ (b) $5y$ (c) $\frac{1}{2}$ (d) 4 (e) $\frac{x}{2y}$ (f) 2 (g) $\frac{a}{2}$ (h) $\frac{2b}{3}$
2. AG, BE, CH, DF
3. (a) a (b) $\frac{10m}{3}$ (c) $\frac{2xy}{9}$ (d) $\frac{4}{aq}$ (e) $\frac{8}{3}$ (f) $\frac{y^2}{5x}$ (g) $2a$ (h) 1
4. (a) $\frac{a}{5b}$ (b) a (c) $\frac{7}{8}$ (d) $\frac{3}{4-x}$ (e) $\frac{5+2x}{3}$ (f) $\frac{3x+1}{x}$ (g) $\frac{4+5a}{5}$ (h) $\frac{b}{3+2a}$
5. (a) 6 (b) 4 (c) $2a$ **6.** AF, BH, CE, DG
7. (a) $\frac{x+2y}{3xy}$ (b) $\frac{6-b}{2a}$ (c) $\frac{2b+4a}{b}$ (d) $x-2$
8. (a) $\frac{x+2}{x-3}$ (b) $\frac{x}{x+1}$ (c) $\frac{x+4}{2(x-5)}$ (d) $\frac{x+5}{x-2}$ (e) $\frac{x+3}{x+2}$ (f) $\frac{x+5}{x-2}$

Exercise 5N

1. (a) $\frac{3x}{5}$ (b) $\frac{3}{x}$ (c) $\frac{4x}{7}$ (d) $\frac{4}{7x}$ (e) $\frac{7x}{8}$ (f) $\frac{7}{8x}$ (g) $\frac{5x}{6}$ (h) $\frac{5}{6x}$
2. AF, BH, CD, EG
3. (a) $\frac{23x}{20}$ (b) $\frac{23}{20x}$ (c) $\frac{x}{12}$ (d) $\frac{1}{12x}$ (e) $\frac{5x+2}{6}$ (f) $\frac{7x+2}{12}$

4. (a) $\dfrac{1-2x}{12}$ (b) $\dfrac{2x-9}{15}$ (c) $\dfrac{3x+12}{14}$ 5. (a) $\dfrac{x}{10}$ (b) $3x$ (c) 11

6. (a) $\dfrac{3x+1}{x(x+1)}$ (b) $\dfrac{7x-8}{x(x-2)}$ (c) $\dfrac{8x+9}{(x-2)(x+3)}$

(d) $\dfrac{4x+11}{(x+1)(x+2)}$ (e) $\dfrac{^-3x-17}{(x+3)(x-1)}$ (f) $\dfrac{11-x}{(x+1)(x-2)}$

7. $\tfrac{1}{4}$

Exercise 5O

3. $(0, {}^-2), (7, 0)$
4. (a) $A'({}^-2, {}^-1), \ B'(0, {}^-3), \ C'(2, 0)$
   (b) $A'(0, 1), \ B'(2, 3), \ C'(4, 0)$
   (c) $A'({}^-1, 1), \ B'(0, 3), \ C'(1, 0)$

Exercise 5P

9. (b) Stretch parallel to the x-axis, scale factor 2
10. (a) $(1, 5)$ (b) $(5, {}^-4)$
11. (a) $y = x^2 + 3x + 5$ (b) $y = x^2 - x - 2$ (c) $y = {}^-(x^2 + 3x)$
12. $a = 2, b = 3$
13. (a) $A'(2, 4)$ (b) $B'(0, \tfrac{1}{5})$, $C'(3, 1)$, $D'({}^-3, 1)$

Exercise 5Q

1. $x = 0, y = 4; \ x = \tfrac{16}{5}, y = {}-\tfrac{12}{5}$ 2. $x = 0, y = 5; \ x = 3, y = {}^-4$
3. $x = 2, y = 1; \ x = \tfrac{41}{19}, y = {}-\tfrac{28}{19}$ 4. $x = {}^-3, x = {}^-2; \ x = 2, y = 3$

Check out AS5

1. (a) 8 (b) $^-1$
2. 10
3. (a) $(x-1)(x-5)$ (b) $(2x+1)(x-2)$
   (c) $(2x+3)(3x+1)$ (d) $x(x-2)(x+2)$
4. (a) 2, 3 (b) $^-0.85, 2.35$ (c) $1.05, 2.95$
5. $5, -\tfrac{3}{13}$ 6. (a) Translate 3 left (b) stretch parallel to x-axis, sf $\tfrac{1}{2}$ (c) Translate up 4
7. $x = {}-\tfrac{9}{17}, y = {}-\tfrac{53}{17}; \ x = 1, y = 1$

Revision exercise AS5

1. $\dfrac{x}{(x+2)}$ 2. (a) (i) $y = (x-1)^2$ (ii) $y = {}^-x^2$ (b) (i) $y = 3\cos x$ (ii) $y = \cos x + 1$
3. (a) $\tfrac{x}{2}$ (b) 3
4. (a) $0.72, 2.78$ (b) $x = 1.75$ 5. (a) $\tfrac{1}{2x}$ (b) $\dfrac{y+2}{2y-1}$

Module 5 Practice calculator test

1. 7.6
2. (a) $25.1\,\text{cm}^2$ (b) $20.6\,\text{cm}$
3. (a) $^-25$ (b) $^-4$
4. (a) $x \leqslant 1$ (b) (i) $x \geqslant {}-\tfrac{7}{5}$ (ii) $^-1$

5. 39 cm
6. (a) 2·1 m (b) 4·8 m (c) 5·6 m
7. (a) 11 cm (b) 7·5 cm
8. 42·5°
9. 0·23, 1·43
10. (b) $2\frac{1}{2}\mathbf{a} + \frac{1}{2}\mathbf{b}$
11. (a) $2(x - 5y)(x + 5y)$ (b) $x = {}^-1·8, y = {}^-7·4; x = 3, y = 7$
12. (a) 15·4 cm (b) 67·7 cm^2 (c) 7·8 cm

Module 5 Practice non-calculator test

1. (a) $2n + 1$ (b) $2a(3a - b)$ (c) $4x^2 + 11x - 3$
2. 9
3. (a) p^7 (b) t^8 (c) y^2
4. (a) (0, 5), (10, 0) (b) (ii) $x = 2, y = 4$ (c) $x = \dfrac{y - b}{a}$
5. (a) B (c) cm^3
6. 56 cm^2 7. $4x^8y^5$
8. (a) (i) 160° (ii) 10° (iii) 40° (iv) 40°
9. $x = 2, y = 1$ 10. (b) $x = 20$
11. (a) 0 (b) $^-2$ (c) $\frac{1}{5}$ (d) $^-4$
12. (a) $y = 4x + 5$ (b) $\dfrac{7(q + 3)}{3 - 2q}$
13. (b) 228·6, 311·4
14. (b) (ii) (0, p) (c) (^-p, 0)
15. $\dfrac{88}{505}, \dfrac{1932}{505}$

Index

acceleration 414–18
addition
   algebraic fractions 431–2
   vectors 375
addition rule 71
adjacent sides 351
algebra 202–51, 317–43, 409–51
   using 458
algebraic fractions 430–3
algebraic indices 339–41
algebraic methods 179–96
algebraic proofs 184
AND rule 75
angle bisectors 264
angles 253–61
   in circles 395–8
   of depression and elevation
      356–8
   exterior 258–9
   finding unknown 354–5
   interior 260–1
   in polygons 258–61
   straight lines 253
   in triangles 253
approximations 111
arcs 283–7
   length 364
area 273–7
   circles 278
   similar shapes 305
   under curves 412–14
arithmetic 99–104
asymptotes 383
averages 2–13
   moving 6–13

back-to-back stem plots 21
bar charts 15 477
bearings 356
bias 38, 472–3
bisectors
   angle 264
   perpendicular 264
bounds 163–72
   lower 163, 167–8
   upper 163, 167–8
box and whisker diagrams 40–1, 475
brackets
   equations 215
   expanding 206–7
   simplifying 204–5

calculators 34, 91–2, 137–43, 148
censuses 32
centres of enlargement 369

charts
   bar 15
   pie 14–15
chords 288
circles 264, 278–90
   angles in 395–8
   area 278
   circumference 278
   inscribed 268
   theorems 395–401
circumference 278
column vectors 369
combined events 67
completing the square 425–6
compound interest 134
compound measures 159–64
computers 480–1
conditional probability 84–6
cones
   surface area 298, 364
   volume 293, 364
congruent shapes 368
congruent triangles 261–3
conjectures 179–82
convenience sampling 471
correlation 23–4
cosine ratios 351, 381–6
cosine rule 389
coursework
   investigative 452–67
   statistical 467–82
cube numbers 94
cubes
   nets 347
   surface area 364
   volume 364
cubic graphs 331–2
cuboids volume 290
cumulative frequency 40–7
cumulative frequency curves 41,
   475
currency 171–2
curved graphs 331–9
curves
   area under 412–14
   cumulative frequency 41, 475
   frequency 16
   gradients 410–11
   transformations 433–9
cylinders
   surface area 298, 364
   volume 290–1, 309, 364

data
   collecting 28, 468, 472–3

   discrete 10–11
   grouped 11, 16
   handling 1–59
   interpretation 468–9, 477
   presentation 14–22
   raw 14
   rounded 15
   secondary 475
decimals 99–101
   and percentages 127
   recurring 151–4
density 162
depression angles of 356–8
diagrams
   box and whisker 40–1, 475
   mapping 223
   scatter 23–7
   stem and leaf 20–1, 475
   tree 78
   see also charts; graphs
diameter 278
difference of two squares 420–1
dimensions in formulae 364–7
direct proportion 115, 186–9
discrete data 10–11
distance 159–62, 414–18
distance-time graphs 237–9, 414–18
division
   long 101
   negative numbers 118
double percentages 136–7

elevation angles of 356–8
elimination method 244–5
enlargements 369–71
equal vectors 374–5
equations 208
   with brackets 215
   solving graphically 192–5
   of straight lines 230, 234
   see also linear equations;
      quadratic equations
errors 168
   percentage 171
   rounding 139–40
estimation 107–12
events
   combined 67
   exclusive 71–4
   independent 75–84
   mutually exclusive 71
   non-exclusive 72
   probability of 63–4
examinations gender differences
   476–9

exclusive events 71–4
expectation 65
exponential graphs 332
expressions 155, 208
    factorising 207–8, 419–22
exterior angles 258–9

factorising
    expressions 207–8
    quadratic expressions 419–22
factors 94
    scale 369
Fermat's Last Theorem 181
formulae 158, 208
    changing subject of 318–23
    dimensions in 364–7
    and fractions 319–20
    quadratic equations 422–5
fractions 104–7
    algebraic 430–3
    and formulae 319–20
    and percentages 127
frequency curves 16
frequency polygons 15
frequency tables 10–13, 14
functions 208

gender differences in examinations
    476–7, 478–9
general terms 223
generalisations 457–60, 466
geometric proofs 402–3
gradients
    of curves 410–11
    linear graphs 230
graphs
    cubic 331–2
    curved 331–9
    exponential 332
    of motion 237–9, 414–18
    quadratic 331
    real-life 237–41
    reciprocal 332
    scatter 478–9, 481
    sketch 239–41
    travel 237–9
    see also diagrams; linear graphs
greater than 323
grouped data 11, 16
guessing game 474–5, 479–80

hexagons 260
highest common factor (HCF) 95
histograms 47–54
hypotenuse 269, 351
hypotheses 27–32
hypothesis testing 31, 469–80

identities 208
independent events 75–84
indices 142–7
    algebraic 339–41
    rules 142
inequalities 323–31

solving 325–8
symbols 323
inexact answers 216–17
intercepts linear graphs 232
interest and percentages 134
interior angles 260–1
interpretation 468–9, 477
interquartile range 40, 41
interviewing 472
inverse operations 98
inverse proportion 115, 189–91
investigative coursework 452–67
irrational numbers 150, 278
isometric drawings 345–6

justifications 460–1, 467

kites 256

laws finding 235
less than 323
letters
    arranging 455–6, 462–3
    as quantities 202–8
like terms collecting 203
line segments 229
line symmetry 349
linear equations 209–15
    problem solving 211–14
    solving 318
    see also simultaneous linear
        equations
linear graphs 227–37
    gradients 230
    intercepts 232
linear scale factors 305
linear/quadratic simultaneous
    equations 247–8, 440
lines of best fit 23
loci 264–8
logical arguments 183
long division 101
long multiplication 101
losses 131
lower bounds 163, 167–8
lowest common multiple (LCM) 96

mapping diagrams 223
maps scales 114
mathematical language 208
mathematical reasoning 453–4
mean 2
measurements 163–72
measures compound 159–64
median 2
mirror lines 367
mode 2
moving averages 6–13
    five-point 6
    six-point 7
multiples 94
multiplication
    long 101
    negative numbers 118
multiplication rule 75

multipliers 135
mutually exclusive events 71

negative correlation 24
negative numbers 116–19
    dividing 118
    multiplying 118
nets 347–8
newspapers, hypothesis testing
    469–73
non-exclusive events 72
numbers 94–178
    cube 94
    irrational 150, 278
    prime 94, 180
    properties 94–9
    random 33–4
    rational 150
    square 94, 321
    see also negative numbers

observations 472
octagons 258
operations 98
operator squares 124
opposite sides 351
OR rule 71

parabolas 268
parallelograms 256
    area 274
patterns triangles 454–5, 462
pentagons 260
percentage errors 171
percentages 127–37
    and decimals 127
    double 136–7
    and fractions 127
    and interest 134
    reverse 1323
perfect tiles 459–60, 463–4
perpendicular bisectors 264
pi (π) 278
pie charts 14–15
pilot surveys 31
planes of symmetry 350
polygons
    angles 258–61
    frequency 15
positive correlation 24
powers see indices
prime factors 94
prime numbers 94, 180
prisms volume 290
probability 60–90
    calculating 63–71
    conditional 84–6
    see also events
problem solving
    linear equations 211–14
    quadratic equations 427–30
    simultaneous linear equations 245–7
profits 131
proofs 184–6
    geometric 402–3

proportion
    direct 115, 186–9
    inverse 115, 189–91
    and ratios 112–16
pyramids
    nets 347
    volume 293
Pythagoras' theorem 186, 269–73

quadratic equations 419–30
    problem solving 427–30
    solving
        completing the square 425–6
        factors 421–2
        formula 422–5
quadratic expressions factorising 419–22
quadratic graphs 331
quadrilaterals 256
    similar 300
quantities, letters as 202–8
quartiles 40
questionnaires 27–32, 472
questions designing 28–30

random numbers 33–4
random sampling 33–4, 471
    stratified 37
range 3
rational numbers 150
ratios
    and proportion 112–16
    trigonometric 351, 381–6
raw data 14
reciprocal graphs 332
rectangles 256
    area 273
    similar 300, 305
recurring decimals 151–4
reflections 367–9
regions 329–31
regular polygons 258
relationships finding 457–60
relative frequency 60–1
reverse percentages 132–3
rhombuses 256
right-angled triangles 269–73, 351–5
rotational symmetry 256, 349
rotations 367–9
rounded data 15
rounding errors 139–40
rules
    addition 71
    finding 219–27
    indices 142
    sequences 223–4
    trigonometric 387–9

sampling 32–9
    methods 471
    selection techniques 33–4
scalars 375–7
scale factors 369
scales maps 114
scatter diagrams 23–7
scatter graphs 478–9, 481
secondary data 475
sectors 283–7
    area 364
segments 288–90, 399
    line 229
    major 288
    minor 288
semicircles angles in 397
sequences
    nth term 223–4
    rules 223–4
shapes
    congruent 368
    similar 300–11
    three-dimensional 345–50
similar shapes 300–11
    area 305
similar solids volume 309–11
simultaneous equations linear/quadratic 247–8, 440
simultaneous linear equations
    algebraic solution 243–5
    graphical solution 241–3
    problem solving 245–7
    solving 440
sine ratios 351, 381–6
sine rule 387–8
sketch graphs 239–41
Spearman's rank correlation 479–80
speed 159–62
speed–time graphs 237–9, 414–18
spheres
    surface area 298, 364
    volume 293, 309, 364
square numbers 94, 321
square roots 321
square sea shells 464–6
squares 256
standard form 146–50
statistical coursework 467–82
stem and leaf diagrams 20–1, 475
straight lines
    angles 253
    equations of 230, 234
straight-line graphs see linear graphs
stratified random sampling 37
stratified sampling 471

substitution 155–9
    method 243–4
subtraction, algebraic fractions 431–2
surds 153–5
surface area 298–300
symmetry 349–50
    planes of 350
    quadrilaterals 256
systematic sampling 471

tables 455
    frequency 10–13, 14
tally charts 14
tangent ratios 351, 381–6
tangents 398–400
terms
    collecting 203, 321–2
    general 223
    like 203
    nth 223–4
    simplifying 204–5
three-dimensional problems 361–3
three-dimensional shapes 345–50
time 159–62
time series 6
transformations 367–74
    curves 433–9
translations 369–71
trapeziums 256
    area 274
travel graphs 237–9
tree diagrams 78
trend lines 6
trial and improvement method 216–19
triangles
    angles 253
    area 273, 364
    congruent 261–3
    patterns 454–5, 462
    right-angled 269–73, 351–5
    similar 300
trigonometric ratios 351, 381–6
trigonometry 351–63

upper bounds 163, 167–8

vectors 374–80
    adding 375
    column 369
    equal 374–5
    scalar multiplication 375–7
velocity–time graphs 414–18
volume 290–7
    similar solids 309–11